Praise for *Atlantic*

"History is rarely as charming and entertaining as when it's told by Simon Winchester. There are fabulous set pieces in *Atlantic*—on piracy, on packet ships, on transatlantic cables and the speeding up of information, on codfish, on sea bass, on plankton."

—*New York Times Book Review*

"Simon Winchester is one of those maddeningly gifted British writers who could probably write the history of mud and make it fascinating. In fact, he sort of did. . . . A rollicking ride. . . . No one tells a better yarn than Winchester."

—*Washington Post*

"Winchester vividly describes how the Atlantic Ocean was born about 190 million years ago, continues to spread at the rate of about an inch a year, and could well disappear as we know it in just another 180 million years. . . . Delightful."

—*USA Today*

"Convincing. . . . A fascinating look at a long sweep of history."

—*Los Angeles Times*

"Simon Winchester . . . is . . . brilliant. To read Winchester is to share the excitement of his travels and adventures. . . . *Atlantic* is a mine of fascinating information and ideas both small and colossal."

—Brian Urquhart, *New York Review of Books*

"*Atlantic* . . . is at once satisying, suspenseful (no mean trick . . . e subject of an ocean), thought-provoking, occasion-'' always absorbing. So big a subject as the A+' certain largeness of spirit and am~'' Simon Winchester is gloriously

"Mr. Winchester—a trained geolog . . . . . .tter— is well suited to tell the story. A . . . .n the sort of panache that he has brought to pre . . . .oks, such as *Krakatoa*, about the volcanic disaster of 1883, and *The Professor and the Madman*, about the creation of the Oxford English Dictionary. . . . His lively, lyrical telling of the ocean's story does much to sharpen our appreciation."

—*Wall Street Journal*

"Telling the story of 'the classic ocean of our imaginings' is a huge undertaking, but Simon Winchester manages it with aplomb."

—*The Economist*

"A formidable writer and storyteller."   —*Entertainment Weekly*

"Fascinating. . . . Simon Winchester's storytelling abilities shine, with personal anecdotes and research expertly woven."

—*The Guardian* (London)

"Mr. Winchester's latest work of nonfiction is, like the thirty-six-pound cod my father caught forty years ago in the deep sea two hours out of Plymouth, Massachusetts, 'a keepuh' (New England parlance). . . . It's one of those you'd like to keep on a shelf for another read sometime in the future."   —*Baltimore Sun*

"Refreshing. . . . A work of 'high specific gravity' and an outstanding example of popular historical scholarship."

—*Times Literary Supplement* (London)

"Simon Winchester has an uncanny ability to connect with readers, holding forth with the erudite charm of the fascinating dinner guest who is on everybody's invitation list."   —*Philadelphia Inquirer*

"Wonderful, encyclopedic. . . . Enthralling. . . . Winchester brings us down to sea level and makes us realize what we owe to the Atlantic."

—*The Telegraph* (London)

"Inspired."   —*San Francisco Chronicle*

"[An] epic new book. . . . With his excellent research and engrossing anecdotes about the ocean as a 'living thing,' Winchester spotlights its inspiration on poets, painters, and writers in its majestic beauty. . . . Winchester's sea saga is necessary reading for those who want to understand the planet better."   —*Publishers Weekly*

"Winchester waxes lyrical, rhetorical, and historical in this broadly focused but eloquently detailed account of the great ocean and mankind's relation to it."   —*Kansas City Star*

"In this detailed, engrossing, and beautifully written book, Simon Winchester shows how the familiar sea at our front door is far more significant to the world than we ever supposed."

—Virginian-Pilot (Norfolk, Virginia)

"Fascinating. . . . Ingenious. . . . Exhilarating, absorbing. . . . Executed with verve and panache. . . . The time is not too far off, it seems to me, when books like this will seem to be infused with a poignant nostalgia, the kind of nostalgia some readers experience nowadays with Joseph Conrad's tales of an empire on which the sun never set."

—Sydney Morning Herald (Australia)

"Enjoyable and richly informative." *—The Telegraph* (London)

"[Simon Winchester] colors the narrative with the history of every human endeavor related to it across the centuries, from the seventh-century moment when Phoenician sailors ventured past the Pillars of Hercules on to the ages of exploration, colonization, and beyond. Winchester has, it seems, stopped at nothing in order to give life to this ocean's story." *—Courier-Mail* (Australia)

"A fine yarn. . . . The pages bustle with fascinating tales and a wealth of unexpected scientific and cultural detail. . . . Winchester works to reinvent the Atlantic with the sense of wonder that it held for generations prior to ours." *—The Age* (Australia)

"Of all of Winchester's amazingly educational and entertaining books . . . his latest one is perhaps the most unique and the most creative in its approach. . . . As we learn from one of the most wondrous facts presented here, oceans actually do have life spans. . . . Lively and extensive. . . . Winchester's latest is bound to follow his previous books onto bestseller lists, and this one should be promoted as one of his best." *—Booklist*

"Poignant. . . . *Atlantic* is a reminder that this most storied ocean will outlive us handily, and is still worthy of our interest, our care, and our respect." *—Winnipeg Free Press* (Canada)

ATLANTIC

ATLANTIC

GREAT SEA BATTLES, HEROIC DISCOVERIES, TITANIC STORMS, AND A VAST OCEAN OF A MILLION STORIES

Simon Winchester

HARPER ● PERENNIAL

NEW YORK ● LONDON ● TORONTO ● SYDNEY ● NEW DELHI ● AUCKLAND

HARPER ● PERENNIAL

A hardcover edition of this book was published in 2010 by HarperCollins Publishers.

P.S.™ is a trademark of HarperCollins Publishers.

HarperCollins books may be purchased for educational, business, or sales promotional use. For information, please e-mail the Special Markets Department at SPsales@harpercollins.com.

Political, physical, exploration, and commerce maps that occur on pages viii, ix, 113, and 319 were created by Nick Springer/Springer Cartographics LLC.

Pangaea and Future Pangaea maps that occur on pages 41 and 446 were created by C. R. Scotese, PALEOMAP Project (www.scotese.com).

All interior photographs, unless otherwise noted, are from the author's private collection. For those photographs and endpaper images that are the exception, grateful acknowledgment is made to the following: front and back endpapers: Fox Photos/Getty Images; page 3: Canadian Pacific Archives; page 6: photograph by Richard Webb; page 30: U.S. Naval and Heritage Command; page 58: photograph by Curis Marean, Institute of Human Origins; page 71: Andrew Vaughan/Associated Press; page 82: photograph by Gregory Howard; pages 94, 117, 129, 175 (New York), 231, 300, 308: courtesy of Library of Congress Prints and Photographs Division; page 155: photograph by Kim Wilkins; page 175 (Liverpool): photograph by Chris Howells; page 175 (Cadíz): photograph by Daniel Sancho; page 175 (Jamestown): courtesy of Wikitravel; page 224: The Granger Collection, New York; page 229: Clement N'Taye/Associated Press; page 254: Associated Press; page 260: STR/Getty Images; page 279: George F. Mobley/Getty Images; page 294: Kean Collection/Getty Images; page 324 (Andrea Doria): U.S. Coast Guard/Associated Press; page 324 (Stockholm): Yale Joel/Getty Images; page 326: Keystone/Getty; page 336: Hulton Archive/Getty Images; page 339: Jani Patokallio/OpenFlights.org; page 354: Alfred Eisenstaedt/Getty Images; page 369: Haywood Magee/Getty Images; page 385: Reuters/Corbis; pages 400, 422: courtesy of NASA; page 429: photograph by William K. Li and Frédéric Partensky, Bedford Institute of Oceanography.

Grateful acknowledgment is made to Charles Tomlinson for the poem on page 205.

FIRST HARPER PERENNIAL EDITION PUBLISHED 2011.

Designed by Leah Carlson-Stanisic

Library of Congress Cataloging-in-Publication Data has been applied for.

ISBN 978-0-06-170262-4 (pbk.)

19 20 OV/LSC 10 9

THIS BOOK IS FOR

Setsuko

AND IN MEMORY OF

Angus Campbell Macintyre

FIRST MATE OF THE SOUTH AFRICAN HARBOUR BOARD TUG

SIR CHARLES ELLIOTT

WHO DIED IN 1942, TRYING TO SAVE LIVES

AND WHOSE BODY LIES

UNFOUND

SOMEWHERE IN THE ATLANTIC OCEAN

ATLANTIC OCEAN: POLITICAL

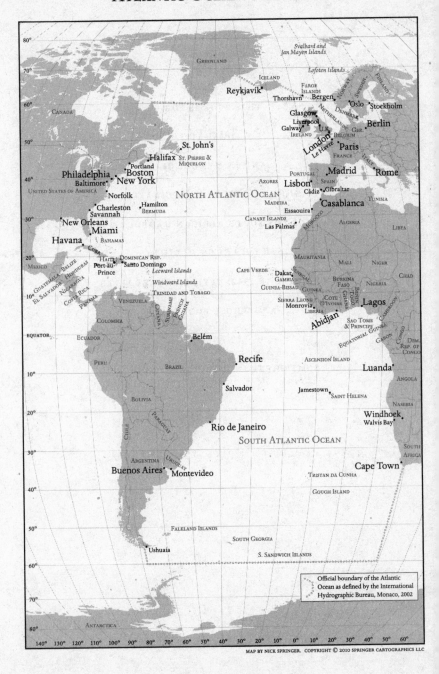

Official boundary of the Atlantic Ocean as defined by the International Hydrographic Bureau, Monaco, 2002

ATLANTIC OCEAN: PHYSICAL

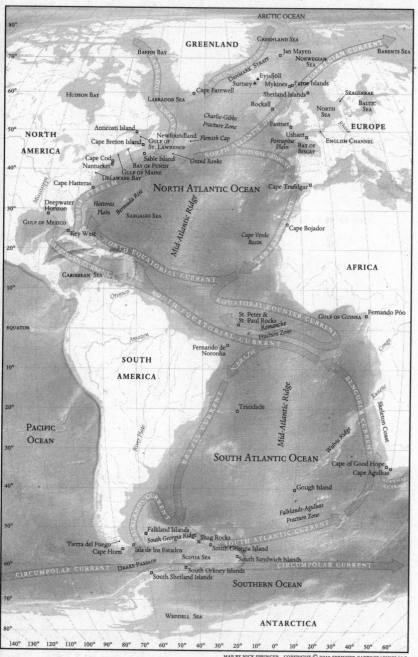

Men might as well project a voyage to the Moon,
as attempt to employ steam navigation against the
stormy North Atlantic Ocean.

DIONYSIUS LARDNER, IRISH SCIENTIFIC WRITER AND LECTURER, 1838

CONTENTS

~~~~•~~~~

MAPS AND ILLUSTRATIONS

PREFACE:
THE LEAVING OF LIVERPOOL

The ocean romance that lies at the heart of this book was primed for me by an unanticipated but unforgettable small incident. It was a clear cool dawn on Sunday, May 5, 1963, and I was eighteen years old. I was alone, on passage aboard a great ocean liner, the *Empress of Britain*, and we were unexpectedly stopped in a remote corner of the northern seas to the east of the Grand Banks of Newfoundland. We were floating quietly above a small submarine plateau some miles off the first headlands of America, an area known to oceanographers and fishermen as the Flemish Cap.

It was there that something rather curious happened.

We were five days out from Liverpool. The voyage had begun on the previous Tuesday afternoon, a wild and blustery day that had sudden gusts chasing the River Mersey's waters with filigrees of spindrift. This was when I first spotted the ship on which I would make this first-ever crossing of the Atlantic Ocean.

It was her flanks that were most noticeable, looming massive and blinding white—the Canadian Pacific's three sister ships were known collectively as the White Empresses—at the end of the lanes running down to the Liverpool waterfront. She was fastened securely to the Pier Head, just beside the old Princes Dock, a dozen manila ropes as thick as a man's arm keeping her quite still, aloof to the weather. But from the bustle of last-minute activity around her and the smoke being torn urgently from her single yellow funnel, it was clear she was already straining at the leash: with her twenty-five thousand tons of staunchly riveted Clydeside steel, the *Empress* was readying herself to sail three thousand miles westward, across the Atlantic Ocean, and I had a ticket to board her.

It had taken six months for me to earn enough to buy it. I must have been on slave wages, because passage all the way to Canada had not cost much more than a hundred dollars, provided I was willing to settle for one of four bunks in a windowless cabin on a deck situated so far below the waterline one could almost hear the slopping in the bilges. But though it was to be an economical crossing, one step up from steerage, in the Canadian Pacific offices off Trafalgar Square—more cathedral than bureau, all teak, marble, and hush, and with scale models of famous ocean liners from the old days illuminated in the windows—even this most modest of transactions was handled with dignity and circumstance.

Maybe time and schoolboy memory have distorted things a little, but I like to fancy that the clerk who took my savings, in frock coat and pince-nez and wearing a company badge embossed with pine trees, polar bears, and beavers, had written out the ticket in longhand, dipping his pen into an inkwell and blotting it with a roller of pink paper. *Liverpool to Montreal, Voyage No. 115*, it had read, and I clearly do remember spending many subsequent moments turning this precious talisman over and over, examining the engravings, the intaglio, the watermarks. It came in a scarlet and white cardboard wallet, thick and stiff and with a pocket to hold luggage tags with waxed string ties, NOT WANTED ON VOYAGE stickers, immigration forms and customs guides, and vague suggestions as to the coming maritime routine—"11 A.M.: Bouillon on the Boat Deck" was the one that stuck most firmly in my mind.

I think I developed a rather unhealthy attachment to this ticket, freighted as it was with so much symbolism—freedom, the New World, adventure, the Atlantic Ocean—and when I handed it in at the top of the gangplank that spring afternoon and saw how the purser took it with only a studied casualness, I must have looked dismayed, for he smiled and handed it back to me. "First time?" he asked, in a kindly way. "Keep it, then. This is a very grand ocean— and you're on a White Empress crossing her. Nothing finer! You should keep your first souvenir of going across."

This 26,000-ton liner, the third to carry the name Empress of Britain, *was launched on the Clyde by Queen Elizabeth in 1955. It was one of the three "White Empress" warhorses that took passengers between Liverpool and Montreal until 1963, when competition from airlines forced her from service.*

By sailing time a watery sun had appeared and was lowering itself toward the horizon. *All ashore who's going ashore!* came the familiar announcement, on cue. The Tannoy speaker carried the call to "ease springs" (sailorspeak for "let go the ropes"); there was shouting from shore, the crackling of radios on the bridgewing and the foredeck—and one by one the heavy iron-bound nooses of the hawsers splashed into the oily gap between hull and pier, the oily gap began to widen, and the dripping ropes were hauled in slowly by capstans that growled at the strain. A pair of well-worn tugs appeared, yelping and snorting, nosing us out into the tidestream, and then turning us, nudging our bows to the northwest.

The famous George clock on the Royal Liver Building struck five. I could see my father down on the quayside checking his wristwatch. He and my mother pointed upward in relief—they had found me at last, among the crowds of passengers lining the taffrail—and as they waved there came the three departure blasts from our steam-horn, echoing and re-echoing along the ship-crowded waterfront. Our decks started to vibrate and rumble as the engines engaged and the propellers began to thrash the waters astern.

I checked my own watch: it was nine minutes past the hour, the moment when the voyage officially got under way. The tugs let go. The *Empress of Britain* was at last under her own power, free of her hawsers and bollards and tugs, free of the shore and free of England, beginning to steam firmly and unstoppably away, bound for the deep ocean and the promise of tomorrow. Some of the passengers, emigrants to Canada probably, looked briefly distraught, waving through tears. I was excited, apprehensive, nervous. I watched my parents as they started to walk back to our tiny tan Ford Prefect, their heads bowed.

Darkness began to fall swiftly, and soon the lights of Liverpool and Birkenhead became a loom of orange, like a damped-down fire astern. At the famous floating lighthouse known as the Bar Light Vessel, somewhere off Crosby, the pilot boat came alongside, and a middle-aged man in a brown pullover and a stained white cap stepped nimbly down onto its afterdeck: he waved up at us, and if he mouthed something like "Take care! Have a good crossing!" his words were whipped away by the breeze. Within the hour he and his wife, I thought, would be dozing in front of his television set, the cat asleep by the fire.

We spooled up the engines once he had cleared our wake, and soon the turbines were pushing us along at a good clip, twenty knots, maybe more, and what little rain was left stung the face like needles. Soon we were positively hissing over the sea, ignoring the waves from a storm that, to judge by that last glimpse of the sunset, was now dying away. I stayed on the foredeck to watch for other vessels: there was a bustle of Fleetwood trawlers scuttling home, an inbound freighter or two, and then the outline of what looked like a warship of some sort, maybe a destroyer heading north with us, but faster and quite silent.

Ocean Passages for the World, the long-haul mariners' route-planning bible, seems frequently eccentric in its suggestions for scheming a voyage. A map will show the obvious: Montreal is some eight degrees of latitude south of Liverpool, and one would think the best way for a ship to travel to the Canadian city from Merseyside would be to turn southward past the coast of Wales, head down through the St. George's Channel, and, keeping Cork and the light on Fastnet Rock* well off to starboard, enter the Atlantic on a direct heading for the St. Lawrence estuary. But the

---

* This lighthouse-capped outcrop, best known today as the outer marker for a dangerous annual yachting race from southern England, is called "Ireland's teardrop" by the sentimentally inclined, since it was invariably the last fragment of the motherland to be glimpsed by emigrants on their way to Ellis Island.

blue-backed bible says otherwise: vessels headed from Liverpool to Canadian ports in springtime, as we were, would find it much more navigationally prudent to make a heading not to the south of Ireland but to the north of it, and only after clearing the coast of Donegal near Bloody Foreland make a very much longer southward swoop for Canada. "Although heavy weather is frequently experienced," *Ocean Passages* offers in its very detailed advice for sailing vessels, "the winds are generally more favorable and the currents from the Arctic assist in the latter part of the voyage."

We were a large, very modern, steel-hulled, and well-found power vessel, with the strength to ignore such bagatelles as winds and storms and currents from the Arctic. Our schedule

Emigrants bound for the American ports gazed wistfully at Fastnet Light at the southwest tip of Ireland, watching the final sight of their homeland fade into nothing. Such sweet sadness prompted its nickname: Ireland's teardrop.

called for us to pick up additional passengers and freight from Greenock, on the Clyde—and so that evening we headed not south but north out of the Mersey into the Irish Sea. Around midnight we saw the flash of the light off the Calf of Man, and later still, spotted a flurry of lights on Galloway on our starboard side and the forbidding basalt cliffs of County Antrim to port.

As dawn came up—and it was raining and blowing again—we were passing Ailsa Craig, a tiny islet made of a fine-grained granite from which are fashioned the world's best stones for use in the wintertime sport of curling. We passed to the east of the Isle of Arran—there was still late snow on the summit of Goat Fell—and by eleven, the promised bouillon time (though none was on offer that day), we were moored off Greenock. A flotilla of small craft brought out the scattering of passengers—two were children with measles, and there was some slight quarantine-related delay until our captain, an evidently compassionate man named Thorburn, decided to take them—and by lunchtime we were back on our way, making down the Clyde for the sea. As we emerged back into salt water, we altered our heading to starboard and set a westerly course to steer safely around the notoriously rough waters north of Rathlin Island.

Now, and at last, we were making steadily for the open ocean, and as we did so, the Atlantic swells began steadily and dramatically to increase. Great rollers began to buffet the bows, big thudding monster waves driven by the springtime westerly gales that blew ceaselessly at the approaches to the British Isles.

Dinner, to no one's surprise given the pitching of the ship, was a thinly attended affair. Those few of us about on that rain-soaked evening could see the tiny island of Inishtrahull through the scudding clouds, three miles off to port, and between it and us the tiny archipelago of the Tor Rocks, Ireland's northernmost possessions. Inishtrahull—the Island of the Empty Beach—

marks one of the beginnings, or one of the ends, of a North Atlantic crossing. Through glasses we could see dimly a scattering of ruined houses and untidy lines of old stone walls, and then the slender pencil of its famous lighthouse, already winking through the gathering gloom, and which has been flashing its welcomes and farewells to thousands of transatlantic vessels for almost two centuries.

From here onward the sea yawned open wide and featureless, and soon took on the character that is generally true of all great oceans—being unmarked, unclaimed, largely unknowable, and in very large measure unknown. Our track was designed to bring us in a great, slow curve almost two thousand miles to a waypoint that hinted at the land of the New World ahead—the notorious and shallowly submerged Virgin Rocks, off Newfoundland. I remembered the Virgins from English literature classes: Kipling had written about the fishing there in *Captains Courageous*—the cod *in legions*, he wrote, *marching over the leathery kelp*, and all usually easily visible in the shallows.

If all went according to plan, and if we kept to the cruising speed of twenty knots that our engines could supply with ease, we should make the Virgin Rocks by Monday night, should soon thereafter sight the lighthouse on Cape Race on the southern corner of Newfoundland, and after threading our way along the St. Lawrence River be safely landed in Canada on Tuesday in time for dinner ashore.

And so it turned out. For the men up on the bridge, Voyage No. 115 was basically just another routine crossing. For me, a rank newcomer to the ocean, the crossing was at first memorable simply for being a crossing of this great ocean. We had what for me were some nail-biting moments of great spectacle and storm; we spent our time almost entirely alone on the sea—encountering just one other vessel en route, despite being on a recognized shipping lane—and that sense of pressing solitude I found more than a little intimidating; and when we passed over the Virgin Rocks we did so in darkness and I never got to see the codfish. But there was nothing desperately unusual—until the single interruption, the one small moment that I remember more vividly than perhaps it deserves, and which took place while we were lying stopped in the shallow Atlantic waters off Flemish Cap.

· · ·

It was just after dawn, and bitterly cold. The season still being early spring, this being *Titanic* waters and with the Arctic ice-fields perilously close by, our crewmen were on alert for icebergs and growlers and other similar hazards. None had yet been seen: the voyage, so far as the navigating officers were concerned, had been entirely plain sailing. Nor were there any of the fogs for which this stretch of ocean is notorious: the Labrador Current and the Gulf Stream collide softly and unseen near here, and the sudden blending of tropical and Arctic waters can thicken the air above into gray pea soup for days at a time. Not this day, however, for which many had reason to be thankful.

I had risen early and, muffled to the ears, was out before breakfast, strolling the length of the boat deck. All was normal: we were hissing along nicely, dawn behind us, darkness ahead. Suddenly, however, bells started to clang, crewmen started running up and down the companionways and the decks, the ship's engines unexpectedly stopped churning, the vessel lost way, and then it swiftly fell silent. We drifted steadily to a halt, our smooth westbound progress replaced by a heavy and ungainly rolling. The gale of the previous night had now all but blown itself out, but a stiff westerly breeze was still whistling through the aerials and gantries up above. Before long, I thought, we would be blown backward.

The ocean here, on the very outer edge of the American continental shelf, appeared quite empty, with not a bird or any marine life in sight. It was quite rough, and though the ship herself had become smothered by an overwhelming deadness, the sea was evidently very much alive, the waves and the swell slapping ferociously against the hull.

After a few moments, though, there came an unexpected sound, from directly ahead. At first it was just a low-frequency sigh, then a hum—then recognizable as the faint sound of a motor. An airplane motor. Up on the bridgewings, I could see the officers of the watch, acting as one, training their binoculars to westward, toward the direction of the sound, and peering anxiously into a still half-dark sky. Soon there came a cry—the aircraft had been spotted. A few minutes later we all saw it: first a single pinprick of light, then two, and finally the outline of propeller plane, its nose glinting in the weak sun. As it approached us it came in low and fast, a large, two-engined machine that roared and smoked as it turned above us and dipped its wings, the roundels of the Royal Canadian Air Force clearly visible on the fuselage.

Events then began to happen fast. From near the stern of the boat deck came a clank of pivots and rusty levers, and then a hard splash as the ship's motorboat was launched. It sped out onto the ocean and came to a stop a mile or so away from us. Once it was holding position the aircraft swooped and turned, opened its cargo doors, and slowed to pass directly over the tiny craft, as it did so dropping something that floated down onto the sea on a small orange parachute. A sailor from the boat's crew swept it up with a billhook and the steersman, giving a thumbs-up, headed the launch back home. The aircraft rose back up into the sky, dipped its wings again in farewell, and headed to its faraway base, becoming a smoke-trailed speck, then vanishing within moments.

The motorboat was winched up, the package—which turned out to be emergency medicine for an elderly woman passenger in distress in our liner's hospital—was duly delivered, and within the hour our engines had throbbed back into life and we were heading back onto our original course once again.

A trivial maritime incident, occasioning no more than a negligible delay in our arrival in Montreal two days later. But it was an event that has remained with me ever since. There was something uncanny about the sudden silence, the emptiness, the realization of the enormous depths below us and the limitless heights above, the universal grayness of the scene, the very evident and potentially terrifying power of the rough seas and the wind, and the fact that despite our puny human powerlessness and insignificance, invisible radio beams and Morse code signals had summoned readily offered help from somewhere far away. It was an augury of sorts, I have come to think in the years since, that this entire small drama had taken place on the first voyage that I ever took across the seas.

The captain's log for the closing moment of Voyage No. 115 is entirely laconic, almost dismissive: "Pilots exchanged at Three Rivers. Fine weather continued all the way up St. Lawrence. Clock tower passed at 1813hrs. Canted into berth with aid two tugs. All fast No. 8 shed at 1853hrs. Finished With Engines." We had crossed the ocean in seven days, six hours, and seven minutes, and despite our mid-ocean rendezvous were just fifty-four minutes late. British railway trains of the day seldom did much better.

. . .

Unknown to all of us aboard that week, and quite by coincidence, forces unseen and unseemly were hard at work. They were the dark forces of economics. As it turned out, the *Empress of Britain* was to make only eight more scheduled crossings of the Atlantic in her life. Just six months later, in October, a peremptory announcement was made that the barely seven-year-old flagship, launched with great fanfare by Queen Elizabeth in 1955, had been withdrawn from Atlantic service and would be sold. Her new owners, Greeks from Piraeus, would instead steam vacationers gently around the Caribbean, in a hurry no more.

The economics of large passenger liners suddenly made no sense. BOAC and Pan American had both begun air service between London's Heathrow and New York's Idlewild (later JFK) airports five years before, in 1958. The first flights were obliged to make refueling stops at Gander, in Newfoundland, but then as the planes became more powerful, both airlines began to cross the ocean nonstop, and scores of other carriers soon began to do the same. One by one the great passenger liners vanished from the ocean trade, and such ships as survived began to cruise instead, helping to inaugurate what would become an entirely different maritime industry.*

So it was tellingly symbolic that I came back from America six months later by air, and did so on what turned out to be the very same week that the stunned crew of the *Empress* was making its final voyage with the much-loved liner. Had I known of the droll coincidence I daresay I might have looked down and seen her plowing her last white eastbound furrow home. But my flight had its distracting moments anyway: it was aboard a Lockheed Constellation, a four-engine, triple-tailed machine designed first as a long-range bomber and then as a troop transport, and operated in this case by a somewhat dubious charter company

---

* My first liner enjoyed many second winds. She was reborn, for different owners and different purposes, as the *Queen Anna Maria*, the *Carnivale*, the *Fiesta Marina*, the *Olympic*, and the *Topaz*. Japanese owners employed her as a floating emissary for peace before finally having her towed to be broken up outside Mumbai, in 2008, fifty-three years after the queen launched her on the Clyde.

known as Capitol Airways of Nashville, Tennessee. We took off from New York, landed four hours later at Gander, then (by the skin of our teeth, the pilot later confessed, as the fuel was alarmingly low) made Shannon in the west of Ireland, but proceeded to discover that for some technical and legal reason we had no permission to land in London and were diverted to Brussels instead. Eventually, and testily, I found a flight to Manchester and made the rest of my way home by railway.

．　．　．

Almost half a century has passed since I made those two crossings—fifty-odd years during which I must have traversed this particular body of water five hundred times, at least. And though I have ventured out from a variety of other ports in both the North and South Atlantic, to cross in other directions, by rhumb lines or diagonally or along the lines of longitude or in huge looping curves, or to make expeditions out to the various islands that are scattered across the sea, it seems to me that the simple and most familiar route, the track from the major British ports to their major equivalents in eastern Canada or the United States, distills one aspect of what this book is about—humankind's evolving attitude toward and relationship with this enormous body of water.

And even in my lifetime, this is a relationship that has changed, and profoundly so.

In the early 1960s it was still something of a rarity to travel across the Atlantic by ship, or by any other means, for that matter. A scattering of the broke still went one-way, westbound, as migrants; a rather larger number of the wealthy and leisured traveled out and back on the great steamers with no care for time or cost. A handful of businessmen, not a few politicians, and clubby aggregations of diplomats went too, but most of them in propeller-driven aircraft rather than propeller-driven ships, for their crossings were said to be more urgent. For those who made the journey, it was still an adventure that could be daunting, exciting, memorable, suffused with romance, or cursed by the travails of *mal de mer.* What it most certainly was not was *routine.*

The same can hardly be said today. Yes, for a while it certainly was an excitement to cross the ocean by air—but for only a very short while. It must have been a considerable thrill, for instance, to take a Pan Am Clipper flying-boat service from the Solent to the Hudson, with stops in the harbors of such strange-sounding and long-forgotten coastal way stations as Foynes, Botwood, and Shediac. It must have seemed the height of style to stretch out in a bed on a double-decker Stratocruiser while the seas unrolled silently below. It was surely memorable—and foolhardy, given the plane's dismal safety record—to fly aboard one of those first BOAC Comet services, and even in the smoky old Boeing 707s when Pan Am and TWA began to fly them nonstop. I remember well taking some of the early Concorde test flights, and being naïvely astonished at just how *fast* they were when, only half-way through the Arts section of the *New York Times*, I was told that we were decelerating over the Bristol Channel and would be in London directly and so would I return my tray table and seat to where they had been when I eased myself aboard just a few moments before. Air travel across the great ocean was for a brief time almost as romantic and memorable as travel by sea. But it all soon changed.

For me it was marked by a small semantic shift. It began some time in the 1980s, when the pilots of aircraft crossing between Heathrow and Kennedy would slip almost casually into their welcoming announcement that "our track today will take us over Iceland"—with a slight emphasis on the word *today,* as if yesterday the flight was much the same except that it had passed over Greenland, or the Faroes. Or else they told the passengers that "the 177" or whatever the flight number might be, and so sounding studiedly casual, would be passing "a little farther north than usual, due to strong headwinds, and we'll make our landfall over Labrador and then head down over the state of Maine."

It seemed to me a shame—as though the flight deck were telling its charges that there was nothing much to get excited about anymore: today's transit was much like yesterday's, or last week's, and the crossing of what had become called "the pond"* (the terminology demoting the great ocean to a body of water almost without significance) would invariably be much as was generally expected at this time of year. Ho-hum, in other words.

And we passengers scarcely noticed. Having made good our nest of books and blankets, having made obligatory noises of good cheer to our stranger-neighbor, having glanced at the menu and wondered idly if it was too early to order a drink, we settled down and barely noticed a takeoff that would perhaps have enthralled us twenty years before. The same was true when it came to our landing six or seven hours later. Maybe there was a little more curiosity—since home was close and one wanted to sense and maybe spot a hint of it. Generally speaking, though, whether we could see six miles beneath us the forests of Labrador or those on Anticosti Island, or whether our first solid encounter with North America was Cape Breton Island or the sand spits of Sandy Hook or Cape Cod, it made little difference: all we really cared about was that we got in on time, that the border formalities weren't too irksome, and that we could get onto dry land and begin at once what we had journeyed to achieve. The gray-green vastness of undifferentiated ocean over which we had perforce to travel was really of no consequence whatever.

. . .

---

* Though the expression sounds modish and modern, it was in fact first used in 1612, and Victorian sailors would often refer to having *crossed the pond*, using the phrase in self-effacing understatement.

That for years was very much the case for me—until one recent
summer's afternoon, as I was crossing to New York on a British
Airways 777, companionless, conversationless, and bored, pinioned uncomfortably into a starboard window seat. Lunch was
long since done. I had finished the paper and my only book. The
entertainment was as much as I could bear. There were three
more hours to run, and I was daydreaming. I looked idly out of
the plexiglass porthole. It was quite cloudless, and miles below
us was the sea, as deep blue as the sky, not smooth but vaguely
crinkled, like dull aluminum foil, or pewter, or hammered steel,
and seeming to inch its way slowly backward from beneath the
wing.

I had been gazing for maybe fifteen minutes at the blue sea
emerging from beneath the gray flaps. Blue, blue, blue . . . and
then as I gazed down, I fancied I saw the water surface unexpectedly and subtly change color, becoming first rather paler, and
within what can have been no more than a couple of moments, or
miles, transmuting itself into a shade of light aquamarine. Seldom had I seen such a thing from this altitude: I supposed that if
it was real, and not imagined, then it must have had something
to do with the angle of the sun, which since I had taken a midday
flight, was higher in the sky than usual.

I glanced at the sky map in the seatback in front. The chart
was large-scale and poor, but the position it showed offered the
obvious reason for the alteration: we had crossed the edge of the
continental shelf. The deep mid-ocean abyss over which we had
been passing since crossing the Porcupine Bank, which marks
the western end of the European shelf and is usually reached
about half an hour off the Irish coast, had now lifted itself up at
last to become the faint submarine stirrings of the North American mainland.

Except that a few moments later, and even more unusually,

the water became dark blue once again, though this time only for a brief interval, before lightening yet again. It was as though the aircraft had passed over a deep river in the ocean, a cleft between two high underwater plains. I squinted as far under the wing as my vision allowed: from where the plain resumed it appeared to stretch away to the west, uninterrupted. And then I remembered, from what I knew of the undersea geography of this part of the North Atlantic: I had long been fascinated by the geography of the Gulf Stream, and as I remembered, it flowed nearby. What I recalled suggested to me that the uninterrupted plain I could now see marked the beginning of the Grand Banks of Newfoundland. The dark blue underwater channel was known as Flemish Pass. And the first patch of green I had spotted was, I realized, the very place where we had stopped all those years before to rendezvous with the Canadian rescue plane: the well-remembered shallows known as the Flemish Cap.

. . .

Nearly half a century has gone by since I first saw the Flemish Cap and watched, captivated, as that Canadian Air Force plane swept in. Back then—I was still a youngster, to be sure, and more easily awed than today—I had savored every detail of what seemed to me a fascinating small moment. And in the hours after our ship had started up and begun to move off westward, I had learned of other historical grace notes to the saga: a friendly deck officer on the *Empress* had told me that the emergency signals we had tapped out the night before had been picked up on Newfoundland by the American coast guard base in a place called Argentia—and they had taught us at school that it was at Argentia, back in 1941, that Winston Churchill and Franklin Roosevelt met aboard the battleship *Prince of Wales* and declared the Atlantic Charter, which so famously delineated the working of the postwar world. That I had just been hove-to, so far from all

mankind, at the mercy of the sea—and yet linked by radio with so
significant a piece of history: that made the moment even more
special and helped burn the memory of this fragment of water-
way ever more firmly into my mind.

Today that same piece of marine geography, spotted briefly
from an overflying aircraft, had been no more than a faraway
patch of mottled and discolored water serving inconveniently
to keep me from the timely arrival at my destination. How sad,
I thought, that so vividly remembered a place should have so
quickly transmuted itself into something little more than an in-
commoding parcel of *distance*.

In August 1941 Roosevelt and Churchill met aboard the battleship HMS
Prince of Wales *to agree on structuring the postwar world. Eventually they
signed the Atlantic Charter, which signaled a changing of the guard, with
America taking over from Britain as titular leader of the Western world.*

But wait—was that not how the world at large had come to think of the ocean as a whole? Wasn't the ocean just distance for most people these days? Didn't we all now take for granted a body of water that, so relatively recently—no more than five hundred years before, at most—was viewed by mariners who had not yet dared attempt to cross it with a mixture of awe, terror, and amazement? Had not a sea that had once seemed an impassable barrier to somewhere—to Japan? the Indies? the Spice Islands? the East?—transmuted itself with dispatch into a mere bridge of convenience to the wealth and miracles of the New World? Had our regard for this ocean not switched from the intimidation of the unknown and the frightening to the indifference with which we now greet the ordinary?

And yet had not this change taken place in some kind of inverse relation to the ocean's ever-growing importance? For hadn't the Atlantic become over the centuries much more than a mere bridge? It had surely also become a focal point, an axis, a fulcrum, around which the power and influence of the modern world has long been distributed. One might say that if the Mediterranean had long been the inland sea of the classical civilization, then the Atlantic Ocean had in time replaced it by becoming the inland sea of Western civilization. D. W. Meinig, the historical geographer, wrote in 1986 of this new perceived role of the Atlantic: the ocean, he wrote, was unique in having "the old seats of culture on the east, a great frontier for expansion on the west, and a long and integral African shore." The Atlantic existed in equipoise between the blocs of power and cultural influence that have shaped the modern world. It is an entity that links them, unites them, and in some indescribable way also defines them.

It was Walter Lippmann, in 1917, who first advanced the notion of the Atlantic Community. In a famous essay in the *New Republic*, he wrote of it as the core of "the profound web of interest which joins together the western world." And though today we recognize what this community is and whom it fully embraces (and even if we do not fully comprehend it), it is clear that despite the coming claims of India and China and Japan, it is a grouping of countries and civilizations that, for now at least, still manages to direct the principal doings of the planet.

It is a community of sorts, a kind of pan-Atlantic civilization, if you will, that at its beginning involved simply the northern countries on the Atlantic shores, with the nations of Western Europe on the one hand and the United States and Canada on the other. More recently both Latin America and the nations of western and central Africa have been incorporated into the mix. Brazil and Senegal, Guyana and Liberia, Uruguay and Mauritania are now every bit as much in and of the Atlantic Community, just as for scores of years have been the peoples of more obviously Atlantic nations such as Iceland and Greenland, Nigeria, Portugal, Ireland, France, and Britain. The community is indeed much larger and more comprehensive than that, as what follows will explain.

And yet the body of water that ties these millions of people and myriad cultures and civilizations together—the S-shaped body of water covering 33 million square miles, which in the Western Hemisphere is called the Atlantic Ocean, and which on the eastern side of the world is generally known as the Great West Sea—suffers the fate of the overlooked. It is an ocean that can fairly be described as hidden in plain sight—something that is quite obviously there, but in so many ways is just not obvious at all.

It is undeniably very visible. "Even if we hang a satellite station in space," wrote the American historian Leonard Outhwaite in 1957, when the first Sputnik was launched, "or if we reach the moon, the Atlantic Ocean will still be the center of the human world."

. . .

Not all bodies of water are so very evidently alive as the Atlantic. Some inland seas that are large, topographically important, navigationally complex, and historically crucial manage somehow to seem strangely still, starved of any readily apparent vitality. The Black Sea, for one, has the feel of a rather moribund, lifeless body of water; the Red Sea also, bathed in its ocher fog of desert sand, seems perpetually half dead; even the Coral Sea and the Sea of Japan, beautiful and placid though they may be, are somehow stripped of any true kind of oceanic liveliness and come off as strangely dulled.

But the Atlantic Ocean is surely a living thing—furiously and demonstrably so. It is an ocean that moves, impressively and ceaselessly. It generates all kinds of noise—it is forever roaring, thundering, boiling, crashing, swelling, lapping. It is easy to imagine it trying to draw breath—perhaps not so noticeably out in mid-ocean, but where it encounters land, its waters sifting up and down a gravel beach, it mimics nearly perfectly the steady inspirations and exhalations of a living creature. It crawls with symbiotic existences, too: unimaginable quantities of monsters, minute and massive alike, churn within its depths in a kind of maritime harmony, giving to the waters a feeling of vibration, a kind of suboceanic pulse. And it has a psychology. It has moods: sometimes dour and sullen, on rare occasions cunning and playful; always it is pondering and powerful.

It also has a quite predictable span of life. Geologists believe that when all is done the Atlantic Ocean will have lived for a grand total of about 370 million years. It first split open and filled with water and started to achieve properly oceanic dimensions about 190 million years ago. Currently it is enjoying a sedate and rather settled middle age, growing just a little wider each year, and with a few volcanoes sputtering away in its mid-region, but generally not having to suffer any particularly trying geological convulsions. But in due course, these will come.

Before what geologists like to think is too much longer, the Atlantic will begin to change its aspect and size very dramatically. Eventually, as the continents around it shudder and slide off in different directions, it will start to change shape, its coasts (according to the currently most favored scenario) will move inward and become welded together again, and the sea will eventually squeeze itself dry and vanish into itself. Planetary forecasters estimate this will take place in about 180 million years.

That is no mean life span. Assume for the sake of argument that the world's total existence, from the postmolten Hadean to the cool meadows of today's Holocene, encompasses some 4.6 billion years. Once tallied up, the Atlantic's 370 million years of existence as a separate body of water within that world will have made up something like 8 percent of the planet's total life. Most other oceans that have come and gone have existed for rather shorter periods: so far as other competing claims for longevity are concerned, the Atlantic will probably turn out to be one of the world's longest-lived, a potential old-timer, a highly respectable record breaker.

It is both possible and reasonable, then, to tell the Atlantic Ocean's story as biography. It is a living thing; it has a geological story of birth and expansion and evolution to its present middle-aged shape and size; and then it has a well-predicted end story of contraction, decay, and death. Distilled to its essence it is a rather simple tale to tell, a biography of a living entity with a definable beginning, a self-evident middle, and a likely end.

But then there is very much more to the tale than that. For we cannot forget the human aspect of the story.

Humans have lived around the Atlantic's peripheries and on its islands, and have crossed and recrossed it, plundered it and fought on it, seized it and surveyed it and despoiled it, and in doing so have made it quite central to our own evolving lives. That is a story, too—a story quite different from, and very much shorter than, that of the making and unmaking of the ocean itself, but one that is yet vastly more important to us as human beings.

Humans were not there when the ocean formed. We will not be there when it ceases to be. But for a definable period, poised almost in the midlife of the ocean itself, we humans arrived, we developed, and—or so we like to think—we promptly changed everything. Only by telling this second story, the kernel within the main shell of the first, can we recount in full the life of the Atlantic Ocean. The physical ocean's history of opening and closing then becomes the context, the frame, for the history of humanity's intimate involvement with and within it.

That human story began when man first settled on the Atlantic's shores. As it happens, mankind spilled down to the sea most probably in southern Africa, and he did so quite possibly (and most fortuitously for this account) very close to Africa's southern Atlantic shores. What follows from that moment is every bit as complicated and multidimensional as one might imagine: the human story of the ocean swiftly becomes a saga of a mélange of peoples and parallels, of diverging languages and customs, of mixtures of acts and events, achievements and discoveries, of confusions and contests. It is a tricky tale to tell. Simple chronology might suit very well the story of the making of the physical sea itself—but the details of the human experience are scarcely so amenable.

For how would it be possible to knit together the experience of, say, a Liberian fisherman with that of an atomic submariner on patrol off Iceland? Or to link the life of an amethyst miner on the shores of Namibia with that of the American director of *Man of Aran*; to write of the captain of a British Airways Boeing and of an ice-patrol ship off the coast of South Georgia; or to connect the long-dead sea-painter Winslow Homer with a wide-eyed Guantánamo detainee from western China, swimming for the first time in the Atlantic Ocean off Bermuda? How best create a sensible structure from all this strange and multicolored variety?

For a long time that remained a puzzle. I wanted so much to write the story of the ocean. But what and where was the structure? I was, as they say, *all at sea.*

Except that one day, gazing down at the rolling waters, I thought: if the ocean had a life, might not mankind's relationship with it have some kind of a life about it also? After all, fossils and finds from digs show that this relationship had a particular moment of birth. It will have a likely moment of death, as well—even the most determined optimist will have to admit that an end to human existence is in sight, that in a few thousand, or maybe a few tens of thousands of years, humanity will be finished, and this aspect of the story will be over, too.

So yes, to corral the life of this human relationship with the sea, and place it within the context of the much more straightforward life of the ocean itself—this might indeed be possible as biography, too. But then there were the details, churning and daunting and devilish. The tide of human history was so filled with facts and incidents and characters and tones of subtle shading that it might be near impossible to swim against it.

But in the end, and out of the blue, I was tossed a quite unexpected lifebelt—and by that most nonmaritime of rescuers, William Shakespeare.

. . .

For many years I had carried with me on tedious plane journeys (and indeed had with me as we passed above the waters of Flemish Cap that recent time) a well-thumbed copy of *Seven Ages*, an anthology of poetry that was assembled in the early 1990s by a former British foreign secretary, David Owen. He had arranged his chosen poems in seven discrete sections, to illustrate each of the seven stages of man's life that are listed so famously in the "All the world's a stage . . ." speech in *As You Like It*. And I was reading Owen's book one day when I realized that this very same structure also happened to offer me just what I needed for this human aspect of the Atlantic story: a proper framework for the book I planned to write, a stage setting that would transmute all the themes of ocean life into players, progressing from infancy to senescence, so that all could be permitted to play their parts in turn.

The Ages are those we remember, if scantily, from childhood, and are listed in Jacques's all too famously gloomy monologue:

At first the infant,
Mewling and puking in the nurse's arms;
And then the whining school-boy, with his satchel,
And shining morning face, creeping like snail
Unwillingly to school. And then the lover,
Sighing like furnace, with a woeful ballad
Made to his mistress' eyebrow. Then a soldier,
Full of strange oaths, and bearded like the pard,
Jealous in honour, sudden and quick in quarrel,
Seeking the bubble reputation
Even in the cannon's mouth. And then the justice,
In fair round belly with good capon lin'd,
With eyes severe and beard of formal cut,
Full of wise saws and modern instances;
And so he plays his part. The sixth age shifts
Into the lean and slipper'd pantaloon,

With spectacles on nose and pouch on side;
His youthful hose, well sav'd, a world too wide
For his shrunk shank; and his big manly voice,
Turning again toward childish treble, pipes
And whistles in his sound. Last scene of all,
That ends this strange eventful history,
Is second childishness and mere oblivion;
Sans teeth, sans eyes, sans taste, sans everything.

Infant; School-boy; Lover; Soldier; Justice; Slipper'd Panta-loon; and Second Childishness. It seemed, all of a sudden, just about the ideal. Pinioned within these seven categories, the stages of our relationship with the ocean could be made quite manageable.

I could examine in the First Age, for example, the stirrings of humankind's initial childlike interest in the sea. In the Second, I could examine how that initial curiosity evolved into the schol-arly disciplines, of exploration, education, and learning—and in this as in all the other Ages I could explore the history of that learning, so that each Age would become a chronology in and of itself. I could then become captivated in the Third Age—that of the lover—by the story of humankind's love affairs, by way of the art, poetry, architecture, or prose that this sea has inspired over the centuries.

In the Fourth Age—that of the soldier—I could tell of the argu-ments and conflicts that have so often roiled the ocean, of how the force of arms over the years has compelled migration or fos-tered seaborne crime, of how national navies have reacted, how individual battles have been fought, and how Atlantic heroes have been born.

In the Fifth Age—that of the well-fed Justice—I could describe how the sea eventually became a sea of laws and commerce, and how tramp steamers and liners and submarine cables and jet aircraft then crossed and recrossed it in an infinite patchwork designed for the attainment of profit and comfort. In the Sixth Age, that dominated by the fatigue and tedium of the pantaloon, I could reflect upon the ways that man has recently wearied of the great sea, has come to take it for granted, to become careless of its special needs and to deal with it improvidently. And in the Seventh and final Age—the Age that ends with Shakespeare's immemorial *sans teeth, sans eyes, sans taste, sans everything*—I could imagine the ways by which this much-overlooked and perhaps vengeful ocean might one day strike back, reverting to type, reverting to the primal nature of what it always was.

Alluring as this all might seem, however, there was something else. First I had to make the frame, to construct the proscenium arch, to attempt to place the long human drama within the very much longer physical context. Only when that had been achieved, and by leave of the enormous natural forces that had made the ocean in the first place, could I try to begin to unveil and recount the human stories. Only then could I attempt to tell something of the ocean's hundreds of millions of years of life, and of the scores of thousands of its middle years during which the men and women who made up its community would eventually go out onstage, and by their own lights, each perform their unique, and uniquely Atlantic, roles.

First—just how was the ocean made? How did it all begin?

PROLOGUE:
THE BEGINNINGS OF
ITS GOINGS ON

All the world's a stage
And all the men and women merely players:
They have their exits and their entrances;
And one man in his time plays many parts,
His acts being seven ages.

A big ocean—and the Atlantic is a very big ocean indeed—has the appearance of a settled permanence. Stand anywhere beside it, and stare across its swells toward the distant horizon, and you are swiftly lulled into the belief that it has been there forever. All who like the sea—and surely there can be precious few who do not—have a favored place in which to stand and stare: for me it has long been the Faroe Islands, up in the far north Atlantic, where all is cold and wet and bleak. In its own challenging way, it is entirely beautiful.

Eighteen islands, each one a sliver of black basalt frosted with gale-blown salt grass and tilted up alarmingly from east to west, make up this Atlantic outpost of the Kingdom of Denmark. Fifty-odd thousand Faroese fishermen and sheep farmers cling

there in ancient and determined remoteness, like the Vikings
from whom they descend and whose vestiges of language they
still speak. Rain, wind, and fog mark out these islanders' days—
although from time to time, and on almost every afternoon in
high summer, the mists suddenly swirl away and are replaced by
a sky of a clarity and blue brilliance that seems to be known only
in the world's high latitudes.

It was on just a day like this that I chose to sail, across a lumpy
and capricious sea, to the westernmost member of the archipelago, the island of Mykines. It is an island much favored by
artists, who come for its wild solitude and its total subordination
to the nature that so entirely surrounds it. And going there left
a deep impression: in all my wanderings around the Atlantic, I
can think of no place that ever gave me so great an impression of
perching *on the world's edge,* no better place to absorb and begin
to comprehend the awful majesty of this enormous ocean.

*The westernmost of the eighteen Faroe Islands, Mykines rises abruptly
from the Atlantic Ocean, buffeted by wind and waves, or else socked in
by thick fogs for most of the year. Puffins, whales, and sheep—the word
faroe is Viking for sheep—support a total Faroese population of fewer
than 50,000 people, all of whom are citizens of Denmark.*

The landing on Mykines was exceptionally tricky. The boat surfed in on the green breaking top of an ocean roller into the tiny harbor, its skipper tying up for just enough time to let me clamber out onto a cement quay lethal with slippery eelgrass. A staircase of rough stones rose up to the skyline, and I scrambled upward, only too well aware of the deep chasm filled with boiling surf far below beside me. But I made it. Up on top there were a scattering of houses, a church, a shop, and a tiny inn, its sitting room heavy with the smell of pipe smoke and warm wet sweater wool. A sudden furious blast of wind had driven away the morning fog, and the sun revealed a long steep slope of grass that stretched right up the island tilt, clear up to the western sky.

There was a grassy pathway leading up to this high horizon, and a skein of islanders was moving slowly up it, like a line of ants. I joined them, out of curiosity. To my great surprise most were dressed in Faroese finery—the men in dark blue and scarlet jackets, with high necks and rows of silver buttons, knee breeches and silver-buckled shoes; the women in wide-striped long skirts, blue vests fastened with an elaborate cat's cradle of chains, and fringed scarves. And though a few men had anoraks with folded felt snoods, none wore hats: the incessant wind would have whipped them away. The children, dressed just as their parents, whooped and skirled and slid on the wet grass, their elders tutting them to keep their boots clean and to be careful not to fall.

It took thirty minutes to do the climb, and none of the islanders seemed to break a sweat. They all gathered at a site by the cliff top, where the grass was flattened. There was a memorial stone here, a basalt cross incised with the names, I was told, of the fishermen who had died in the Icelandic fishing grounds off to the west. The crowd, perhaps a hundred in all, arranged themselves beside the summit marker, a cairn of basalt boulders, waiting.

After a few minutes a white-haired man of sixty or so, puffing a little from exercise, appeared at the top of the path. He was dressed in a long black surplice with a ruffled high collar that made him look like he had stepped from the pages of a medieval chapbook. He was a Lutheran pastor, from the Faroese capital town of Thorshavn. He proceeded to lead a service, helped by two churchwardens who played accordions and one island lad with a guitar. A pair of pretty young blond children handed around some damp hymn sheets, and the villagers' high voices set to singing old Norse holy songs, the thin music instantly swept away to sea by the gale, as it was designed to be.

The islanders said the small religious ceremony was quite without precedent: in the past it had always been a visiting pastor from Denmark, a thousand miles south, who would come here to bless the islands' long-drowned sailors; but today made history, it was explained, because for the first time ever the minister was Faroese. In its own gentle and respectful way the dedication service, with its prayers offered in the local tongue, offered an indication of just how these remote mid-ocean islands had drawn themselves steadily away from the benign invigilation of their European motherland. They had gone their own way at last: an *island* way, remarked one of the congregants. *An Atlantic way.*

After the service was finally over, I strolled behind the dispersing crowd—and without warning suddenly and terrifyingly reached the cliff edge. The grass cut off as with a blade, and in its place there was just a huge hollow emptiness of wind and space, the black wet walls of a hurtling precipice of basalt cliffs with, crawling almost half a mile below, the tides and currents and spume of the open sea. Hundreds of puffins stood in nooks in the cliff edge, some no more than an arm's length away, and all quite careless of my presence. They looked like ridiculous, stubby creatures, with that mask-face, chubby cheeks, and a colored bill that was often crammed full with a clutch of tiny fish. But every so often one took to the air and soared off into the sky with an easy and contented grace, ridiculous no more.

I must have sat at the edge for a long, long time, staring, gazing, mesmerized. The gale had finally stopped its roaring, and the sun had come out and was edging its way into the afternoon. I was sitting on the cliff edge, my legs dangling over half a mile of emptiness. I was facing due west. Just below me were clouds of seabirds, the gannets and fulmars, kittiwakes and storm petrels, and beside me were the chattering congregations of puffins. Ahead of me there was just nothing—just an endless crawling sea, hammered like copper in the warm sunshine and stretching far, fifty miles, a hundred—from up this high I felt I could have been looking out on five hundred miles and more. There was an endless vacancy that at this latitude, 62 degrees north or so, I knew would be interrupted only by the basalt cliffs of Greenland, more than a thousand miles away. There were no ships' wakes on the sea, no aircraft trails on the sky—just the cool incessant wind, the cries of the birds, and the imagined edge of the known world set down somewhere, far beyond my range of sight.

And it is very much the same on any Atlantic headland, whether in Africa or the Americas, in the Arctic or from the dozens of other oceanic islands like these, places from where the views are limitless, the horizons finely curved with distance. The view is enough to give the viewer pause: it is just so stupefying, so haunting, the impressions welling up, one after another.

How eternal the ocean appears, and how immense. It is anything but trite to keep reminding oneself how incalculably large the Atlantic happens to be. The big seas are so big that after just a little contemplation of this ocean you understand why it was once perfectly fitting of someone—in this case Arthur C. Clarke, who knew a thing or two about immensity—to remark on *how inappropriate it is to call this planet Earth, when clearly it is Sea.*

Then again, above all the dominant color of this ocean is gray. It is gray, and it is slow-moving, and it is heavy with a steady heaving. The Atlantic is in most places not at all like the Pacific or the Indian oceans—it is not dominated by the color blue, nor is it overwhelmingly fringed with leaning palm trees and coral reefs. It is a gray and heaving sea, not infrequently storm-bound, ponderous with swells, a sea that in the mind's eye is thick with trawlers lurching, bows up, then crashing down through great white curtains of spume, tankers wallowing across the swells, its weather so often on the verge of gales, and all the while its waters moving with an air of settled purpose, simultaneously displaying incalculable power and inspiring by this display perpetual admiration, respect, caution, and fear.

The Atlantic is the classic ocean of our imaginings, an industrial ocean of cold and iron and salt, a purposeful ocean of sea-lanes and docksides and fisheries, an ocean alive with squadrons of steadily moving ships above, with unimaginable volumes of mysterious marine abundance below. It is also an entity that seems to be somehow *interminable.* Year in and year out, night and day, warm and cold, century after century, the ocean is always there, an eternal presence in the collective minds of those who live beside it. Derek Walcott, the Nobel laureate poet, wrote in his famous epic work *Omeros* of his fisherman-hero Achilles walking finally and wearily up the shingled slope of an Atlantic beach. He has turned his back on the sea at last, but he knows that even without his seeing it, it is behind him all the while and simply, ponderously, magnificently, ominously, continuing to be the sea. The Ocean is, quite simply, "still going on."

Three thousand years ago Homer introduced the poetic idea of *Oceanus*—the son of Uranus and of Gaia, the husband of Tethys and father of a score of river gods. The word itself signified a vast globe-encircling river, which the ancients imagined to be rimmed by both the Elysian fields and Hades. To Homer the ocean was a river that rose far away where the sun sets. It was something totally daunting for Mediterranean sailors who spied its great grayness crashing and storming outside the Pillars of Hercules, at the Strait of Gibraltar. It was known as the Great Outer Sea, and it was a thing hugely to be feared, a world of crashing waters inhabited by terror-inspiring monsters like Gorgons and Hecatonchires, or by bizarrely unfamiliar humans like Cimmerians, Ethiopians, and pygmies. And forever, *always going on.*

This poetic notion of the sea's ceaseless activity is one that manages to be at once familiar, comforting, and mildly unsettling. One has a sense that the sea, whatever else it may be, however gray or immense or distempered or powerful, is a permanent presence in the world, whether it is rumbling or calm, storming or drowning. We think of it as an immutable living being, ceaselessly occupied in its unfinishable business of washing and waiting.

Yet strictly speaking, this is hardly true at all. Oceans have their beginnings and their endings, too. Not in the human imagination, perhaps, but in a physical sense, most certainly. Oceans are born, and oceans die. And the Atlantic, the once much feared Great Outer Sea, the most carefully studied and considered of them all, was not always there, and it will not remain either where it is or what it is.

For an ocean to begin, a planet must have two elemental essentials. One is water. The other is land. The enormous tonnage of water* that presently exists has not always been there, of course—but recent research suggests that it came into existence fairly soon after the earth was first coalesced out of clouds of space-borne planetesimals, almost five billion years ago. Studies of zircon crystals found near an iron ore mine in Western Australia indicate that liquid water was on earth just a few hundred million years after the planet was formed. It was extremely hot water, and it had all manner of noxious and corrosive dissolved gases in it; but it was liquid, it sloshed about, it could (and did) erode things that it poured over, and most important of all, it was the undeniable aqueous ancestor material of all of our present seas.

The ocean I gazed down on from the puffin cliffs of Mykines is in essence the selfsame water that was created all those years ago; the principal difference is that while the Hadean sea was hot and acid and incapable of supporting anything but the most primitive of thermophilic cyanobacteria, the Faroese Sea was cold and clean, had been purified and well salted by millions of years of evaporation and condensation and recycling, was rich in chemical ions from all over, and was vibrant with life of great complexity and beauty. In all other respects the frigid waters off the North Atlantic islands and the steaming acid waters of our early and territorially undifferentiated planet of long ago were more or less the same.

Territorially undifferentiated though that early planet may have been, it would not remain so for long. Solid, habitable earth was being manufactured in the cooling planet at about the same time, too.

At first this land was represented by little more than the ap-

---

*   The water weighs 1.3 billion billion tons, give or take—on a planet that is calculated to weigh 6,000 billion billion tons total.

pearance of countless huge supervolcanoes, each separated from the other so that their clusterings might have looked from the air like the chimneys of a planet-sized industrial complex, giant marine mountains that belched out choking clouds of smoke and spewed thousand-mile-long puddles of thick black lava. Eventually these isolated volcanoes managed to vomit out so much new rock that they started to coalesce, and some of these coagulating masses became more or less stable, such that they could be thought of in aggregate as *landmasses*. Some long while later, these landmasses formed into even larger bodies of land that could fairly be described as *protocontinents*. And thus did the defining present-day characteristic of our planet—an entity formed of continents and seas—have its beginnings, although the process of reaching a configuration that looked anything like today's world was infinitely slow and involved a fantastic complexity. The making and unmaking of a multidimensional topography is only now beginning to be understood.

The earth in its early days may have been both water and land, but it was a scalding and wretched place. It spun on its axis much more rapidly than today: once every five hours the sun would rise, though had any inhabitants been around they would not likely have seen it through the vast clouds of ash and smoke and fire and noxious gas. If the skies ever cleared, the planet below would have been scourged by unfiltered pulses of ultraviolet radiation and gamma rays, making the surface hostile to almost everything. And the newly made moon was still so close that each time it swept around in orbit, it raised great acid tides that would inundate and further corrode such continents as existed.

But some continents most certainly did exist. Today's geological record contains the relict remains of half a dozen or so identifiable former bodies mighty enough to be as continents. Their remains have been dispersed by billions of years of planetary restlessness: no longer is any one of these early bodies intact. All that is left is a collection of stratal shards and sunderings that can be dated from at least three billion years ago, and which are now scattered to places as otherwise unconnected as present-day Australia (where parts of this earliest of continents are to be found) and Madagascar, Sri Lanka, South Africa, Antarctica, and India.

The detective work needed to piece together the original continents is prodigiously difficult. Yet it has become possible, by looking carefully at the ages and structures of such rocks, to come up with at least an approximate sequence of events that led to the formation of today's Atlantic Ocean and the continents that now border it.

It is a sequence featuring the dozen or so continents and seas that have come into existence, briefly or for aeons, over the planet's life. The lineage commences with the arrival of the world's first continental body: a mighty, two-thousand-mile long landmass shaped much like the silhouette of a monstrous albatross, which formed itself and hoisted itself above the boiling seas some three billion years ago. Today's geological community has given it a suitably sonorous and memorable name: it is known, in honor of the Chaldean birthplace of Abraham, as the supercontinent of Ur.

The remains of other ancient continents have been discovered since the finding of Ur, and they have been given names reflecting either the national pride of those living where they lie, the classical education of the explorers who discovered them, or the realities of modern global politics. They are names mostly unfamiliar beyond the sodalities of geology: Vaalbara, Kenorland, Arctica, Nena, Baltica, Rodinia, Pannotia, Laurentia. They are names that define bodies either as small as present-day Greenland, or as immense as present-day Asia. They were bodies constantly in motion, constantly changing their shape, topography, and position.

Over immense stretches of time, during periods of scourging heat and colossal physical forces, they all shifted themselves slowly and in stately fashion around the surface skin of the planet. Sometime they collided with one another, creating what are now ancient and much-flattened mountain chains. More often than not, they broke apart in a series of slow-motion explosions, events that took millions of years to play out. The shards of their ruin then banged and ricocheted their way around the earth, reordering themselves and occasionally recombining with one another, as though the planet's surface were covered with the pieces of some enormous jigsaw puzzle that was being operated by an unseen and none-too-bright giant. And all the while, the spaces between the continental bodies were filled with the seas—being constantly shape-shifted and divided up and redivided and configured into bodies of water that were each recognizable, from about one billion years ago, as true and proper oceans.

By Cambrian times, some 540 million years ago, one of these oceans was starting to have a familiar look to it. When it first appeared, its shape was inconsequential—it was merely very big. But during the Ordovician period, it started to become fairly narrow, vaguely sinuous, no more than a thousand miles wide, like a great river coursing across the world from northeast to southwest. That is to say—*it was in appearance not altogether unlike the North Atlantic to be.*

And because it washed the shores of what would in time become the east of North America and the northwest of Europe, so this supposed Ordovician sea was given the name that it should by rights bear. It was called Iapetus, for the mythical figure known by the ancient Greeks as the father of Atlas. The Iapetus Ocean, long since dry, and now seen at its spectacular best in the sandstones and deepwater gray limestones in northern Newfoundland that memorialize its existence, was the precursor, the father or mother, of the true and eventual Atlantic Ocean.

. . .

The modern and recognizable world began to come about some 250 million years later—250 million years ago, indeed—during the end of the Permian and the beginning of the Triassic eras. It was a process that got under way when four of the original protocontinental jigsaw pieces collided and formed themselves into the one supercontinent that has since managed to achieve wide familiarity: the great body known as Pangaea. This vast entity contained every piece of Permian real estate that then existed on the globe. Its name alone says this was one land that comprised all of the world's land, and it was surrounded by one sea—Panthalassa—that was all of the world's sea.

Out of these two bodies—one water, the other land—today's Atlantic Ocean would be made. The process began with a long era of spectacular volcanic violence, one of the planet's most violent episodes in its entire recent history. Soon thereafter there was a mass extinction of life-forms, both at sea and on the land; and then finally Pangaea started to break apart and the new ocean started to form. The extent to which these three events were connected has been debated at length—especially over whether the vigorous volcanic activity caused both the extinction and the breakup—but these events did occur, and within relatively short order.

The volcanic period was so comprehensively and terrifyingly violent, so generous in its extent and so profound in its consequences that it must have felt as though the entire world were ripping itself apart. A gigantic series of explosions started to cannonade around the central core of Pangaea. Thousands of mighty volcanoes, first thousands of Heclas, and then in time thousands of Krakatoas, or Etnas or Strombolis or Popocatépetls, pushed themselves up and out of the countryside and started to spew fire and magma thousands of feet into the air. A ceaseless round of unbearably huge earthquakes began to shake

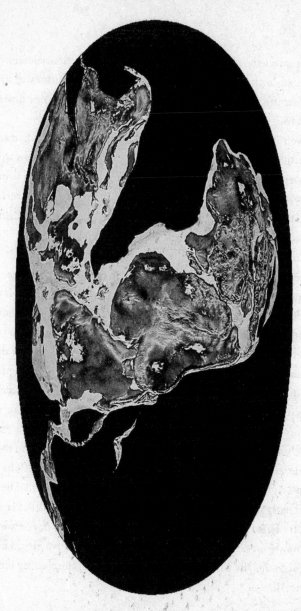

Some 195 million years ago, Pangaea began to break up, and the first tonguelike extension of the Panthalassan Ocean (center) began to seep into the narrow but widening gap between America and Europe, and in time between Africa and South America too. The Atlantic was being born: it would exist for 440 million more years.

and shatter the planet, trending along a roughly delineated line that ran for hundreds of miles northward and southward, and splintering and smashing the earth for scores of miles downward into the crust.

Even if the immense universal continent of Pangaea had not yet broken up, it certainly had started to weaken and groan with the weight and weariness of its own long existence. The world was witnessing the beginnings of a brief and yet merciless series of spasms of tectonic mayhem that started tearing the world's one stretch of land into pieces, from end to end.

And water began to seep into the growing gap between the two halves of Pangaea that were beginning to form. The tiny weasel-tongue of water that laid down sediments that are found in today's Greece turned into an almighty spigot: trillions upon trillions of tons of seawater started to rush inward from it and from the feeder-waters of the surrounding Panthalassan Ocean. In doing so—by beginning the process of prying apart, levering open, wielding a tectonic crowbar—this potent combination of volcanoes, earthquakes, and lots and lots of water started the making of a brand-new ocean. It opened up only a crack, like a door cautiously ajar: but this was a process that would continue, and then accelerate and proceed without let-up, for scores of millions of years, right up to the present day. The resulting ocean had been paternally prefigured by the Iapetus two hundred million years before. This tiny filigree of seawater that was fast rising between the newly made volcanic cliffs of what are now Nova Scotia and Morocco was the first small-scale indication of the coming birth of the Atlantic.

. . .

The volcanoes lasted for only a few score thousand years (though some say as much as two million), but their pulses were so violent and the amount of magma they disgorged was so prodigious that the cliffs and mountain ranges that today stand as memorial are awesomely impressive.

I took a family vacation in 1975 on the Canadian island of Grand Manan in New Brunswick, a short distance from where Roosevelt took his summer's ease on Campobello. We spent happy afternoons investigating the tide pools at Southwest Head, a high cape from where only the Atlantic could be seen, misty and cold, endlessly stretching to the south. Afterward we walked home to watch the huge Fundy tides at Seal Cove, and on the way passed by a curious assortment of pure white boulders that sat incongruously at the top of a cliff composed of sheer columns of a dark brown rock. The boulders, deposited by glaciers, were called the Flock of Sheep. But it was the brown rock below them, a columnar basalt, that has most intrigued geologists— ever since, in the late 1980s, it was realized that they were quite similar in appearance and probable age to another huge pile of basalts, in a mountain range in Morocco.

I went to these mountains, the High Atlas, when I was researching a different aspect of this book. I had no idea then of their connection with the Grand Manan rocks, nor did I know until I started to ask around. For although Morocco is known for its Paleozoic as well as its Jurassic and Cretaceous fossils, the Atlas mountains have large outcrops of basalt, too—layers of volcanic rocks sandwiched between the sedimentary rocks, which, it was realized by researchers in 1988, were of exactly the same age as the rocks in places like Grand Manan, in eastern Canada. This discovery, which I was told about while sitting sunning myself in a rooftop bar in the coastal town of Essaouira, led geologists on a huge Easter egg hunt around other Atlantic coastal countries for more basalts of the same antiquity. A series of expeditions in the 1990s found scores of outcrops—sills, dikes, flood basalt sequences—all in enormous abundance, which showed almost certainly just what had been going on a little over two hundred million years ago.

The outcrops were all over—four million square miles of lavas, covering parts of what in time would become four continents: in North America they ranged along the Appalachians from Alabama to Maine, and then well beyond up into Canada and along the shores of the Bay of Fundy; in South America they were found in Guyana, Surinam, French Guiana, and, most impressively, throughout the Amazon basin of Brazil; in southern Europe they were detected in France; and in Africa there were swarms of sills and dikes found not only in Morocco but in Algeria, Mauritania, Guinea, and Liberia. And all these puzzle pieces had alignments and ages and proximities that positively shouted their intimate geological connections and their probable common origin.

The average age of their deposition eventually came in with some accuracy: most of the basalts had been laid down or extruded or blown into the sky 201.27 million years ago, a figure computed with an error either way of only perhaps three hundred thousand years. Some discrepancy exists between the age of the basalts on what would be the eastern side of the region—in North Africa, especially—and those in what would become North America: the American basalts seem older. This discrepancy has led to an impassioned debate over whether the volcanoes led to the extinction of so much of the flora and fauna, since that massive wiping-out—when huge numbers of amphibian species vanished, leaving environmental niches perfectly suited for the arrival of scores of Jurassic dinosaur types—occurred around 199.6 million years ago. Would volcanoes, however almighty, have their principal biological effect almost *two million years* later? It seems a little improbable—but some laboratories are still trying to link the two events, not least because it makes for a more dramatic, and anthropomorphically comprehensible, story.

The great continent unzipped, though not like a fly on a pair of pants. It was an inelegant, jerky process, rather like watching a camel getting to its feet, with one part of the ocean opening, then another far away, then a portion of the middle, then another section in the distance, and then back to the middle again. The first waves of water washed the shores of eastern Canada and northwest Africa as they pushed apart from each other, almost at the very beginning of the Jurassic, 195 million years ago. This was the first true moment of the Atlantic Ocean's life.

Twenty million years on, the process of sea-floor spreading got under way in earnest, in the middle of the sea—like two unrolling carpets, or two unspooling conveyor belts running away from each other from a vague submarine midpoint. The bottom of the sea started to split open, and its two halves began to diverge, the continents on either side shifting steadily apart. West Africa shifted itself about three hundred miles away from South Carolina; Mali moved a couple of hundred miles off Florida; there was a large stretch of wide-open ocean around where the Windward Islands would eventually be, and then a gap of almost a thousand miles opened between Liberia and Venezuela. In this midsection a body of seawater as large as today's Mediterranean was created, and yet unlike the rather stable-sized Mediterranean, this body only continued to get bigger.

By 150 million years, continuing a Canaveral-style countdown, Greenland* had begun to pull away from Norway, and

---

* In 1965 I was part of an expedition to determine, by measuring fossil magnetism in basalts collected from nunataks high on the East Greenland ice cap, how much the island had drifted in the fifty million years since the rocks had been laid down. We found that Greenland had drifted about 15 degrees westward—an impeccable example of the kind of movement confirming the tectonic plate theory just then being advanced.

Iceland began to be built up from deep down in the sea. (The spectacular eruption that began in the spring of 2010 from Eyjafjoll, an Icelandic volcano that had been quiet for the previous two centuries, and which disrupted air traffic across northern Europe with its immense swaths of high-altitude volcanic dust, is part of the process of building up. Surtsey, an entirely new island born just a few miles away in 1963, may have provided somewhat clearer evidence of the steady swelling of Iceland, but Eyjafjoll produced much more lava, even if most of it was blasted high into the sky.)

At the same time the shallow waters off the northern parts of the British Isles had deepened, and serious wave-tossed oceanic expanses now separated Ireland from Labrador. By ten million years later, Guinea, the Gambia, Senegal, and Sierra Leone had pulled relentlessly away from the coastlines of the putative Guyana, Surinam, and French Guiana, which would occupy a similar dependent position in South America. Hitherto they were in the same place: five hundred miles of ocean now separated them.

By the early Cretaceous, 120 million years back in time, the conveyor-belt-unrolling-carpet mechanism that was now evidently driving the entire process—for there was to be no further dramatic volcanism to complicate matters—had an apparent source: the Mid-Atlantic Ridge had been formed. This linear bulge in the seabed, its center fissured and faulted and alive with submarine volcanoes, would play a vital role for the rest of the ocean's history. It was the place where new crustal material would be belched out of the inner earth, where the ocean floor to the east and west of it would spread out and away, and where islands— a long string of them, the Azores, the Canaries, St. Helena, Fernando do Noronha, Tristan da Cunha, a jagged line stretching from Jan Mayen in the far north to Bouvet Island, 9,200 sea-miles away to the south*—occasionally poked their peaks above water level, only to be pushed away in their turn to end up, remote and mostly unpeopled, in the new ocean's farther reaches.

And still the opening went on. Fifty million years more, and the north and middle portions now began to create and separate the southern coasts of Africa and South America. There was at first another sudden outbreak of volcanic activity—floodplains of basalt poured from numberless vents. But then separation began down here, too, though it is still not clear if this was connected with the volcanic spasm. And here the process did indeed look like the unzipping of a fly, and it was accomplished with similar speed. It was an opening up that rippled southward, one coastline following hard on another. Nigeria stripped itself away from Brazil. The valleys that would one day house the Congo on one side and the Amazon on the other snapped apart. The flood basalts of the southern edge of Pangaea separated into two: on one side the enormous Etendeka Traps, which would come to lie in southern Africa—and over the edge of which the Victoria Falls now cascade—and on the other the Paraná basalts of Argentina, currently home to the sprawling spray curtains of the falls the Guarani called *big waters*, the Iguazu.

And then in a final protracted frenzy of tearing, all of eastern Patagonia wrenched itself away from Angola, and the flatlands that were then off Cape Horn freed themselves from their geological embrace of what is now Namibia and the South African cape, and swept away to become the foothills of the southern Andes.

This was all accomplished at a remarkable speed, for though

---

* Both islands are Norwegian possessions, giving Norway a unique perspective on the ridge from its ownership of both ends. Jan Mayen, fogbound and miserable, has an airstrip and a manned weather station; Bouvet, a jumble of cliffs and Southern Ocean ice, had its weather station destroyed by an avalanche, is uninhabited, and is classed as the most remote island in the world.

in the north matters unfolded in a somewhat leisurely fashion, down south they raced almost breathlessly. The Atlantic coastlines that had once been welded together between the bulge of Brazil and the armpit of Africa—the apparently natural fit that led nineteenth-century figures like Alfred Wegener to think out loud that continents might once have moved apart, thoughts that condemned him to live in near-universal and near-perpetual ridicule—had managed in a scant forty million years to spring five thousand miles apart from one another. The sea in these parts must have opened up at rather more than four inches a year—infinitely more rapidly than the separation that took place up in the brisk waters of the North Atlantic, and more than three times the rate at which the ocean continues to spread wider today.

And that movement has never ended. The outline of the Atlantic Ocean that we know today was fixed perhaps ten million years ago, and though to us and our cartographers it appears to have retained its boundaries, its coastlines, and its "look" ever since the days of Columbus and Vespucci and the great German map of Martin Waldseemüller that first defined it, it has been changing, subtly and slightly, all the time. Coastlines in the east continue to advance, those in the west to retreat. *Things fall apart: the center cannot hold.* The Mid-Atlantic Ridge continues to disgorge untold tonnages of new ocean floor; some of it appears above the water's surface and creates new islands and reefs. And the islands that do exist continue to move, slowly and slightly, away from the sea's center.

By ten million years ago the great split was done, and the Atlantic was fully born. At some time in the distant future—but not the unknown future, as we shall see—the rocks that opened will close and the sea will be forced to go elsewhere, and it will find another home. The vast earth-ocean, with its essentially and eternally constant volume of seawater, will be obliged by continental movement to reconfigure itself, and in time other shapes and sizes of its constituent water bodies will appear. The Atlantic that was born will in due course also die.

But that will not be for a very long while. In the meantime, the Atlantic Ocean, *Mare Atlanticus*, the Great West Sea, is like an enormous stage set. It was ten million years ago just as it is today: a sinuous snakelike river of an ocean, stretching thousands of miles from the Stygian fogs of the north to the Roaring Forties in the south, riven with deeps in its western chasms, dangerous with shallows in eastern plains, a place of cod and flying fish, of basking sharks and blue-finned tuna, of gyres of Sargasso weed and gyres of unborn hurricanes, a place of icebergs and tides, whirlpools and sandbanks, submarine canyons and deep-sea black smokers and ridges and seamounts, of capes and rises and fracture zones, of currents hot, cold, torrential, and languorous, of underwater volcanoes and earthquakes, of stromatolites and cyanobacteria and horseshoe crabs, of seabird colonies, of penguins and polar bears and manta rays, of giant squid and jellyfish and their slow-and-steady southern majesties, the great and glorious wandering albatrosses.

The stage, now so amply furnished with all this magic and mystery, has been prepared for a very long while. The supporting cast of players, all the beasts and plants, have now mostly made their entrances. The Atlantic Ocean is open wide, its physical condition fully set, and all is ready for the appearance on stage of the creature that will give full force to the human idea of the great sea.

For what promises or threatens to be in relative time just the briefest moment only, the central character is set to step into the light. Mankind is finally about to confront the gray-heaving reality of all these mighty waters. To see at last, just what is going on.

Chapter One

FROM THE PURPLE ISLES OF MOGADOR

At first the infant
Mewling and puking in the nurse's arms

1. ALLUREMENTS

The Kingdom of Morocco has on its most widely used currency bill neither a camel nor a minaret nor a Touareg in desert blue, but the representation of the shell of a very large snail. The shell of this shore-living marine beast—a carnivore that uses its tongue to rasp holes in other creatures' shells and sip out the goodness—is reddish brown, slender, and spiny, with a long spire and an earlike opening. It is in all ways rather beautiful, the kind of shell not to be idly thrown away by anyone lucky enough to find one.

Yet it was not the shell's curvilinear elegance that many years ago persuaded directors of Morocco's Central Bank in Rabat to place its image on the back of its 200-dirham bills. The reason for their choice, more befitting a currency note, had all to do with money and profit. For it was this curious sea creature that formed the basis of Morocco's fortunes, long before Morocco was an organized nation.

The shell of Haustellum brandaris, *or dye murex, on the 200-dirham Moroccan banknote underscores its importance to the North African economy three thousand years ago. Phoenician traders harvested the shells from the Atlantic coast and from the mollusk's hypobranchial gland extracted purple dye, which they sold at ports around the Mediterranean.*

The Berbers of the desert were not mariners, nor were they especially interested in harvesting these snails and putting them to good use. It fell instead to sailors from far away, who came thousands of miles from the Levantine coast of the eastern Mediterranean, to realize the potential for using these gastropods to make a fortune. The great challenge would come in the business of acquiring them.

For the sea in which these elegantly fashioned beasts lived so abundantly was a body of water very different in character from the placid waters of the Mediterranean. The gastropods were generally to be found, thanks to complex reasons of biology and evolutionary magic, clinging stubbornly to the rocks and reefs in the great and terrifying unknown of the Atlantic Ocean, well outside the known maritime world, and in a place where traditional navigational skills, honed in the Mediterranean, were unlikely to be of much value. To harvest these creatures, any mariners of sufficient boldness and foolhardiness would be obliged to grit their collective teeth and venture out into the deep waters of the greatest body of water it was then possible to imagine.

But they did so, in the seventh century B.C. They did so by blithely setting sail past the Pillars of Hercules, the exit gates of their own comfortable sea, and out into the great gray immensity of the limitless unknown beyond. The sailors who performed this remarkable feat, and yet did so so casually, were Phoenicians; they used sailing ships that were built initially only to withstand the waves of their familiar inland short sea but now had to tackle the much more daunting water of an unknown long sea beyond. There must have been something remarkable about the sailors, yes; but there must also have been something even more so about these particular North African snails that made them worth so much risk.

And indeed there was. First, however, it seems appropriate to set the snails to one side and consider the lengthy and necessarily complex human journey that brought the Phoenicians to Morocco in the first place.

2. ORIGINS

Early man's march down to the ocean began in remarkably short order. Just what impelled him to move so far and so fast—curiosity, perhaps; or hunger; or a need for space and living room—remains an enigma. But the fact remains that a mere thirty thousand years after the fossil record shows him to have been foraging the grasslands of Ethiopia and Kenya—hunting for elephants and hippos, gazelles and hyenas, building shelters and capturing and controlling lightning-strike fires—he began to trek southward through Africa, a lumbering progress toward the continent's edge, toward the southern coastlines and a set of topographical phenomena the existence of which he had no inkling.

The weather was becoming cooler as he went: the world was entering a major period of glaciation, and even Africa, astride the equator, was briefly (and before it became very cold indeed) more climatically equable, more covered with grassland, less wild with jungle. So trekking down south along the Rift was perhaps the least complicated of early man's explorations—the mountain ranges on each side offering him a kind of protection, the rolling grassy countryside more benign than the jungles of before, the rivers less ferocious and more crossable. And so in due course, and after long centuries of a steady southern migration, man did reach the terminal cliffs, and he did find the sea.

He would have been astonished to reach what no doubt seemed to be the edge of his known world, at the sudden sight of a yawning gap between what he knew and what knew nothing about. At the same time, and from the safety of his high and grass-capped cliff top, he saw far down below him a boiling and seemingly endless expanse of water, thrashing and thundering and roaring an endless assault against the rocks that marked the margin of his habitat. Quite probably he was profoundly shaken, terrified by the sight of something so huge and utterly unlike anything he had known before.

Yet he didn't run yelping back to the safety of the savannah. All of the recently discovered evidence suggests he and his kin stopped where they were and made shelter on the shore. They chose to do so in a large cave that was protected from the waves below by its location well above the high-tide levels. Then— whether timidly or boldly or apprehensively we will never know—he eventually clambered down and made it onto the beach proper. Then, while keeping himself well away from the thunder of the breakers, he knelt first to investigate, just as a child might do today, the magical mysteries of the seashore tide pools.

With the cliffs of land on one side and the brute majesty of the water crashing on the other, he was briefly captivated by the entirely new world of the pools. He gazed into their depths, which were crystal clear, yet fronded with green and furtive with darting movement. He dipped his finger into the water, withdrew it, tasted—it was very different from all he had known before, not sour and brackish like the poorer of the desert wells, but not fresh and sweet, probably not good for him to drink.

But nonetheless, it supported some kinds of life. The pool, as he looked at it more closely, was furiously alive with creatures— crabs, small fish, beasts with shells, weeds, anemones. And so by the same process of trial and error that had dominated his feeding and foraging habits on land in the millennia before, he eventually discovered in the pools an abundance of food for himself and his family. It was, moreover, good and nutritious food, and it was of a kind that he could hunt without running, that he could eat without cooking, that he could gather without putting his life at risk. Moreover, and mysteriously, it was a food that was somehow magically replenished with every one of the twice-daily refillings of the small watery world that lay before him.

Inevitably, man's fascination with this strange new aqueous universe brought him to settle by the sea. He had come at last to Pinnacle Point.

3. COASTING

The Western Cape, the administrative name for the most southerly corner of South Africa, is where the waters of the Indian Ocean blend into the cold chill of the South Atlantic. It is a fearfully dangerous coastline, heaped with shipwrecks: oil tankers too big for Suez pass south of Cape Agulhas and then hug the coastline on their way to or from the wellheads, and seem to collide with each other with a dismaying frequency, resulting usually in the spillage of much unattractive cargo and the deaths of scores of African penguins.

I have sailed in these waters and know them to be very trying. Almost all vessels like to keep close to shore to avoid the notoriously extreme waves of the deeper sea, and there are few harbors to which one can run in the event of bad weather. The combination of crowded sea-lanes (all littered with local fishing vessels, too), of cold and rough waters, and a forbidding and havenless rockbound coastline is one that few mariners—and certainly few quasi-inept strangers, as I used to be—care to experience.

I still have my old *South Africa Pilot*, the blue-backed sailing directions I used on the yacht. The rather beautiful Vlees Bay it notes as a landmark lies between two rocky headlands—Vlees Point to the south and, nine miles to its north, Pinnacle Point.

Back when the *Pilot* was written, its hydrographer-authors noted the presence near Pinnacle Point of "a group of white holiday bungalows," noting them not for aesthetic reasons, but because they would have provided visible markers for ships making their way along the coast.

This gathering of second homes has transmogrified over the thirty years since the *Pilot* was written into an almighty village—the Pinnacle Point Beach and Golf Resort—devoted to costly seafront hedonism. The peddlers of the resort like to say that the sea air, the Mediterranean climate, the white water, and the peculiar local floral kingdom known as fynbos, which here renders their ironbound coastal landscape unusually attractive, all combine to transform the place into "a new Garden of Eden."

Little do they know how apposite their slogan is. Pinnacle Point may be on the verge of enjoying fame among professional golfers and retired businessmen—but it has been known for very much longer to archaeologists consumed by the history of early man. For Pinnacle Point appears to be the place where the very first human beings ever settled down by the sea. Specifically there is a cave, known to the archaeological community as PP13B, situated a few score feet above the wave line (though not quite out of earshot of the ninth tee), where evidence has been found showing that the humans who first sheltered there did such things as eat shellfish, hone blades, and daub themselves or their surroundings with scratchings of ocher. And they did so, moreover, 164,000 years ago, almost exactly.

An American researcher, Curtis Marean, from the School of Evolution and Social Change at Arizona State University, was among the first to realize the importance of the cave, in 1999. He had long suspected, from what he knew of Africa's chilly and inhospitable climate during the last major glaciation, that such humans as existed would probably travel to cluster along the southern coast, where the ocean currents brought warmer water down from near Madagascar, and where food was available both on the land and in the sea. He decided that these humans probably sheltered in caves—and so he looked along the coastline for caves that were sufficiently close to the then sea level* to allow the humans to get to the water, yet sufficiently elevated that their contents weren't washed away by storms and high tides. Eventually he found PP13B and had a local ostrich farmer build him a complicated wooden staircase so his graduate students didn't fall to their deaths clambering to the cave mouth—and began his meticulous research. Marean's paper, published in *Nature* eight years later, drily recorded a quite remarkable find.

This large sea cave in the South African coast appears to be one of the places where humankind first reached the shores of the sea. Researchers have found evidence that these humans also ate their first seafood here: shell fragments include oysters, mussels, and limpets.

---

*   Lower than today, because the glaciation had locked up so much of the ocean as polar ice.

There was ash, showing the inhabitants lit fires to keep them-selves warm. There were sixty-four small pieces of rock fash-ioned into blades. There were fifty-seven chunks of red ocher, of which twelve showed signs of having been used to paint red lines on something—whether walls, faces, or bodies. And there were the shells of fifteen kinds of marine invertebrates, all likely to have been found in tide pools—there were shore barnacles, brown mussels, whelks, chitons, limpets, a giant periwinkle, and a single whale barnacle that the Arizonans believe came at-tached to a whale skin found washed up on the beach.

How the community decided to dine on shellfish remains open to conjecture. Most probably the inhabitants saw seabirds picking up the various shells, cracking them on the rock shelves, and gorging themselves on the flesh within. Disregarding the so far unsaid assertion* that it was *a brave man who ate the first oyster,* the cave dwellers swarmed down to the oceanside and promptly wolfed down as many mollusks as they could find—eventually repeating what must have been this very welcome gastronomic adventure on every subsequent occasion that the tides gener-ously provided them with more.

The experience had a signal effect on this small colony, and on humankind in general—which makes it all the more remarkable that the financiers of the local golf course chose "Garden of Eden" as their slogan. The effect was of far greater significance than might be suggested by a mere change of diet, from buffaloes to barnacles, from lions to limpets. The lim-itless abundance of nourishing food meant the settlers could now do what it had never occurred to them to do before—*they could settle down.*

They could at last begin to consider the rules of settlement—

---

* Attributed variously to Jonathan Swift, H. G. Wells, and G. K. Chesterton.

which included the eventual introduction of both agriculture and animal husbandry and, in good time, civilization.

Moreover, their ocher colorings suggests that for the first time these cave colonists began to employ symbols—signs perhaps of warning or greeting, information or suggestion, pleasure or pain, simple forms of communication that would have the most enduring of consequences. An early seaside human might go down to a certain crab-rich tide pool and merely expect or hope that others would follow him. But then he might decide to create a sign, to use his recently discovered color-making stick to mark this same tide pool with an indelible ocher blaze—ensuring at a stroke that all his cave colleagues would now be able to identify the pool on any subsequent occasion, whether its initial finder went there or not. Thus was communication initiated—and from such symbolic message making would eventually emerge language—one of the many kinds of mental sophistication that distinguish modern man.

4. DEPARTURES

The Atlantic, at its beginnings, was a very one-sided ocean, with many peoples distributed along its eastern coasts and yet for many thousands of years no one—no human or humanoid—on its western side. Moreover, its populated coasts were settled initially by newcomers from the continental heartlands, who had little experience with or aptitude for the ways of the sea. Not surprisingly it took a long time for sailors to venture any distance from the coastline; it took thousands of years for the islands within the Atlantic to be explored; and it took an inordinately long time for anyone to cross the ocean. It was to remain a barrier of water, terrifying and impassable, for tens of thousands of years.

Today's research, which permits this kind of certainty, is hugely different from the archaeological diggings and probing that went on since before Victorian times. The unraveling of the human genome in 2000 made it possible to find out who in antiquity settled where, and when, simply by examining in great detail the DNA of the present-day inhabitants. The romance of finding potsherds and pieces of decorative artwork remains, of course, but for speedy determination of the spread of humankind, no better way can be devised than the computerized parsing of the genetic record.

Communities were already forming on the Atlantic's east side while native newcomers were still nervously pushing their way through the woodlands in the west. The first Neolithic peoples in the Levant had already created their world's first town, Jericho. By now all the world's peoples were *Homo sapiens*; no other human kind had made it beyond the end of the Paleolithic epoch—and their advances, seen from this end of the telescope of time, seem to have taken place at an almost exponential rate. When Jericho was first founded—and this is when the western Atlantic was still essentially unpopulated—its inhabitants were busy carving stones and raising millet, sorghum, and einkorn wheat. Just a few thousand years later, when the first skin-wearing and shivering Ojibwe and Cree and Eskimos were doing their artless best to create the first hardscrabble settlements in the American north, men in the Fertile Crescent and beyond, in places as far away as Ireland, were already throwing pots, were raising dogs, pigs, and sheep, had created from stone the adze and the sickle blade, had built tombs and henges, had used salt to preserve their food, and were on the verge of smelting metals.

. . .

Moreover, these easterners had also made their first boats. Very early settlers in Holland and France had first carved or scorched the interior of fallen trees as much as ten thousand years ago, producing dugouts that they used to navigate their rivers and swamps and cross some of the less formidable estuaries. But these craft were really just canoes, at once both unstable and elephantine, and without keels, sails, rudders, or the kind of freeboard necessary for even the most limited push into the sea. It was to be the Crescent, once again, where the first major advance occurred: in Kuwait, two thousand years later, there appeared a proper sailing craft, made of rushes and reeds and lacquered with bitumen, that was capable of journeying at least through the tricky and unpredictable waters of the Red Sea and perhaps beyond.

Oman also had such a boat, and in 2005 a very eager Omani sultan sponsored a crew of half a dozen to pilot a replica from Muscat to the Indian coast of Gujarat. The journey was to have been 360 miles, but the bitumen must have leaked, because the reeds in the hull became waterlogged three miles off the Arabian coast. The tiny craft promptly sank and everyone had to be rescued by a ship from the Royal Oman Navy.

5. SAILINGS

The Phoenicians were the first to build proper ships and to brave the rough waters of the Atlantic.

To be sure, the Minoans before them traded with great vigor and defended their Mediterranean trade routes with swift and vicious naval force. Their ships—built with tools of sharp-edged bronze—were elegant and strong: they were made of cypress trees, sawn in half and lapped together, with white-painted and sized linen stretched across the planks, and with a sail suspended from a mast of oak, and oars to supplement their speed.

But they worked only by day, and they voyaged only between the islands within a few days' sailing of Crete; never once did any Minoan dare venture beyond the Pillars of Hercules, into the crashing waves of the Sea of Perpetual Gloom.

The Minoans, like most of their rival thalassocracies, accepted without demur the legends that enfolded the Atlantic, the stories and the sagas that conspired to keep even the boldest away. The waters beyond the Pillars, beyond the known world, beyond what the Greeks called the *oekumen*, the inhabited earth, were simply too fantastic and frightful to even think of braving. There might have been some engaging marvels: close inshore, the Gardens of the Hesperides, and somewhat farther beyond, that greatest of all Greek philosophical wonderlands, Atlantis. But otherwise the ocean was a place wreathed in terror: *I can find no way whatever of getting out of this gray surf,* Odysseus might well have complained, *no way out of this gray sea.* The winds howled too fiercely, the storms blew up without warning, the waves were of a scale and ferocity never seen in the Mediterranean.

Nevertheless, the relatively peaceable inland sea of the classical world was to prove a training ground, a nursery school, for those sailors who in time, and as an inevitable part of human progress, would prove infinitely more daring and commercially ambitious than the Minoans. At just about the time that Santorini erupted and, as many believe, gave the final fatal blow to Minoan ambitions, so the more mercantile of the Levantines awoke. From their sliver of coastal land—a sliver that, in time, would become Lebanon, Palestine, and Israel, and can be described as a land with an innate tendency toward ambition—the big Phoenician ships ventured out and sailed westward, trading, battling, dominating.

When they came to the Pillars of Hercules, some time around

the seventh century B.C., they, unlike all of their predecessors, decided not to stop. Their captains, no doubt bold men and true, decided to sail right through, into the onrushing waves and storms, and see before all other men just what lay beyond.

The men from the port of Tyre appear to have been the first to do so. Their boats, broad-beamed, sickle-shaped "round ships" or *galloi*—so called because of the sinuous fat curves of the hulls, and often with two sails suspended from hefty masts, one at midships and one close to the forepeak—were made of locally felled and surprisingly skillfully machined cedar planks, fixed throughout with mortise and tenon joints and sealed with tar. Most of the long-haul vessels from Tyre, Byblos, and Sidon had oarsmen, too—seven on each side for the smaller trading vessels, double banks of thirteen on either side of the larger ships, which gave them a formidable accelerative edge. Their decorations were grand and often deliberately intimidating—enormous painted eyes on the prow, many-toothed dragons and roaring tigers tipped with metal ram-blades, in contrast to the ample-bosomed wenches later beloved by Western sailors.

Phoenician ships were built for business. The famous Bronze Age wreck discovered at Uluburun in southern Turkey by a sponge diver in 1982 (and which, while not definitely Phoenician, was certainly typical of the period) displayed both the magnificent choice of trade goods available in the Mediterranean and the vast range of journeys to be undertaken. The crew on this particular voyage had evidently taken her to Egypt, to Cyprus, to Crete, to the mainland of Greece, and possibly even as far as Spain. When they sank, presumably when the cargo shifted in a sudden storm, the holds of the forty-five-foot-long *galloi* contained a bewildering and fatally heavy amassment of delights, far more than John Masefield could ever have fancied.* There were ingots of copper and tin, blue glass and ebony, amber, ostrich eggs, an Italian sword, a Bulgarian axe, figs, pomegranates, a gold scarab with the image of Nefertiti, a set of bronze tools that most probably belonged to the ship's carpenter, a ton of terebinth resin, hosts of jugs and vases and Greek storage jars known as *pithoi*, silver and gold earrings, innumerable lamps, and a large cache of hippopotamus ivory.

The possibility that the Uluburun ship journeyed as far as Spain suggests the traders' ultimate navigational ambitions. The forty ingots of tin included in the cargo hints at their commercial motive. Tin was an essential component of bronze, and since the introduction of metal coinage in the seventh century B.C., the demand for it had vastly increased. It was known anecdotally to the Levantines that alluvial tin was to be found in several of the rivers that cascaded down from the hills of central southern Spain—the Guadalquivir and the Guadalete most notably, but also the Tinto, the Odiel, and the Guadiana—and so the Phoenicians, at around this time, decided to move, and disregard the legendary warnings. For them, with the limited knowledge they had and the jeremiads on daily offer from the seers and priests, it was as audacious as attempting to travel into outer space: full of risk, and with uncertain rewards.

And so, traveling in convoy for safety and comfort, the first brave sailors passed beneath the wrathful brows of the rock pillars—Gibraltar to the north and Jebel Musa to the south—made their halting way, without apparent incident, along the Iberian coastline, and finding matters more congenial than they imagined—for they were in sight of land all the time, and did not venture into the farther deep—they then set up the oceanic trad-

---

* His famously imagined quinquireme, homebound to Palestine in *Cargoes*, carried ivory, apes, peacocks, sweet white wine, and sandalwood, together with plenty of cedarwood, presumably as dunnage.

ing stations they would occupy for the next four centuries. The first was at Gades, today's Cádiz; the second was Tartessus, long lost today, possibly mentioned in the Bible as Tharshish,* and by Aristophanes for the quality of the local lampreys, but believed to be a little farther north than Gades, along the Spanish Atlantic coast at Huelva.

It was from these two stations that the sailors of the Phoenician merchant marine began to perfect their big-ocean sailing techniques. It was from here that they first embarked on the long and dangerous voyages that would become precedents for the following two thousand years of the oceanic exploration of these parts.

They came first for the tin. But while this trade flourished, prompting the merchantmen to sail to Brittany and Cornwall and even perhaps beyond, it was their discovery of the beautiful murex snails that took them far beyond the shores of their imagination.

The magic of murex had been discovered seven hundred years before, by the Minoans, who discerned that, with time and trouble, the mollusks could be made to secrete large quantities of a rich and indelible purple-crimson dye—of a color so memorable the Minoan aristocracy promptly decided to dress in clothes colored with it. The color was costly, and there were laws that banned its use by the lower classes. The murex dye swiftly became—for the Minoans, for the Phoenicians, and most notably of all, for the Romans—the most prized color of imperial authority. One was *born to the purple:* only one so clad could be part of the vast engine work of Roman rule, or as the *Oxford English Dictionary* has it, of the "emperors, senior magistrates, senators and members of the equestrian class of Ancient Rome."

---

* In 1 Kings 22: "For the king had at sea a navy of Tharshish with the navy of Hiram: once in three years came the navy of Tharshish, bringing gold, and silver, ivory, and apes, and peacocks."

By the seventh century B.C., the seaborne Phoenicians were venturing out from their two Spanish entrepôts, searching for the mollusks that excreted this dye. They found little evidence of it in their searches to the north, along the Spanish coast; but once they headed southward, hugging the low sandy cliffs of the northern corner of Africa, and as the waters warmed, they found murex colonies in abundance. As they explored, so they sheltered their ships in likely-looking harbors along the way—first in a town they built and called Lixus, close to Tangier and in the foothills of the Rif: there remains a poorly maintained mosaic there of the sea god Oceanus, apparently laid by the Greeks.

Then they moved on south and found goods to trade in an estuary close to today's Rabat. They left soldiers and encampments at still-flourishing coastal towns like Azemmour, and then, in boats with high and exaggerated prows and sterns, decorated with horses' heads and known as *hippoi*, they pressed farther and farther from home, coming eventually to the islands that would be named Mogador. Here the gastropods were to be found in suitably vast quantities. And so this pair of islands, sheltering the estuary of the river named the Oued Ksob, is probably as far south as they went,* and this is where their murex trade commenced with a dominating vengeance.

What are now known as Les Îles Purpuraires, bound inside a foaming vortex of tide rips, lie in the middle of the harbor of what is now the tidy Moroccan jewel of Essaouira. This town is now best known for its gigantic eighteenth-century seaside ramparts, properly fortified with breastworks and embrasures, spiked bastions, and rows of black cannon, and which enclose a

---

* With one caveat—a claim by Herodotus that in about 600 B.C., on the orders of the Egyptian pharaoh Necho II, a party of Phoenician sailors made *a three-year circumnavigation of Africa*. Necho—who built an early version of the Suez Canal—was an ambitious and imaginative leader, and may have ordered such an expedition, though there is much skepticism.

handsome cloistered medina. The walkways on top of the curtain walls are the perfect place to watch the ever-crashing surf
from the Atlantic rollers, especially as the sun goes down over
the sea. The Phoenicians found that the snails gathered in the
thousands there, in rock crevices, and they scooped them up in
weighted and baited baskets. Extracting the dye—known chemi-
cally as 6.6'-dibromoindigo, and released by the animals as a
defense mechanism—was rather less easy, the process always
kept secret. The animal's tincture vein had to be removed and
boiled up in lead basins, and it would take many thousands of
snails to produce sufficient purple to dye a single garment. It was
traded, and the trade was tightly controlled, from the home port
of the sailors who harvested it: Tyre. For a thousand years, genuine Tyrian purple was worth, ounce for ounce, as much as twenty
times the price of gold.

· · · ·

The Phoenicians' now-proven aptitude for sailing the North African coast was to be the key that unlocked the Atlantic for all
time. The fear of the great unknown waters beyond the Pillars of
Hercules swiftly dissipated. Before long a viewer perched high
on the limestone crags of Gibraltar or Jebel Musa would be able
spy other craft, from other nations, European or North African
or Levantine, passing from the still blue waters of the Mediterranean into the gray waves of the Atlantic—timidly at first maybe,
but soon bold and undaunted, just as the Phoenicians had been.

"Multi pertransibunt, et augebitur scientia" was a phrase
from the Book of Daniel that would be inscribed beneath a fanciful illustration, engraved on the title page of a book by Sir
Francis Bacon, of a galleon passing outbound, between the Pillars, shattering the comforts and securities of old. "Many will
pass through, and their knowledge will become ever greater,"
it is probably best translated—and it was thanks to the purple-

veined gastropods and the Phoenicians who were brave enough
to seek them out that such a sentiment, with its implication that
learning comes only from the taking of chances and risk, would
become steadily more true. It was a sentiment born at the entrance to the Atlantic Ocean.

6. WESTERINGS

The Phoenicians eventually vanished from the scene in the
fourth century B.C., vanquished in battle, their country absorbed
by neighbors and plunderers. And as their own powers waned,
so other mariners in other parts of the world would begin to
press the challenge of the newfound Atlantic ever more firmly.
There was Himilco the Carthaginian (who lost the Second Pu-
nic War to the Romans, despite his fleet of forty quinquiremes),
and there was Pytheas from Marseilles (who sailed up to and circumnavigated Britain, and gave it its name, then pressed on up
to Norway, encountered ice floes, gave us the name Thule, and
found the Baltic).

Then came the Romans—a martial people never especially
maritime in their mind-set, and perhaps as a consequence
somewhat nervous sailors at the beginning. According to the
Roman historian Cassius Dio, some of the legionnaires involved
in the Claudian invasion of Britain in 43 A.D. were so terrified at
having to cross even so mild a body of Atlantic water as the Strait
of Dover that they rebelled, sat on their spears, and refused to
march, protesting that crossing the sea was "as if they had to
fight beyond the inhabited earth." In the end they did embark on
their warships, and they did allow themselves to be transported
to the beaches of Kent, and the empire did expand—but even at
its greatest extent in 117 A.D., it was an empire firmly bounded by

the Atlantic coast, from the Solway Firth in the north to the old Phoenician city of Lixus in Morocco to the south. They may have cast off and kept to the shallows for coastal trade, but otherwise the Romans kept a respectful distance from the real Atlantic, never to be as bold as their predecessors.

Nor as bold as their eventual successors. For after a lengthy and puzzling period of mid-Atlantic coastal inactivity, the Arabs—sailing in the eighth century A.D. from their newly acquired fiefdom in Andalusia—and later the Genoese from northern Italy began trading in the North African Atlantic. Records show that they went as far south as the coast off Wadi Nun, close to the former Spanish possession (and a philatelist's favorite) of Ifni, where the sailors met desert caravans from Nigeria and Senegal laden with all manner of African exotica to be hurried back to customers in Barcelona and the cities of Liguria.

Yet navigational advancement and casual fearlessness were not to be a monopoly of the Mediterranean sailors. Long before the voyages of the Arabs and the Genoese—though long after the Phoenicians, whose efforts trumped those of everyone else—northern men had launched their boats into the much colder and rougher waters of the northern Atlantic. Their motives were different: curiosity, rather than commerce, tended to drive the northerners out into the oceans. Curiosity, and to a lesser extent, Empire and God. Two groups of sailors dominated, at least in the first millennium: the Vikings most famously, but initially and often half forgotten in the fog of history, the Irish.

There could hardly be vessels more different than the products of the first millennium's Scandinavian and Irish boatyards. The Vikings, men who to this day are renowned as having essentially conceived the tradition of freebooting violence, and who kept mostly to the coast, sailed out in their famous longships, bent on pillage and sack; the Norsemen, which is today's preferred name for the more congenial and numerous of the early Atlantic's Scandinavian traders and explorers, used slightly chubbier, more stolid vessels known (in the plural) as *knarrer*.

Both were clinker-built wooden craft, high-prowed and in the case of the more menacing longships, more than a hundred feet from bow to stern, made of oak and with a figure high on the bow. Both kinds of craft had an enormous square sail, maybe thirty feet across, and weighing tons, needed a crew of at least twenty-five who, with a following wind, could manage fifteen knots on a smooth sea.

The Vikings, who became notorious for sack and pillage in Europe, generally sailed in longships; those Norsemen who, more peaceably, sailed west to Iceland, Greenland, and North America used smaller craft like this knarr.

The Irish, by contrast, sailed into the wild waters of their western seas in boats they still insist on calling, with typical Celtic self-deprecation, *canoes*. A curragh, the proper Gaelic

name for today's still-used descendant, is a small and stubby
boat, round and squat where the longship and the *knarr* were
sleek and fast. It required few crew, had a single sail and a single
steering oar, and was made of a cage of ashwood laths covered
with ox leather that had been soaked in a solution of oak bark
and marinated in lanolin; the whole was stitched together with
flax thread and leather thongs. Tim Severin, the noted Irish
sailor-explorer who would later construct and sail one, asked
a noted curragh maker from County Cork if such a tiny and
fragile-looking craft could make it all the way to America.

"Well now," the builder replied, "the boat will do, just so long
as the crew's good enough."

. . .

Legend has it that the wandering Irish abbot, St. Brendan, was
the first to make a sustained journey through the waters of the
North Atlantic. Whether he was guided on the voyage by much
more than blind faith in what he supposed to be a kindly god is
unknown. Most imagine he carried with him the only Atlantic
map then known—not that it would have been much use: it was
an illustration—drawn in the first century, in Egypt, from Ptolemy's biblically authoritative book, *Geographica*, which in later
copied versions had the Atlantic as a mere sliver on the western
edge of the sheet, and had named it either *Oceanus Occidentalis* or,
more ominously to the north, *Mare Glaciale*.

The beginning of the great Irish-Scottish missionary expeditions, all bent on exporting Christianity to the more remote
nooks of the northern world, is generally dated with some precision at 563 A.D., the year that St. Columba brought knowledge of
the Trinity to Iona, in Argyll. According to the rollicking yarns
found in the medieval *Navigatio Sancti Brendanis Abbatis*, Brendan's voyage was taken somewhat before this time; together with
perhaps as many as sixty brother monks, he sailed from a small

estuary on the Dingle peninsula of far southwestern Ireland, first north to the Hebrides, then on farther to the Faroe Islands and Iceland, and finally westward, maybe even to Newfoundland, the Promised Land of the Saints.

. . .

It is unknown who brought Christianity to the Faroes, but its legacy survives and is still in robust good health. When Brendan and his monastic brethren landed, after beating their way up through two hundred miles of gale-wracked waters from Butt of Lewis, the northern tip of the Hebrides, they were impressed by the islands' innumerable sheep, by the extraordinary number and variety of seabirds and an almost equally abundant variety of fish, as well as by the rain, the sheer and eternally dripping rock faces, and the deep green of the omnipresent tussock grass.

Little has changed in almost fifteen hundred years. It was on a blustery spring day that I first sailed in the Faroes, and just as St. Brendan is supposed to have done, crossing the strait between the two most westerly Faroese islands of Vagar and Mykines. I was in a little boat that bounced merrily across the swells, passing under basalt cliffs that were quite sheer and black and so high that they quite vanished into the swirling clouds above.

But on close inspection the cliffs were not entirely black. Blotches of green grass stood out, edged by cascades of running water after each blustery shower passed through; and on each patch of grass, which must have been angled seventy, eighty degrees, such that a man could not stand upright for fear of falling hundreds of feet down into a bottomless sea of the purest indigo, were sheep.

Young island men had placed them there as lambs, early in the spring. The island shepherds had climbed up the cliffs—fixed ropes could be seen, spun between a network of pitons and carabiners that glinted against the rocks when the sun was right—

and men in Faroese rowing boats would hand up the lambs to them, one by one, and each climber would sling a mewing lamb across his shoulders and then heave himself, hand over hand, boots sliding on the wet rock face, up to the tiny and precipitous pasture.

With one hand he would hold on to the rope, and with the other unclasp the frightened and warm-wet animal from around his neck and place it as firmly as possible on solid ground. A thousand feet below, the boat looked tiny, the occupants barely visible, just craning faces gazing up to make certain everything was still all right. The young sheep would stagger for a moment, bewildered, would then sniff the air and look with amazement at the drop—and finally would realize how best to stand, four-square, in order to survive. The animal, by now more calm, would tuck its nose into the rich grass that had been long fertilized with the guano from the whirling puffins, and remain there, nervously content, for the rest of the year.

From down below I could see them, hundreds of wool-white dots, shifting slowly behind their noses and always *ohmygod* about to fall, but never apparently doing so, even in the gales and when the rains made the grass as slick as oilskin and blubber.

St. Brendan, if he had voyaged to the Faroes at all, sailed almost due northward from the Hebrides. But after his visit (where his *Navigatio* reports encounters with others similar to him, suggesting he was not the first Irishman to get there), the prospect of continuing north was quite bleak: to do so would mean cold, and then intense cold, and then ice. Eastward, too, was no picnic: the expedition would have ended on the known and dangerously rockbound coast of Norway. So westward was the only way to go; but the small boat had to brave seas and storms and winds and currents possibly entirely beyond the competence of even the most navigationally expert of this group of innocent and most likely discalced Clonfert friars.

When Tim Severin took a replica boat across the Atlantic in the summers of 1976 and 1977 (arguing that since it had taken St. Brendan fully seven seasons to cross the ocean, he could legitimately take two), he made landfall in the Faroes, in Iceland, and eventually, after weathering ferocious storms in the Denmark Strait, in Newfoundland. His expedition proved it was entirely possible to cross the Atlantic in a leather boat, providing (as the Irish curragh builder had earlier told him) the crew was good enough. But while he showed that such a journey could have been made, Severin did not prove that such a journey *had* been made, nor that Irish monks had ever done such a thing, had been to any of these three countries at the time suggested. Nor has any firm evidence ever been adduced that suggests either that Irishmen visited or settled, nor, more crucially, that they ever completed a crossing of the ocean. No early Irish artifact has ever been found in North America.

So the Irish were almost certainly not the linear antecedents of Christopher Columbus. Moreover, even though many Italians claim to this day that Columbus had no predecessors at all, and that 1492 was a true historical watershed for transoceanic contact, a discovery in the middle of the twentieth century changed everything. An archaeological find in northern Newfoundland in 1961 proved that the first ocean crossing had been made four hundred years *after* the supposed evangelizing mission of the Irish, and fully four hundred years *before* the commercial expedition of Columbus, but by neither an Irishman nor a Genoese.

The first European to cross the Atlantic and reach the New World was a Norseman, a Viking, and probably from a family born in the fjord lands south of the coastal towns of Bergen and Stavanger, in Norway.

7. ARRIVALS

Four years before these archaeologists announced their discovery, a group of antiquarian booksellers piqued public interest in the possibility that Columbus had been completely outdone.

In 1957 a young dealer in New Haven, Connecticut, Laurence Witten, approached Yale University with an extraordinary offer: he had bought, by way of a dealer in Italy, what appeared to be a fifteenth-century map of the known world, but with one crucial feature that had never before been seen: the presence of a large island, with two elongated indentations on its east coast, situated on the left side of the map to the west of Greenland. The island was identified on the map as *Vinlanda*, and the rubric above it, written in Latin, said that it had been visited in the eleventh century, first by "companions Bjarni and Leif Eriksson," and later by a legate from the Apostolic See.

It was eight years before the discovery of the map was announced—mainly because Paul Mellon, the banking millionaire who had eventually acquired it from Witten and offered it as a gift for his alma mater, decided he would hand it over only once it had been authenticated. Eight years of tests later, a team of British Museum specialists finally declared it to be genuine, and Mellon allowed Yale to release the news. It sparked a sensation—as if it had been an arm of the True Cross, a fresh revelation about the Shroud of Turin or the rudder from Noah's Ark. It was "the most exciting cartographic discovery of the century," said the university's curator of maps, "the most exciting single acquisition in modern times," said the head of the Beinecke Library, "exceeding in significance even the Gutenberg Bible." It made the front pages everywhere.

What thrilled the world—or at least most Americans (though

not Italian-Americans) and all Norwegians—was that the map appeared to be final cartographic confirmation that the "Vinland" famously mentioned in two of the best-known thirteenth-century Icelandic sagas was in North America. The map appeared to prove once and for all that Leif Eriksson—Erik the Red's peripatetic Iceland-born son—had indeed, and in the very precisely remembered year 1001 A.D., landed somewhere on the American continent.

Here was documentary confirmation of something all red-blooded Italians had long feared—that it was not Columbus who had first crossed the Atlantic, but an eleventh-century Norseman. Adding insult to this injury to Genoese pride was the fact that Yale, with magnificently self-evident cheek, chose to display its Viking treasure by throwing a lavish celebration dinner featuring a Viking longboat carved from ice and the normally owlish university librarian wearing an iron helmet flown over by the King of Norway—and held the party on Tuesday, October 12, that year's Columbus Day. It was hardly the most appropriate moment to suggest that a Norwegian made the first American landfall, and it caused much huffing.

"Twenty-one million Americans will resent this great insult," said the then president of the Italian American Historical Society.

The only problem was that the fragile and yellowing little parchment document, eleven inches by sixteen, has turned out to be a tissue freighted with all manner of uncertainties and bitter argument. The book dealer had lied about where and how it had come to be his. The Italian (an irony not lost) who had sold it to him (for $3,500) and who had previously tried in vain to sell it to the British Museum turned out to have been both a fascist and a convicted thief. Tests on the map's ink showed high levels of chemicals that had not been invented at the time the map was said to have been made—and although the parchment itself was proved to be fifteenth century, it appeared to have been coated with an oil made in the 1950s. The fold down the map's middle turned out not to be a fold at all, but a splice, with traces of curious chemicals at its edges. And the map's Latin text was peppered with the æ ligature, a lexical form seldom used at the time of the map's supposed creation.

It all became too much for Yale, and in 1974 the exasperated librarian declared their costly treasure to be a forgery. This was not, however, to be the end of the story. Further tests were conducted in the mid-1980s, and these suggested that the tests of the previous decade had been botched—and so in 1987 Yale changed its mind once more, said it now had confidence in the document and had it insured for $25 million, just in case. At the time of writing the skeptics and the believers were still endlessly trading the ascendancy: more chemical and spectroscopic and subatomic tests have raised ever more complicated doubts, and the name of a curious anti-Nazi forger[*] who might have had a powerful though complicated motive for forging such a map has come to light, even though the most senior of all Danish conservators was as recently as 2009 still insisting that the map was true.

In any case, there is a further irony, a further puzzle. The ink on the map is now fading to the point of near invisibility, despite Yale's best efforts at conservation. Just why such deterioration might suddenly accelerate, nine hundred years after the map was supposedly drawn, remains unexplained. If this is all an elaborate ruse, then this fading has produced an ironic coda

---

[*] A German-Austrian Jesuit priest, Josef Fischer, an expert on medieval cartography, is thought by some to have had the unique combination of opportunity, motive, and sufficient free time to create the map—to twit the Nazis in their belief in Nordic world supremacy. His rubric note referring to the Vinland visit of the papal legate supports a belief that the Catholic Church was involved in the transatlantic mission—something impossible to square with the Nazi ideal. Fischer died in 1944, long before the controversy broke.

to the story: just like the Cheshire Cat's smile, the Vinland Map seems to be drifting away into nothingness.

However, despite all the brouhaha surrounding Yale's document, the finding in Scandinavian libraries of a series of other (and this time undoubtedly genuine) charts, and a further discovery in 1960 that was brought about by what those maps had drawn onto them, finally dashed any further claim for the primacy of Columbus. The other maps were all fair copies of a much less sensational but in the end far more useful document that is known today as the Skálholt Map. It was drawn in Iceland in 1570 by a schoolmaster named Sigurd Stefansson, simply as an exercise to show from his readings of a variety of Icelandic texts just where the Nordic explorers and traders had landed on the various shores of the North Atlantic Ocean.

The original is long gone; but the copies that exist all show the same thing: an Atlantic—here called *Mare Glaciale*, the icebound sea, with islands such as the Faroes, Iceland, Shetland, and Orkney all in their more or less accurate relative positions—bordered by an almost wholly connected skein of landmasses. There was Norway, of course; then Gronlandia, then Helleland, Markland and Skralingeland (which Nordic scholars suggest—as *flagstone land, forest land,* and *land of the savages*—to be portions of Labrador); and then finally, jutting from the southwest of the chart, a slender, north-pointing peninsula—marked simply as *Promonterium Vinlandiae*, the Peninsula of Vinland.

This was the clue that concluded a decades-long search. Ever since the Icelandic sagas had mentioned Vinland, Americans, and Canadians, mainly in the northeast, had been scouring their properties and their neighborhoods for anything that might suggest a onetime Norse settlement—for who would not wish to know that European feet had first been placed on their front garden, or that Nordic sailors had walked first on their own village beach? Runestones—most of them fakes—popped up in unlikely places like Minnesota and Oklahoma; a Nordic statue was uncovered beside Thoreau's Merrimack River; the unusual color and height of the Narragansett Indians of Rhode Island was bruited as evidence that Norsemen had once set up a colony near Providence; and a wealthy Harvard chemistry professor named Eben Horsford* claimed to have found the site of Leif Eriksson's house in Cambridge, no less, beside a traffic light near Mount Auburn Hospital. He, along with a violinist named Ole Bull, raised funds to have a statue raised to the Nordic settler on Commonwealth Avenue in Boston. It still stands.

Despite all this nonsense, in the mid-1950s a Norwegian scholar of Viking history, Helge Ingstad, felt certain from his studies of the Skálholt Map that he knew where Leif Eriksson's Vinland must have been. It was, he surmised, in the Canadian province of Newfoundland, and somewhere on great north-running peninsula, under the mountains of the Long Range, which lies on the island's western side. Armed with this educated hunch, he began to make annual excursions to Canada, asking repeated questions of villagers and farmers in the outports between the town of Stephenville Crossing and the tiny inlets nearly three hundred miles north, on the shores of the Strait of Belle Isle.

One day in 1960, he and his daughter Benedicte sailed on their small yacht up to the tiny settlement of L'Anse aux Meadows, at the very northern tip of the island. There Ingstad met a local fisherman, George Decker, and asked him the question he felt he had asked a thousand times. Were there by any chance any ruins nearby, which might have been from a settlement of Norsemen?

---

* He made his fortune from inventing baking powder.

Decker replied with a studied casualness. "Yeah, I know where there are some old ruins. Follow me." He took a stunned Ingstad over to a field of cloudberries, wild iris, and gale-stunted pines, to where there stood almost a dozen very large grass-covered mounds, all set on a slight slope that ran down to Epaves Bay. Decker watched as his visitor gulped with amazement. He was delighted that the Norwegian was so astonished, he said later, but often wondered to himself why it had taken outsiders so long to get around to asking.

In that instant, the world—at least, the world of archaeologists—shifted on its axis. Once the diggings began, history was rewritten, profoundly and at a stroke. L'Anse aux Meadows—the words are a linguistic contortion of the French for *The Bay of the Jellyfish*—became in short order the most famous archaeological site in North America. With barely a cirrus cloud of doubt, this spot is now acknowledged as the base of operations for the Norsemen who settled and lived and created homes for themselves on the far side of the sea. Quite possibly—quite probably, in fact—L'Anse aux Meadows was the Vinland settlement itself. Leif Eriksson, his kin and his kith, provably and finally had now joined that select group of men and women who had been first to cross the Atlantic Ocean. The excavations continued. They were conducted by Ingstad and his wife over the next seven years, the pair reburying the site each winter, for protection against both the ferocious snowstorms and the destructive grinding of icebergs washed up on the beach.

Formal public announcement of the find was made in the pages of the *National Geographic* magazine in November 1964. It was revealed that the Norsemen had built themselves a total of three large stone-and-sod houses and five workshops, one of which clearly was a smithy. Iron nails had been found, and spindles, and a copper pin used for decoration. Specialists working

At the far northern tip of Newfoundland, these sod-covered huts, discovered in 1960 but estimated to date from the beginning of the eleventh
century, offered the first concrete evidence of early Norse settlement in the
Americas.

with the University of Toronto's main subatomic particle accel-
erator and at the Radiological Dating Laboratory in Trondheim
both brought the very latest technologies to bear on the various
samples—mainly charcoal from the smithy furnace—and came
up with agreement that everything at L'Anse aux Meadows had
been created between 975 and 1020 A.D. The sagas' date for the
Vinland settlement—depending how the stories were read—was
1001 A.D. It was like the last piece of a jigsaw puzzle, snapping
itself neatly into place, locked solid.

The excavations continued until 1976, after the Canadian
parks service had taken over from the then quite elderly Ing-
stads. They found cooking pits, bathhouses, and an enclosure
for keeping cattle. Rotted remains of butternuts were also dis-

covered—and since agricultural climatologists are certain that
in the first millennium, no butternut tree could have grown
north of New Brunswick, it is assumed that the visitors must
have taken their *knarrer* and penetrated even farther south. It is
also assumed that they sailed southwestward from their camp,
crossed the notoriously bumpy waters of the St. Lawrence estuary to the American continent, eventually landing on Gaspé or
on Cape Breton Island and then heading either upriver or even
overland to seek richer pastures and tastier crops. (Since New
Brunswick is also the northerly limit for the appearance of wild
grapes, it offers some additional credence to the Norse use of the
name Vinland—Wine Land.)

A second site, smaller in extent than L'Anse aux Meadows,
may have been found more recently: archaeologists working at a
site on the southern end of Baffin Island in 2000 say that the discovery of sod-and-stone walls, a whalebone spade, and a crude
form of a house drainage system all suggest the work of Norsemen. Rival scholars deride this as wishful thinking and insist
it is no more than an indication of the gathering sophistication
of the paleo-Inuit of the Dorset culture who are known to have
occupied this corner of the Canadian subarctic. Those who support the Norse claim say it suggests their Viking-inspired *knarrer* were shuttling between sites in Newfoundland and Labrador
and the Hudson Bay islands for far longer than imagined, and
that the notion that all the Europeans scuttled back to Greenland or Norway and left Canada alone for centuries is myopic and
wrongheaded.

But one further intriguing morsel of history remains. In 1004
a child was born in Vinland. It was a boy, named Snorri, the child
of Gudrid and Thorfin Karlsefni. According to an Icelandic
custom that continues to this day, the lad was given a surname
derived from his father's first name, thus Snorri Thorfinsson.

He was, undoubtedly, the first European child to be born on the American continent. Since he traveled back to Greenland with his parents if and when the L'Anse aux Meadows outpost was eventually closed down, in around 1008, he most probably died there, or in Europe—unaware to the last that he would in time come to be remembered as Canada's first European native son.

8. REPUTATIONS

How should history regard this entire Nordic Atlantic venture, when compared with the ocean crossing much more famously undertaken five centuries later, by Christopher Columbus?

It certainly seems true, so far as archaeology and literature indicate, that no other sailor crossed the ocean with success* in the nearly five centuries between the Erikssons' ventures over the Labrador Sea in 1001 and the self-styled Admiral of the Ocean Sea's six-week journey from Andalusia to San Salvador island in the late summer of 1492. But even though these expeditions had precisely the same initial outcome, that of firmly placing European men onto American shores, there are many differences between the two voyages.

It is 4,500 miles from Bergen to Newfoundland, but Leif Eriksson did not have to travel nearly so far, because he first came to Newfoundland only from Greenland, less than a thousand miles away. Not that this short journey was exactly a picnic.

---

* There are a welter of unproven claims of others having been first across the ocean—based on the supposed finds of, among other things, bones of Portuguese fishermen in Canada, Greek amphorae in Brazil, Roman coins in Indiana, Hebrew lettering on an Indian burial mound in Tennessee, and the relics of the Welsh language spoken in Mobile Bay, Alabama, courtesy of one Prince Madoc, itinerant. Travelers going in the other direction get an opportunity, too: faint chemical traces of nicotine and coca were supposedly found on some ancient Egyptian mummies.

In winter, though the sea does not freeze, the weather is dire, the sea thick with ice, packs and bergs and the much more dangerous half-submerged chunks known unimaginatively as *bergy bits.* The winds Eriksson encountered were exceptionally fierce and almost always came from the west or the northwest, contrary to the Norsemen's intended direction. These gales were so fierce the tiny *knarrer* would have had to lay-to for hours, sometimes for days at a time. Masts broke, sails ripped, everyone aboard was drenched, chill and miserable. Even in summertime matters were little better: drenching fogs and the endless high-latitude daylight—which mightily complicated sleep—made for unhappy navigation.

When the party finally reached land—which was pocketed with settlements of the Dorset people—they made bases for themselves, peaceably and with a recognized veneer of civilized behavior (these were Norsemen, remember—not Vikings). They brought women with them on the comparatively short voyage across the Labrador Sea; they settled into routines of amiable domesticity; and they got on reasonably well, so far as one can tell, with the local peoples, even though their word, *skraelinger,* described them as barbarians (mainly since the natives wore animals skins, not clothes of woven wool, as did the Europeans). The Norsemen refused to give the *skraelinger* any weapons: the sagas say they bartered with them, employing as trade goods not beads or useless trinkets, but milk, which the Eskimos appeared to like.

All told, the Norsemen's brief stay in America seems to have been motivated by curiosity, marked by maritime courage, and sustained with a degree of apparent civility. The much better-known voyage of Columbus, by contrast, was motivated by a combination of commercial ambition, a growing Spanish exasperation at the blockage of land trade routes to the east by the Ottoman Turks (and the thought that this East could be reached instead by heading west and sailing halfway around the world), and the evangelical yearnings of the Church. It turned out to be a voyage carried out in comparative nautical comfort, and it never actually reached the North American mainland, with Columbus to his death believing he had reached the East—the Indies—and in all probability, Japan.

His three small carracks, the *Niña*, the *Pinta*, and the *Santa María*, were cleverly routed to the south of the Canaries (for no one would dispute that Columbus was an exceptionally canny navigator). He then turned right, due west—for he supposed that China and Japan, the cities Marco Polo knew, and the islands where the spices grew were all on the same latitude as the Canaries—and he led his tiny squadron and his ninety crewmen on a relatively pleasant sojourn through sunny seas pressed only by gentle easterly trade winds which sped the vessels toward their destination without significant incident. It was to be a far longer sea voyage without stops than any hitherto known—and since no navigator knew the extent of the sea into which they were heading, it must have been frightening: would they fall off the edge of the world, would they reach an area of impossible storms, were there sea monsters, whirlpools, angry gods?

But by great good fortune the three tiny vessels sashayed their way over the waves quite easily, their logs sometimes recording passages of more than 150 miles a day, cruising at up to eight knots. They did so until that heart-stopping predawn, moonlit moment—on the long-remembered date, October 12, 1492—when the *Pinta*'s lookout, Rodrigo de Triana, spotted a line of white cliffs directly ahead. It was the sudden vision of a new world—or the New World, as it would soon be realized.

This first piece of territory seen almost certainly was one of the outer cays of what are now the Bahamas, most probably the low and sandy windward outpost now known as Watling's Island.

Columbus had his flagship's longboat convey him ashore under the banner of Castile, then kissed the ground, wept grateful tears, annexed the place—as Isabella had given him the contractual right to do—and named it for the Holy Savior, or in Spanish, San Salvador. Rodrigo de Triana was given five thousand maravedis* for being so adept a lookout.

Had this one journey been all, the reputation and worth of Columbus might have remained intact. But of course, his assumptions were wrong: So great a shame that the spice islands were never to be found so close! So great a pity that a jungle-covered landmass—yet still a part of the Indies, the Admiral continued to insist—had somehow so inconveniently settled itself down, blocking easy passage!

But Columbus was not content with this single journey—there were to be three more ventures, all bent on the acquisition of land for Spain, and for periods of annexation and governorship that were marked for their cruelty, tyranny, greed, vindictiveness, and racism. Columbus favored slavery; he had a long record of cruelty toward the native peoples; and he condemned his own followers for various infractions, having their tongues cut out, their nose and ears sliced away, the women given the vilest of public humiliations. On his second voyage he brought a cargo of pigs, which he set free and which bred and provided succeeding seamen-explorers with food (but may also have brought with them some of the diseases that helped decimate the native populations). His third voyage, in 1498, brought him to the American mainland, in Venezuela, where he found the Orinoco

---

* A less impressive sum than it sounds: a maravedi—named for the Berber Almoravids, and so a subtle reminder of Atlantic coastal influences—was valued at just one thirty-fourth of a real, which itself was an eighth (hence "pieces of eight") of a Spanish peso. Maravedi coins were to be the first minted in the New World, on Hispaniola from the start of the sixteenth century.

and assumed it to be one of the rivers mentioned in Genesis; his fourth, in 1502—made while he was still clinging stubbornly to the belief that all his discoveries were unfound parts of the Indies, such that on this particular voyage he might well find the Strait of Malacca—brought him to Honduras. And it was here that he heard whispers of an isthmus, and of a short land passage to another, mysterious ocean.

But the penny never dropped—and the notion that America was a continent, and that the body of water that separated his native home from the lands he was conquering was an ocean, separate from the waters of the east, just never occurred to him. The sea he had crossed was called the Atlantic, true; but in Columbus's mind, the Atlantic was an ocean cunningly joined to the Pacific, quite as seamlessly as if the two seas had long been one.

Christopher Columbus, though he was a courageous and highly skilled sailor, was not the first to cross the Atlantic Ocean. While his voyages introduced Europe to the existence of an entirely new universe across the seas, he himself never reached North America. And in the prosecution of his aims and his duties he behaved not infrequently as a tyrant and a bully, as a slaver, an unrepentant imperialist and a man of immense avarice and self-promotion.

And yet for all of this Americans have adopted and proudly sport as central to their identity the name Columbus, in Columbia; the District of Columbia; the Columbia River; Columbia, South Carolina; Columbia University; Columbus, Ohio—and Columbus Day. For the United States the reputation of the man remains intact despite the best efforts of enlightened teachers. The troubling details of his life, if known, appear in fact to trouble very few.

The calendar still bends before his brand. Ever since 1792, when New Yorkers marked the three hundredth anniversary of his first landing; ever since 1869, when the Italians in the newly founded San Francisco held a similar celebration; ever since 1892, when President Benjamin Harrison urged all Americans to celebrate the four hundredth anniversary; ever since FDR made October 12 a holiday; and ever since 1972, when President Nixon shifted its observance to the second Monday in October, Americans have taken formal and honored notice of Christopher Columbus, with the establishment of a great national holiday in his name. And even if he was somewhat more violent and avaricious than was perhaps necessary, history generally treats him well. By contrast, Leif Eriksson, who almost certainly was the first man to cross the ocean, who probably did make it onto the mainland, and who was a man whose motives seem to have been directed to the general good and who left no legacy of harm, passes largely unremembered, little memorialized. True, there has been since 1964 an annual and presidentially proclaimed Leif Eriksson Day to honor the contributions of Nordic people to the United States. Minnesota and Wisconsin were the first to observe it, by closing some offices, and with some local merchants offering discounts. But in all other respects the American nation remains largely mute and oblivious to the Norsemen. Most Americans prefer pizza, as someone put it, to lutefisk.

It seems a peculiar misreading of history, one that performs a small and nagging injustice to the long story of the Atlantic Ocean. Matters are changing, though slowly. Perhaps in time a wise counselor will accept the inequity and will publicly suggest some measure of right by moving to limit the excesses of the one memorial and to restore to its proper degree the unsung other. But one has to doubt it.

Perhaps the reason for this lies less in Italian chauvinism and Nordic modesty, and more in the one undeniable reality: that though Leif Eriksson got to North America first, he never truly realized he was there. Nor did he suppose that he was anywhere of particular importance. One might argue that he just *didn't get it.* As the historian and Librarian of Congress Daniel Boorstin once put it: "What is remarkable is not that the Vikings actually reached America, but that they reached America and even settled there for a while without discovering America." And so their reputation has suffered for the lack of ambition of their wanderings, *for their lack of vision*, ever since.

. . .

And there is always one further question that niggles away at critics of colonial adventuring and white hegemony. Is it conceivable that the pre-Columbian peoples themselves, the original inhabitants of the Americas, ever tried to head out east across the ocean, to Europe? Could any of these—the Caribs, say, or the indigenous Newfoundlanders or Mexicans—have made the journey that Eriksson and Columbus eventually made, but in reverse?

Circumstantial evidence hints at the possibility, certainly. Tobacco leaves and traces of coca in Egyptian sarcophagi. A sculpted bronze head in the Louvre, said to be Roman of the second century A.D., and which displays features uncannily similar to those of Native Americans. Mosaics from near Pompeii with images of objects that resemble pineapples, chili peppers, and lemons. And the suggestions—made with varying degrees of enthusiasm by a small army of competing translators—that Christopher Columbus encountered a husband and wife from the Americas, in of all places Galway, Ireland, in 1477. Whether he met them socially, or saw their corpses, or merely heard of their existence, remains tantalizingly unclear.

"People from Katayo came to the east," wrote one of the translators of the marginal scribbles Columbus made in a history text that he was known to have read. "We have seen many notable things, and especially in Galway, in Ireland, a man and a woman on some wood dragged by the storm, of admirable form."

But could a couple have survived, in a dugout canoe—for that is the kind of craft most Caribs appear to have been using at the time the Europeans first saw them—for a journey that crossed the entire Atlantic Ocean, from the Americas to Ireland? The Gulf Stream might have carried them—it carries all manner of flotsam with it. But they would have sailed with it at a mere three knots—a total of fifty days of sailing to reach the Irish coast, and with neither food nor fresh water to sustain them. It seems highly doubtful that they came to Ireland by accident; and if it was by planning and design, which is the only possible manner in which any transatlantic voyage could reasonably be made, then one suspects that there would have been others who tried to follow them, that there would have been more artifacts of their journeying found, more evidence that such a voyage had happened.

And none has ever turned up. The champions of the theory that Native Americans reached Europe by sea are as vocal as they are passionate, but thus far the arguments remain thin. The balance of probability suggests that it was Europeans—northern or southern—who sailed across the Atlantic first.

9. REALIZATIONS

In a matter of months after the death of Christopher Columbus in 1506, three men—one a Tuscan from Chianti who at one time or another was a sailor-explorer, a pimp, and a sorcerer; and the two others simply solid German cartographers from Freiburg— put the requisite two and two together and gave formal birth both to a continent that would be called America and to a recognizably self-contained ocean called the Atlantic.

Columbus had found only the vaguest adumbrations of a continent-sized landmass. He had encountered, charted, and colonized hundreds of tropical islands, as well as a subequatorial coastline that sported rivers big enough to suggest that they drained something rather larger. But in all of his voyages he had found no real evidence of a great land that was large enough to block westward passage on all the navigationally available latitudes.

But then toward the turn of the century news started to trickle in from other explorers that hinted that such a body might exist. John Cabot, for instance, had almost certainly landed in eastern Newfoundland in 1497, reporting back to his sponsors in Bristol, England, on the presence of a large landmass. Then two Portuguese brothers, Miguel and Gaspar Côrte-Real, reached a variety of points also on the northern coast, and on their return to Lisbon in the autumn of 1501 suggested—and for the first time by anyone—that the land they had just encountered in what are now the Canadian maritime provinces might well be physically connected to the landmasses already discovered to the south—the body of land we now know as Honduras and Venezuela.

A somewhat inelegant little map had also started to confirm the gathering suspicions of the educated European public. It had been drawn in 1500 by Juan de la Cosa, a Cantabrian pilot who had twice accompanied Columbus and who would make five further voyages to the New World—only be murdered by natives with poisoned arrows in 1509, on the Atlantic coast of Colombia near Cartagena. But his map, held today in Madrid's naval museum, lives on; it was the first ever made that displayed a representation of the New World—an edge-to-edge border of territory on the map that lay far to the west of Europe. It was marked by an enormous concave embayment, with the lands found by Cabot on its northern side, those found by Columbus and company to the south (and all of the territory, thanks to the Treaty of Tordesillas,* supposedly Spanish). But no names were offered on the map, either on the landmass or on the sea.

That was to happen just seven years later, in 1507. It took the German mapmaker Martin Waldseemüller to fasten the name *America* onto what now even more clearly was a newly recognized continent. Waldseemüller and his poetically inclined colleague, Matthias Ringmann, did so despite a welter of confusions, deceptions, and falsehoods that have intrigued scholars and occupied writers for centuries, because of a vastly popular booklet they had lately seen. This slender book, truly more of a pamphlet, was known as the *Mundus Novus*, and together with a subsequent brief document known as the Soderini Letter, was purportedly written by Amerigo Vespucci, the colorful Italian explorer and sorcerer (and in later life the aforesaid pimp) who appears to have been the very first to claim from his own navigational evidence that the great body of land in the west was in fact a separate continent, *the fourth part of the world.*

The *Mundus Novus* is a prolix, flamboyant, and in detail quite unreliable work, thirty-two pages long, printed and written in Latin, addressed initially to his Medici sponsor and then published in 1503 simultaneously, like the opening of a modern feature film, in many cities around Europe. Printers in Paris, Venice, and Antwerp saw to it that Vespucci's graphic descrip-

---

* This agreement, made in 1494, allowed Spain sovereignty over any newfound lands that lay west of a meridian drawn 370 leagues to the west of the Cape Verde Islands, and gave Portugal the rest. Since Brazil lies east of this meridian, it alone in Latin America fell under the rule of Lisbon.

Amerigo Vespucci's first name gave the world America. But this Florentine navigator was also the first to realize that the Americas formed a continent between Europe and Asia—and therefore that the Atlantic was a discrete and separate body of water, an ocean.

tions of his sailing adventures along the coasts of what we now know to be Guyana, Brazil (where he was the first European to enter the mouth of the Amazon), and perhaps even Patagonia enjoyed a massive circulation.

The book was indeed wildly popular—helped no doubt by Vespucci's loving discussions of the cosmetic self-mutilation, anal cleanliness, and sexual practices of the people he met along the way. It was a book that not only gave him personal immortality but also led to the explosion of European interest in the New World and the beginnings of a rolling tide of exploration and immigration that one might fairly say has not abated since.

The crucial sentence in Vespucci's pamphlet stated simply that "[on] this last voyage of mine . . . I have discovered a continent in those southern regions that is inhabited by more numerous peoples than in our Europe, Asia or Africa, and in addition I found a more pleasant and temperate climate than in any other region known to us. . . ." He had found a new continent—or, more precisely, he had identified the land that he had found *as* a new continent, something that Columbus, some years before, had been entirely disinclined to do. To Columbus it was—and it wrongly was—an already existing continent: Asia. To Vespucci it was—and it correctly was—a totally new continent, and at the outset it was a continent without a name.

It fell to the Freiburg mapmakers to give it one. At the time the pair happened to be working in an academic community in the Vosges mountains of eastern France—and it was here that they got finally to christen this great body of land, and to offer it an identity it would then have for all time. Both of the mapmakers had read the *Mundus Novus;* both had read and were taken in by the more evidently forged Soderini Letter. Both agreed that in the preparation of an enormous new world map that had been commissioned from them, they would give, at least to the thinly sinuous southern part of the new continent that would be drawn on their masterpiece, a name. They would give it the feminine form of the Latinized version of Amerigo Vespucci's Christian name: the properly feminine place nouns of Africa, Asia, and Europe would now be joined, quite simply, by a brand-new entity that they would name *America.*

And so, in 1507, when the new map was published, and with images of the two giants of Ptolemy and Vespucci presiding in profile over an entirely fresh cartographic representation of the planet (but with neither Leif Eriksson nor Christopher Columbus in illustrated evidence anywhere), so, in large letters across the southern half of the southern continental discovery, just where Uruguay is situated today, was this single word. *America*. It was written in majuscule script, a tiny bit crooked, curiously out of scale and looking a little last-minute and just a little tentative—but nevertheless and incontrovertibly *there*.

The name caught on. A globe published in Paris in 1515 has the word written on both segments of the continent, north and south. It was published in a Spanish book in 1520; another from Strasbourg five years later listed "America" as one of the world's regions; and finally, in 1538, Mercator, the new arbiter of the planet's geography, placed the phrasal titles "North America" and "South America" squarely on the two halves of the fourth continent. With that the name was fully secure; and it would never be changed again.

And with a new continent in place, so the sea that lay between it and the Old World continents of Europe and Africa—the sea that had variously been named the Ocean Sea, the Ethiopian* Ocean, *Oceanus Occidentalis*, the Great West Sea, the Western Ocean, *Mare Glaciale*, and by Herodotus in *The Histories* in the fifth century B.C., *the Atlantic*—became, at last and with certainty, a discrete and bordered ocean, too.

It was no longer appended to any other sea. It was no longer a part of some larger and more amorphous worldwide body of water. It was a thing—a vast and, back then, almost unimaginable thing, true—but it was a thing nonetheless, with borders, edges, coastlines, a rim, a margin, a fringe, a brink, and a northern, a southern, a western, and an eastern limit.

---

* *Okeanos Aethiopikos* was the name given by the Greeks to that part of the Atlantic south of its narrow neck between Brazil and Liberia, and was still employed on some maps published as recently as Victorian times. Ethiopia itself is not on the ocean, but the name was once given to all Africa—in part perhaps because of the region's perceived importance as the birthplace of humankind. The use of the word to describe the South Atlantic is thus a means of calling it the "African Ocean."

From a simply inexplicable green-gray immensity that stretched without apparent cease beyond the Pinnacle Point tide pools, to an even more frightening turbulence of waves and winds raging beyond the Pillars of Hercules, to a warm sea stained with purple dye or a cold sea choked with ice, and to a body of little self-importance and which was supposedly conjoined with other seas that lay far beyond, the Atlantic Ocean now at last, and from the moment of being given the early-sixteenth-century imprimatur of Mercator, had a proper identity, all of its own.

It remained now to find out just what that identity was, and to set this newfound ocean in its rightful place on the world stage.

The Atlantic had been found. Now it demanded to be *known*.

Chapter Two

ALL THE SHOALS
AND DEEPS WITHIN

⚊⚊❦⚊⚊

Then the whining schoolboy, with his satchel,
And shining morning face, creeping like snail
Unwillingly to school.

1. THE DEFINING AUTHORITY

The Principality of Monaco, that sunny haven for shady cash on
the French Riviera, is not a place sufficiently blessed by a noble
history to be littered with an abundance of grand public statues.
The parks and plazas naturally have plenty of marble representations of members of the Grimaldi family, the Genoese nota-
bles who have run the place since the thirteenth century. There
is a kindly sculpted bust of Hector Berlioz, who is remembered
for once falling over near the Opera House, and there is a dull
bronze rendering of the Argentine speed demon Juan Manuel
Fangio, standing beside the Formula One Mercedes in which he
won so many of his local car races.

But those monuments aside, there is little other statuary of interest—except in the entrance to a rather anonymous-looking modern office building on the Quai Antoine the First, beside a harbor that is permanently jammed gunwale to gunwale with large cruising craft. There stands a striking and rather magnificent statue in polished teak, of the great Greek god of the Sea, Poseidon. He stands there decently naked, full-bearded, and wielding his trident in the stance of a guardian outside the little-known office that, since 1921, has defined, delineated, and approved the official names of all the many oceans and seas, bays and inlets on the surface of the planet.

The International Hydrographic Organization has been in Monaco since 1921, invited to this improbable setting* by the then ruler, Prince Albert I, a man who collected charts and portolans and had a fleet of research vessels, and held great admiration for and great knowledge of the deep-sea fish and marine mammals for which he went exploring. The organization he helped create has as members almost all of the oceanside states of the world—Algeria to Venezuela, by way of Jamaica, Tonga, and Ukraine, and with all the obvious big-sea countries among the founders.

One of its principal mandates is to define—in a de facto rather than a de jure sense—the boundaries of the world's oceans and seas. This turns out to be a most contentious matter. Right from the start there was argument: "Your proposed western limit of the Mediterranean," huffed a Moroccan delegate in the 1920s, when comments about the first proposed boundaries were invited, "makes Tangier a Mediterranean port, which it certainly is not."

---

* Hydrographers—"droggies" in the naval vernacular—are usually seagoing science types, by no means a patrician group. But in Monaco, thanks to Prince Albert's munificence, they work cheek by jowl with those who are, or wish to be, patricians. Fellow academics at the local university, for instance, teach courses in such subjects as Wealth Management, Hedge Funds, Financial Engineering, and the Science of Luxury Goods and Services, while the droggies deal with lighthouses, buoys, and dredging.

The original architects had thought fit to make the boundary of the North Atlantic Ocean pass on the outside of the entrance to the Strait of Gibraltar, a decision that appeared to unsettle everyone. So on instructions from the senior brass, the clerk promptly erased his first boundary line and drew in a second, a mile to the east of Tangier—elevating it at a stroke to the status of an Atlantic Ocean city, and not a mere Mediterranean port—and all were reported to be happy.

The IHO's other important, practical remit is to ensure that all the world's navigation charts look more or less the same. This is not quite as dull as it sounds. It stems from a conference held in Washington, D.C., in 1889, at which grim stories were told about ships' captains who were compelled to use charts made by countries poorly skilled in chartmaking and so came to sudden grief on unmarked shoals or on the approaches to ill-drawn harbors. The only way such marine misadventures could be prevented, said the conferees, was for all charts and all navigational aids to be the same, and for all sailors' maps, whether made in Britain or Burma, the United States or Uruguay, to adhere to exactly the same high standards.

At a navigation conference held in St. Petersburg, Russia, just before the Great War, the world's navies and merchant mariners promptly urged that an international commission be set up to study such problems. Finally in 1921, once the European dust had settled, Monaco's well-regarded Serene Prince offered room and board and a clutch of Monegasque typists (together with one charmingly styled "boy-attendant") to help establish the IHO, which was then formally constituted and guarded by its pet Poseidon where it resides contentedly, if rather obscurely, to this day.

Its most important publication occurred in 1928. Back then, and costing thirty-five American cents, it was a handsome green-covered pamphlet, printed letterpress by the Imprimerie Monégasque of Monte Carlo and titled IHO Special Publication No. S.23, "Limits of Oceans and Seas." In the twenty-four pages of this endearing publication one would find such official pronouncements as the formal description of the limits of the English Channel:

> *On the West: From the coast of Brittany westward along the parallel of the E. extreme of Ushant (Lédènes), through this island to the W. extreme thereof (Le Kainec), thence to the Bishop Rock, the SW extreme of the Scilly Isles, and on a line passing to the Westward of these Isles as far as the N. extreme (Lion Rock) and thence Eastward to the Longships, and on to Lands End.*

The world may not have expanded in the years that followed, but the definitions and denominations of its seas, and the arguments among the countries that lay beside them, most certainly did. In consequence the size of this pamphlet grew, modestly at first, and then prodigiously. The twenty-four pages of the first edition grew to twenty-six pages in the second, and then thirty-eight pages in the third edition—but when the fourth edition was published in 2002, it had ballooned into 244 pages. Seas so obscure that only those who live beside them have ever heard of them now officially exist: there is a Ceram Sea, for example, a Cosmonauts Sea, an Alboran Sea, a Lincoln Sea, a somewhat tautological Sound Sea,* and scores upon scores of others.

Three senior naval officers from member states are elected to preside over the International Hydrographic Organization, usually for five years at a time. Before I traveled to Monaco to see

---

* There are also some highly unfamiliar capes and headlands used to delineate certain of these seas, of which northern Russia's Cape Vagina presents many sailors with particular frisson.

them I had visions of this trio, all splendid in crisp blue uniforms and with coils of gold tassellage, ruling definitively on lofty matters of world navigation—on how best to define the new limits of the Kattegat, on demanding the mapping of where the Arafura Sea abuts against the Gulf of Carpentaria, on determining whether L'Anse aux Meadows was truly washed by the Labrador Sea or the Gulf of St. Lawrence. They would settle these quibbles while quaffing pink gins, smoking pipefuls of rough shag, and carving scrimshaw doodles on the side.

As it happened, two of the officers—from the navies of Greece and Chile—were away when I called one blissful midwinter morning, and the only sailor "on deck," as seamen in offices like to say, was the representative from Australia. He turned out to be a middle-aged, full-bearded Briton in civilian dress, a man who had long ago left the Royal Navy for its Royal Australian counterpart and was now usually based in Melbourne. His driving passion was not so much ships and the sea—they were his job—but the building in his modest apartment in Villefranche of model railway layouts, HO scale.

Officially, however, he and his brother sailors spend a great deal of time wondering about, fulminating over and trying to reverse what they see as a general world ignorance of the oceans. The world's seas may now have more names than ordinary man may care to know—that much seems true, but this is the fault of politicians and a consequence of national pride. What troubles the IHO, which, as mentioned, has as another of its mandates the creation of charts to help ships navigate safely around the world, is just how dangerously unaware most landlubbers are of what goes on beneath the surface of these bodies of water. To illustrate the point, they mention repeatedly one unanticipated statistic: even though mankind now knows the precise altitude of the entire surface both of the moon and Mars at points little more than five feet apart, he knows the altitude of the bottom of the sea only at points that are separated in many cases by as much as *five miles.*

For all the hydrographic surveying that has been done over the years, all the soundings taken and the reefs plotted and the headlands marked, the admirals complain that the current inhabitants of the earth know far too little about their seas, even though oceans cover seven-tenths of their world. This is not for want of trying, however. Europeans especially have been attempting to divine the details of their ocean for the past five hundred years. Ever since Columbus and Vespucci came home, and ever since it became clear that Europeans were inevitably going to trade and fight their way across the Atlantic and all the other seas, there have been great national efforts—in Britain, in Portugal, in Spain, and in time in America and Canada and Brazil and South Africa, too—to survey and chart the waters, to find out the seas' depths and shallows, their tides and currents, their races and whirlpools and the accurate measure of their coastlines, their islands and their reefs, and all the other features that mark them out so peculiarly. Educating the world about the ocean—with the knowing of Atlantic in the very forefront of the effort—was a venture that got under way as early as the fifteenth century, and it has not stopped for a moment since.

To survey an entire ocean required access to all of its farther limits—access that in the case of the Atlantic was for a long while frustrated by more than a few navigational challenges. The severest limit was the existence of a highly inconvenient sandstone headland known as Cape Bojador—a West African cape that the Arab sailors had feared for centuries and knew as *Abu khater* or *the father of danger.*

2. THE ROADBLOCK IN THE WATER

The road into the Sahara south from the old Moroccan seaside fort city of Essaouira happens also to be the main trunk Atlantic coast road into West Africa—it passes on to Mauritania, and then to Senegal, the Gambia, Guinea-Bissau . . . With careful planning, good fortune, decent springs in your car, and a fair amount of time, a determined driver could make it to Cape Town, arriving in time for tea under the jacarandas at the Mount Nelson Hotel.

For most of its early miles, the journey has a steady tedium. After the spectacle of the Atlas Mountains dipping into the sea to which they gave their name, and after passing through the tiny Spanish enclave of Ifni, and after seeing the chains of great French-built lighthouses and the surfers carelessly riding the rollers thundering in from the sea, you drive inland for a few miles and the road becomes flat. The groves of argan-oil trees and goat-busy scrub eventually give way to the stony desert plains of the hammada, and there is a dreary little junction town called Goulmime, where the desert proper begins.

Beyond the dust and chaos of its medina—with blue-robed Touareg still to be seen, and desert-weary cameleers bringing trade goods for the souks—the two-lane highway, oil-dark against the sands of the hammada, winds empty over the horizon, with just the occasional tanker truck whooshing past, and fleets of rickety Mercedes taxis traveling too fast for their own good. The sea rumbles endless to the west, and there is the glint of the high ergs of the Sahara far to the east. The east wind constantly whistles, leaving grit in one's hair and teeth. This until lately was the entranceway to Spanish territory, and one sees it in the landscape and the feel of the place. The north of Morocco possesses a certain silky plenitude, whereas this more south-erly corner of the place has a harshness: dry, dusty, and stained with oil.

The towns are far apart, and generally worth stopping at only for fuel—though one of them has a monument to Antoine de St.-Exupéry, memorializing his time as a pony express pi-lot for the 1930s coast airmail service between Toulouse and Dakar; and there are plenty of fishermen's huts where one can find grouper, swordfish, and sardine, plucked from the sea and grilled over driftwood fires. The coastline itself becomes more interesting, too. Near Tarfaya it turns abruptly out to sea, a fifty-mile change of direction jutting into the ocean and into which, over the years, scores of vessels with captains sleeping, stupid, or drunk have plowed: there are wrecks of fishing boats sitting high and dry and majestic on the rocks, all being slowly chewed away into nothing by the ever-digesting surf.

The seas here have a particular reputation for peril. From high up on the hills above this headland's terminal point, Cape Juby, it is just possible to make out the closest of the Canary Islands, Fuerteventura. Until the ever-gnawing seas did their work, a fa-mous shipwreck also lay here: the great transatlantic star of the fifties, the Virginia-built liner SS *America*, broke loose in a storm while being towed to Thailand in 1994 to be turned into a float-ing hotel. She now lies almost wholly sunk a hundred yards off a Fuerteventura bathing beach, a forlorn memorial to the brief greatness of America's merchant marine.

. . .

It is indeed coastal danger that is most memorable along this stretch of the African shore. At latitude 27 north, some 150 miles south of the wrecks of Cape Juby, there rises a long, low headland, extremely undistinguished in appearance. This is a headland that remains hugely important in the history of Atlantic navigation, though in reality it is disappointingly not at all like those other famous Atlantic capes—Finisterre, Horn, Good Hope, Farewell, St. Vincent, Race—that are the stuff of great poetry and legend. This, more modest in its majesty and its menace, is Cape Bojador.

Though the Portuguese word *bojador* hints at "protrusion," the land that makes up this low hurdle of cliffs is not a protrusion at all; nor does it pose anything but the smallest inconvenience to a vessel passing south along the African coast. But for many centuries no sailing vessel ever dared to pass it, nor was one physically able to do so. *Quem quer passar além do Bojador, Tem que passar além da dor*, wrote the modern Portuguese poet Fernando Pessoa. *He who wants to pass beyond Bojador, must also pass beyond pain.*

Beyond it lay a totally unknown sea—a terror-inspiring, monster-filled wilderness known in all ports as the *Green Sea of Darkness.*

Until the fifteenth century, no sailor—whether Spaniard, Portuguese, or Venetian, whether Dane or Phoenician, and by all existing accounts no African sailors, either—had ever successfully rounded Cape Bojador from the Atlantic Ocean side. All the early navigational academies of Europe regarded the sea beyond Bojador as quite impassable. Its very existence stands as one of the reasons why the central Atlantic Ocean, despite having almost certainly the world's most populous shores, was the last of the great seas to be properly navigated. Polynesian navigators had long before crisscrossed the Pacific; Persians and Gulf Arab sailors had taken their reed-and-creosote sailing craft across upper parts of the Indian Ocean; Chinese sailors knew the intricacies of the eastern Indian Ocean and their various littoral seas; and the Vikings knew the navigational complexities of the far north. But traditional navigation seemed not to work so well or so speedily in the Atlantic as it did elsewhere, and Cape Bojador, so far as the literature records, was one of the reasons why.

The problem with Bojador was created by a unique combination of circumstances—topographic, climatic, and marine. No hint of impending difficulty would be apparent to a southbound sailor, who would perhaps leave from an Iberian port, head past the Strait of Gibraltar with light winds still favorably on his starboard quarter, and dip steadily along the African coast at a comfortable five or six knots. He would mark his passage each day by the sight of the three obvious Moroccan capes: Rhir, Draa, and Juby. He would see the twinkling fires of the settlements of Casablanca, Essaouira, and El Ayoun, taking comfort from their proximity—for like most in those early days he was probably a nervous sailor, and would always be most reluctant to leave the sight of shore, finding a degree of security in his crabwise progress along the edge of land.

And then he would come to Bojador—and in an instant his illusion of comfort would evaporate. An unseen sandbar, stretching twenty miles out from the low cape and reducing the depth under his keel to a mere couple of fathoms, would first compel him to turn to starboard and, against his better judgment, head out into the deep ocean. At the same time the telltales on his mainmast would show that the slow winds off Morocco had suddenly changed to full easterly, and may well have picked up to a steady half gale. (For most of the year the winds veer east just at this point, and modern satellite pictures will show trails of desert sand being wafted each summer out across the Atlantic.*) And third, once clear of the submerged point, a current—the North Equatorial Current—would catch his vessel in its power-

---

* Sands from the hammada around Bojador are blown as far away as Brazil, where they settle on and help fertilize the alluvial Amazon soils. The local soybean farmers are unaware of the debt they owe to the dunes of Morocco.

ful maw and begin to drag it westward, too, for possibly as much as six hundred miles.

The perils of the cape are actually even more conspiratorial than this. For most of the voyage down along the coast a persistent southbound current, called by the Portuguese mariners the Guinea Current (modern mariners call it the Canary Current), helped sailors of old to scurry along the shore, providing only that they kept close to land. This was important, because one characteristic of the Guinea Current was that it became steadily weaker the farther offshore. This then gave a ship's master two equally unpalatable options: remain close inshore and risk being swept into the irresistible arms of the westbound equatorial stream, or sail well away from land and encounter only a fading current and weak winds, and remain motionless at sea, with food and water running out, the vessel trapped in doldrums of your own making.

Small wonder not a single seaman managed to get past the cape—until a seminal moment, seventy years before Vespucci, in 1434. It was the growing intelligent acquaintance with the complexities of the sea that eventually allowed the Bojador problem to be solved—as the early phase of exploration of the Atlantic steadily gave way to a period of rigorous ocean education. *To know the sea* became the working phrase, for only by knowing it could its dangers be avoided and its treasures exploited. The Cape Bojador story is a classic example of this shift in attitude.

It was a young Portuguese navigator named Gil Eannes who is commonly credited with having the maritime intelligence and *feel for the sea* so necessary for blazing a southbound trail. Though most of the papers relating to his voyage were lost in the Lisbon earthquake three centuries later, sufficient anecdotal evidence remains to hint at just how he did it. It was entirely a matter of intelligence: of using such intellectual techniques as observation, forethought, timing, planning, and calculation.

Prior to Eannes, sailors merely set themselves a goal (or had it set it for them by their financial sponsors), victualed their ship, and set off—and in the case of their West African ventures, all were forced to turn back after little more than a thousand miles. These sailors employed the rituals of old—they followed the currents, they sailed before the winds, they followed the paths of seabirds. But what Gil Eannes did next involved an immense amount of planning, and it invoked the growing science of celestial navigation, already known to the Arab merchants after diffusing slowly from the East, where it had already evolved to a fine degree under the Chinese.

Eannes believed it would be possible now to move through the Atlantic and reach places unfavored by winds and currents and migrating birds if a mariner used the new tools fast becoming available: astronomy, timekeeping, a sophisticated knowledge of weather and climatic history, and the geography of the sea. To get around Cape Bojador specifically, or to *double* it, in seamen's jargon, involved the scrupulous measuring of the water speeds and directions, and the measuring in detail of the average directions and strengths of the winds. It involved the development of a technique now known as *current sailing*. Eannes also drew *current triangles* onto his crude but ever-improving charts sets, and used vectors, intelligent tacking, and careful, hour-by-hour timing. Once he knew the currents and the winds, their directions and their speeds, it remained a matter of simple trigonometry to plot a course that would take advantage of them both. However, his planning also involved choosing a season when the winds of one kind would be blowing while those of another would not.

Only after all that information had been digested and calculated and factored in could Eannes set his rudder and trim his sails and point his bowsprit in a direction that might have seemed to his unsuccessful predecessors eccentric—but eccentric in the way that modern Great Circle routes seem oddly counterintuitive when compared with the apparent directional simplicity of a voyage made along a straight line of constant bearing, in other words a rhumb line or loxodrome.

The precise details of his famous voyage remain unclear—there was no surviving diary, no ship's log, not even the name of the ship is recorded. All we know is that Eannes went south on the explicit orders of Henry the Navigator, the princely architect of Portugal's imperial ambitions. Henry had drily remarked that fourteen previous attempts to double the cape had failed. Eannes, who was no more than one of the personal servants in Henry's court, might just as well make an attempt, too.

He did precisely as he was ordered—he sailed southwest to Madeira and the Canaries, then performed all his complicated arithmetical elaborations. He initiated a deep-sea twisting-and-turning voyage that has been known for many years since as the Portuguese *volta*, and by doing so succeeded finally in rounding the dreaded cape. He was then blown by the gusts of a harmattan wind onto the African desert coast some thirty miles south of Bojador and picked a sample of the woody desert plant known as St. Mary's flower or the rose of Jericho to bring back as proof. It didn't work: none of this convinced the skeptical prince Henry, who promptly ordered Gil Eannes to go out to sea once again.

So back he went the following year, 1435, this time with a companion—a man who was also a household servant, though a spare-time sailor—and they took their puny fishing *barca* on almost the exact same plotted route, with its wide westerly diversion south of the Canary Islands. The men landed at almost the same African coastal spot, they named a river, they saw the footprints of men and the hoofprints of camels, and thus realized that the Torrid Zone was peopled, and they came back to a finally credulous Henry the Navigator, and in consequence to a brief period of court rapture followed by a lengthy period of public obscurity.*

And the two ventures did the necessary trick. Within months other expeditions had set out from the Portuguese harbors, and they fanned out along the at-long-last-accessible coast of Africa to explore it, round it, and then eventually to head off east beyond the continent of Africa, to the great treasure grounds of the Indies.

The ships grew steadily in size—from the tiny *barcas* used by Eannes, to the three- and four-masted caravels, and the gigantic *naos* employed on the spice runs of the sixteenth century. The equipment carried on the ships' bridges became more sophisticated: the astrolabe was soon to be perfected, the compass to be employed, sounding wires to be made long enough to deal with exceedingly deep waters, and tide tables and sight reduction tables to be published.

The mariners became ever more adventurous, and history is littered with their names: Bartholomew Diaz, who first rounded the Cape of Storms; Vasco da Gama, who first went to India; Pedro Cabral, the first to land in Brazil; Alfonso d'Albuquerque, first in Malabar, Ceylon, and Malacca; and all those other sailors whose names—Fernando Póo, Tristan da Cunha, Luis vaz de Torres—are memorialized in islands or straits (or as these three are remembered, for a slaving colony off Africa, a dangerous volcano in the far South Atlantic, and a narrow passage between New Guinea and the northern tip of Australia). Perhaps greatest of them all, though claimed by others, was Fernão de

---

* However, there is statue of Eannes on the seafront in Lagos, the ancient town on the windward side of the Algarve where Prince Henry maintained his headquarters, and from where the Bojador expeditions set out.

ATLANTIC OCEAN: ROUTES OF EXPLORERS AND SETTLERS

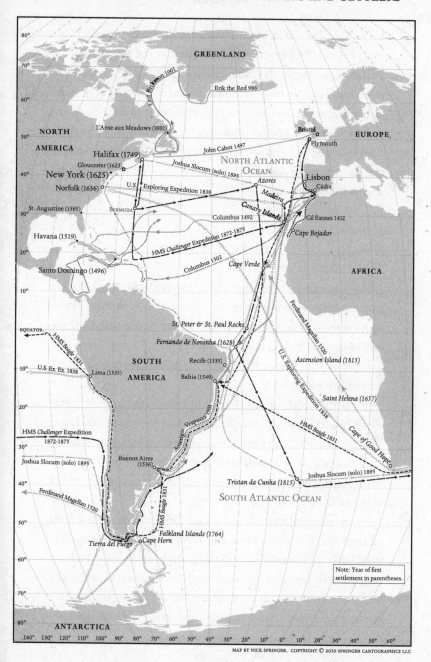

MAP BY NICK SPRINGER. COPYRIGHT © 2010 SPRINGER CARTOGRAPHICS LLC

Magalhães, the would-be circumnavigator who was born in Portugal but sailed for Spain, and died as Ferdinand Magellan in the Philippines in 1521. All of these indefatigable sailors and a score more—who mostly came from a Portugal of which it used to be said *such a tiny land to live in, but the whole world to die in*—came to be legatees of the pioneering sailing techniques of Gil Eannes. They followed in his wake, both literally and figuratively, to begin the organized acquisition of knowledge about the Atlantic and all the other oceans besides.

3. MOVING THE WATERS

It has to be remembered that until Amerigo Vespucci, there was no knowledge—nor even a suspicion or a hint—that the Atlantic was a separate sea. Culturally, this was an ocean that until the end of the fifteenth century was not known to exist. Then, and at a stroke, with Vespucci's voyage, the Atlantic Ocean was born; suddenly it was *there*.

With this realization of a brand-new sea, anchors were weighed and sails unfurled, brass clocks were wound and heaving lines leaded. Scientists were appointed and chartmakers assigned, and bold and fearless skippers in their legions took their little ships out of port and headed off to measure and to mark this new body of water.

At the edges of a sea, it is the daily tides that prove the most obvious features to measure and record. Out in the deeps beyond, beyond the influence of tides, the seaman must look for other things: for the size of waves and the direction of swells, the tenor of storms, the press of fish and birds, the depths beneath the bow. And most important, the unexpected and initially mysterious ways that the waters appear to move.

Since these motions are among the most clearly influential on the passage of any ship—as Gil Eannes experienced off Cape Bojador and then made use of—they were noticed very early on in the exploration of the Atlantic. They seemed like great underwater rivers, or torrents. Currents—from the French, *things that run*—were the first of the ocean's many unseen features to become properly known. And perhaps no stream was more famously so than that immense rushing extension of the North Equatorial Current known, from Florida, where it begins, to the west of Scotland, where it ends (with palm trees growing beside the waters that it so conveniently warms), as the Gulf Stream.

Like many mariners around the world, Columbus noticed the currents—and here the exceptionally strong currents that seemed to him so unusually prevalent in Caribbean Atlantic waters. "I found the sea ran so strangely to the westward," he wrote in the log of his third voyage, describing his passing through the notorious Dragon's Mouth, between Trinidad and the Venezuelan mainland, "that between the hour of Mass, when I weighed anchor, and the hour of Complines, I made sixty-five leagues of four miles each with gentle winds. . . ." There are also accounts by Peter Martyr, the Spanish court historian who, coincidentally, was among the first to recognize the huge potential importance of the Gulf Stream, of a vain attempt by Columbus to take a sounding off the coast of Honduras, only to have the "contrary violence of the waters" force his lead upward and never once allow it to touch bottom.

But Columbus was too far south to experience the power of the Gulf Stream. That happy discovery was left to his successor, Ponce de León,* who found it in 1513 while on his quest for the

---

* There is some slight evidence that John Cabot's doughty little ship the *Matthew* was pushed along by the Gulf Stream between Ireland and Newfoundland, but Cabot didn't seem to recognize it as such—he just accepted its north-bearing nudges as part of the deity's eternal munificence.

fountain of youth—a search that eventually won him the ironic substitute of being the first European to find Florida. He was charting the topography of this new coast—thinking it to be a large island, *the flowered one.*

Ponce made rendezvous with two other ships coming north from Puerto Rico, and the three vessels set themselves to sailing farther south, keeping Florida just in sight on their starboard side. One afternoon, when they were perhaps thirty miles from shore, Ponce de León and his fellow sailors suddenly found themselves swept into and caught up in "a current such that, although they had a great wind, they could not proceed forward, but backward, and it seems that they were proceeding well; at the end it was known that the current was more powerful than the wind." Whatever was the cause, this wide river of water, which he soon found swept northward and in time turned toward the east, had huge and unstoppable power. The Spaniard became swiftly aware of its commercial implications: that however difficult it might be for ships to beat their way westward across the middle reaches of the Atlantic, the power of this submarine river offered the guarantee that anyone who floated onto it would be taken home, in style and with considerable speed. Empty galleons might find the outbound passage a trial, but treasure-laden and stately, they could dip home from the Isthmus of Panama, pushed along by this newfound current, with a very welcome dispatch.

Riding the Gulf Stream home quickly became a kind of navigational sport. The traditional means of return to Spain—though it was barely a tradition, since the passage had only been first opened two decades before—was to use the winds alone, to take advantage of the westerlies that blew for most seasons in the middle latitudes of the ocean. But there was a risk inherent: on a voyage from the Main it was tempting to turn east, to turn for home, too early, and in doing so chance becoming becalmed in the fickle breezes of what is now known as the Bermuda High. Now that the Gulf Stream was known, the solution was simple— though, as with the sharp turn to sea made by Gil Eannes in rounding Cape Bojador, it was also counterintuitive. He headed west to go south; homebound Atlantic skippers needed to head north to go east.

Coming from the Isthmus they would tease out the Gulf Stream's beginnings in the Caribbean and then more properly in the shallow waters off what is now known as Cape Hatteras. Once it was found, a homebound sailor would attempt to slot his ship

When he was America's first postmaster general, the fearsomely poly-mathic Benjamin Franklin heard packet-boat skippers tell of the power and extent of the Gulf Stream, and he drew a map of it. Crude though it may look, it is remarkable for its generally accurate depiction of the current's size, shape, and direction.

neatly into the sixty-mile-wide band of its warm, fast-flowing waters; let the current carry him north at nearly six miles per hour, and then as it turned, head eastward with it too, following its warm blue stream for most of its two-thousand-mile curving, Europe-bound length.

Once this marvel had been discovered, and once its spread and its speed had been mapped and measured, the Gulf Stream swiftly became an object of widespread fascination. Its most resolute early champion was perhaps its most improbable: the polymathic American statesman and founding father Benjamin Franklin. In a most remarkable letter written on board a Falmouth, England—bound packet boat in the summer of 1785, he ruminated with precision and wisdom on *Sundry Circumstances Relating to the Gulph Stream*—a document of such intellectual richness that it is easy to see why this most unforgettable of men went on to invent such wonders as the lightning rod, bifocal lenses, lending libraries, a superior kind of fireplace,[*] and the underlying principle behind the glass harmonica.

The letter, to a French academician and friend named Alphonsus le Roy, is quite stunning, fascinating at every line. The Gulf Stream does not appear until halfway through—and by the time of getting down to writing his thoughts about it, Franklin had already offered to his friend a meandering dissertation on the design of ships' hulls, on the possible use of propellers to steer balloons, on the most common causes of accidents at sea, and on the kinds of foodstuffs best stowed for long ocean voyages (almonds, rusks, lemons, and "Jamaica spirits" foremost among them).

---

[*]   The Franklin stove, long popular in postcolonial American homes, enclosed the fire in a ventilated iron box. Its rival was the shallow, brick-lined Rumford fireplace, invented by an Anglo-German count who also created the coffee percolator, invented a nutritious soup for feeding the poor, gave Munich its biggest beer garden, and, fascinated by the complex physics of heat and cold, made the dessert that is known today as baked Alaska.

But then came the Gulf Stream moment, when Franklin reminded le Roy that a decade before, he had been America's first postmaster general, and colonial postmaster before that, and that this was when he first fully apprehended the North Atlantic's most unusual phenomenon of the time:

About the year 1769 or 70, there was an application made by the board of customs at Boston, to the lords of the treasury in London, complaining that the packets between Falmouth and New York, were generally a fortnight longer in their passages, than merchant ships from London to Rhode-Island. . . . There happened then to be in London, a Nantucket sea-captain of my acquaintance, to whom I communicated the affair. He told me the difference was that the Rhode-Island captains were acquainted with the gulf stream, which those of the English packets were not. We are well acquainted with that stream, says he, because in our pursuit of whales, which keep near the sides of it, but are not to be met with in it. . . . I then observed that it was a pity no notice was taken of this current upon the charts, and requested him to mark it out for me which he readily complied with, adding directions for avoiding it in sailing from Europe to North-America. I procured it to be engraved by order.

This stream is probably generated by the great accumulation of water on the eastern coast of America between the tropics, by the trade winds which constantly blow there. It is known that a large piece of water ten miles broad and generally only three feet deep, has by a strong wind had its waters driven to one side and sustained so as to become six feet deep, while the windward side was laid dry. Having since crossed this stream several times in passing between America and Europe, I have been attentive to sundry circumstances relating to it, by which to know when one is in it; and besides the gulf weed with which it is interspersed,

I find that it is always warmer than the sea on each side of it,
and that it does not sparkle in the night.

Franklin then helpfully drew a map—a map somewhat short
on both accuracy and elegance, but one that heralded a new field
of oceanic cartography and by extension helped inaugurate the
entirely new science of oceanography.

4. WRITING THE SEA

This calling, as its curious name suggests—*oceanography, the
writing of the ocean*—was at least in its early days something of a
fugitive science: for how could it be possible to write of a body of
water, especially deep water beyond land, an entity without visible coasts as reference points and no detectable seabed below?
It was like trying to describe the invisible mass of air in a room—
a task rather beyond the imaginative and descriptive powers of
the time.

It's not surprising that of the graphical sciences, oceanography was so late in being born. Geography and hydrography, the
descriptive analyses of bodies of land and water, were disciplines both created in the sixteenth century; it was not until the
middle of the eighteenth century, two hundred years later, that
there was sufficient confidence within the academic community
to name a similar study that would be called *oceanography.* Matters might have been simpler had the science been called *ocean-
ology,* but it never was, and now only the Russians use the term.

On some levels the study of the sea had obvious features that
were well worthy of study. There were the zoological aspects—
the fish, swimming mammals, and seabirds and other animals
both hugely exotic and vanishingly small to catch and record and

classify. There were matters botanical: the existence of floating and sunken ocean plants—Sargasso weed in immense quantities in the center of the gyre of the North Atlantic, kelp banks around islands in the South, and a thousand other pelagic and benthic pieces of botany besides. There was a unique maritime meteorology, too: there were ocean winds in particular to record, in their variety and persistence—trades wafting steadily from the northeast, westerly gales powering the fierce climatic tantrums of the north, and then the fitful and skittish baffling airs around the equator, which were given the name of the tantrums' literal antithesis, the *doldrums*. There were the dangerous circulations of the wind, too—hurricanes, waterspouts, typhoons, cyclones. There was ice and snow, and floes and tabular icebergs. And there were maritime curiosities—St. Elmo's fire, mermaids, the Bermuda Triangle, sea serpents, giant squids.

There was all of this—but each one turned out to be merely peripheral to the ocean itself, in much the same way that the discovery of a new land mammal would be considered peripheral to geography, and the realization of the ferocity of the harmattan wind incidental to the study of oasis formation in the Sahara. Oceans have their own very peculiar physical attributes—a list of inherences and essentials that at the very least would include such matters as the topography of the sea's invisible underneath, the temperature and chemistry of the water, and the movement of the ocean's currents and its tides. And early scientists did indeed notice and inquire: in the seventeenth century alone we had Robert Boyle writing on the sea's salinity, Isaac Newton offering his views on the causation of tides, and Robert Hooke—the famously ill-tempered polymath and philosopher who is better known for establishing the principles of elasticity, inventing sash windows, championing microscopy, first seeing Jupiter's Great Red Spot, and creating an elegant escapement mechanism for watches—designing a host of devices and methods that might be used for research into the deep seas.

. . .

So scientists did eventually begin to focus their attentions, to fathom the unfathomable, and they did begin to come to grips with the immensity of the challenge posed by such a vastness as the Atlantic. They did so especially in Victorian and Edwardian times, a period of both British and American history when the stupendously difficult often seemed unusually possible; this was a time when unraveling the immensity of an ocean looked only marginally more difficult than, say, the cataloging of all the earth's creatures, or the corralling between hard book covers of all the words of the English language, or the building of a transcontinental railroad, or the construction of a canal between the Atlantic and the Pacific.

Fame in the early days belonged to the explorers, those hunting for land and territory and tangible acquisitions, rather than to students of the ocean itself. Bold adventurers like James Cook, Sir John Ross, the Comte de la Pérouse, Robert Fitzroy, and the Chevalier de Bougainville are still remembered and memorialized in capes and straits and islands around the world—while the very earliest true oceanographers have largely faded from memory. Who now remembers James Rennell, for instance, a young sailor from Devon, England, who first came upon the Atlantic proper on a long-sea trick from military service in Bengal? There is today only his tomb, a scattering of long-forgotten books, and the name of a lecture theater at Britain's National Oceanography Centre in Southampton. Yet he was a properly heroic figure, of the mold of Cook and la Pérouse, the kind of seaman who would do whatever was necessary in the pursuit of his calling. While leading a team in the survey of Bengal he had almost his entire arm sheared off at the shoulder during an attack by saber-wielding tribesmen, and then had his original maps of India stolen by pirates off Calcutta, yet persisted in acquiring new knowledge of the sea, in spite of it all.

Rennell's oceanic achievements began in 1777 when he came home by sea—fathering a daughter who was to be born on the quintessentially Atlantic oceanic island of St. Helena, where Napoléon would later be exiled—and en route became captivated by the Atlantic currents that his vessel was compelled to cross, and then by ocean circulation generally. He then helped to survey portions of the deep ocean, and wrote papers on the Gulf Stream, on the North Atlantic Drift, and on the then mysterious current that somehow compelled transatlantic ships bound for the English Channel to head north of Cornwall and toward the Bristol Channel instead. And all the while he delved painstakingly into historical curiosities: the average speed of Saharan camels, the probable landing place in Britain of Julius Caesar, and the likely site of the shipwreck of St. Paul. He lived and worked until he was nearly ninety, and though he was distinguished enough to be buried alongside other national heroes under the nave of Westminster Abbey, he is otherwise widely overlooked.

5. PLUMBING THE DEEP

Comparing James Rennell's interest in the ocean with Benjamin Franklin's a few years before illustrates to some small degree the diverging European and American motivations behind the science that would tackle the strangely sinister world of the deep. Rennell's fascination verged toward the academic and the conceptual; Franklin, whose interest in the Gulf Stream arose out of the reports that mail packets were being mysteriously delayed, had more of a commercial take on the subject. And for years this divergence continued: Britain looked at the sea as something of great theoretical interest, as well as an entrance-way to its ever-expanding empire; America had its eye on the ocean as an obstacle over which mastery could be won only by practical means—by making ever more efficient the shipping lines, by laying and then expanding the use of submarine communications cables, by adroitly harvesting the sea of its edible and usable creatures.

It was the lobbying of powerful merchants in the East Coast ports that eventually persuaded the U.S. Congress to establish a coastal survey, yet at the very same time scientists in Britain, France, Germany, and the Scandinavian countries were all looking to the ocean as the ultimate source, not of trade or funds or fortune, but of an endlessly diverting cavalcade of unknown animals and plants. To Europeans—the generalization may be as unfair as most, yet has enough truth to it to stand—to win knowledge of the Atlantic was to gain knowledge of the planet; to those on its far side in the nineteenth century, to know the Atlantic was to be the better equipped to make money.

Charles Darwin was among those early nineteenth-century Britons who sailed into the Atlantic for the pleasure of study alone. He was just twenty-two, newly graduated from Cambridge, when he was invited in 1831 to sail "to Tierra del Fuego and home by the East Indies" on the ninety-foot-long, ten-gun naval brig HMS *Beagle*. It was a journey that lasted an unexpected five years, and it was principally a survey mission—there were all manner of new devices on board, including accurate chronometers, lightning conductors, and anemometers specially calibrated to measure the newly created Beaufort wind scale. On the way south, Darwin saw and collected specimens from the Cape Verde Islands, the Peter and Paul Rocks, Brazil, Montevideo and Buenos Aires and the Falkland Islands, and on the way home three years later looked in on St. Helena and Ascension, too. But his interest was mainly in the geology or the wildlife of his various landfalls—the maritime aspects of the enterprise were largely left to the ship's captain, Robert Fitzroy.

Perhaps the most memorable event so far as Darwin was concerned occurred as he was about to leave his home ocean and round the Horn into the Pacific: Fitzroy had aboard the ship three extremely large Fuegian natives, captured two years before as specimens* and brought to London to be taught English, clothed, instructed in basic Christianity, and in other ways "civilized." Now they were being taken back home. Despite their London tailoring, fine manners, and good knowledge of English, Darwin regarded them as little more elevated than animals, and was not entirely surprised when one of them, Jemmy Button (the others were a woman named Fuegia Basket and a man, York Minster; a fourth, who was named Boat Memory, had died of smallpox), reverted to his aboriginal state within days of being dropped near the Horn. Soon after being left, he was reencountered when the storm-savaged ship had to put back into harbor—and to the surprise of the ship's company he appeared as shaggy-haired and near naked as when first found, two years before. He could not be persuaded, despite Darwin's entreaties, to return to the ship and come back to London yet again. Though the finches of the Galapagos Islands would eventually reveal much more, these Patagonian unfortunates offered Darwin lessons for his eventual thoughts

---

* The Fuegians were in a sense similar to Omai, the Tahitian boy brought to London on HMS *Adventure* sixty years before. Imported as an example of "the noble savage," the courteous and affable youngster became the darling of London society and had his portrait painted by Joshua Reynolds. On his return to the Pacific he found it increasingly difficult to fit back into island society, and died unhappily, possibly violently.

on evolution: he could say with some certainty from his knowledge of Jemmy Button that the biblical story of human creation was uncertain, at best—for some kinds of clothed men could always revert to nakedness, whatever Genesis suggested took place in the Garden of Eden.

Two expeditions were landmarks in the winning of Atlantic knowledge: the first was conducted by a flotilla of American vessels that set out from Norfolk, Virginia, in the summer of 1838, and the second was the venture by a single Royal Navy vessel that started out from Portsmouth, Hampshire, in the winter of 1872. The former was known somewhat portentously as the United States Exploring Expedition, and in terms of Atlantic history was made more famous by the absence of one invited member who resigned shortly before sailing. The second expedition has come to be remembered, rather more economically, as simply the voyage of HMS *Challenger.* The convoluted fate of the first is still a matter of discussion to this day; but of the second—in more recent times one of the five American space shuttles was named in honor of the single British ship, which testifies to the success of that pioneering sea voyage undertaken almost exactly a century before.*

The American venture—known more familiarly at the time as the Ex-Ex—was an ill-timed, ill-organized, and ill-accomplished congressional attempt to divine the mysteries of America's two neighbor oceans, especially the Pacific. Commerce was Capitol Hill's driving force: the fast-growing American whaling and fur-sealing industries needed new hunting grounds to exploit,

---

* All five of the space shuttle fleet were named after pioneering surface ships, two of them American, three British. *Columbia* was named to honor the first American vessel to circumnavigate the world, *Atlantis* after a stalwart research vessel of the Woods Hole Oceanographic Institute in Massachusetts (but now renamed and working for the Argentine coast guard). *Discovery* and *Endeavour*, the latter deliberately spelled in the English manner, both carried James Cook on his eighteenth-century global navigations; and *Challenger* was named after the ship of the 1872–76 voyages.

and landlubber traders needed new territories with which to do business. Congress offered funds, and then got itself into the most terrible pickle trying to mediate between competing claims of the scientists and the naval officers from which it had to choose to drive the venture out into the ocean. The figure who because of the endless rows chose not to go—but who would nonetheless become nineteenth-century America's most celebrated oceanographer—was a young naval lieutenant named Matthew Fontaine Maury. His decision to pass on the expedition (he had been invited along as the official astronomer, but decided the organizing civil servant in charge was an "imbecile") turned out to benefit his own reputation: few of those on this expedition would win much kudos.

For when the ill-assorted gaggle of six ships sailed off toward Madeira in the late summer of 1838, it turned out that scientifically incompetent professional sailors had been handed most of the venture's more important knowledge-gathering positions. Not that the professional officers were especially good at sailing, either. One of the ships went down in a river estuary and its crew was saved by an African-American member of another ship's company who plucked them all from the water with one of the locals' canoes. A sailor on the *Vincennes*, one George Porter, was allowed to be caught up by a trailing rope and swept up by his neck to the main topgallant sail, swinging there a hundred feet above the sea, being slowly strangled to death. (He survived, his neck unbroken but his face completely black for want of oxygen. His first demand on opening his eyes was for a glass of grog.)

Then there was an almighty fuss on Fiji when the Americans managed to insult someone, had two of their crewmen murdered by enraged islanders, and then, in an augury of policies to come, retaliated by burning down a village or two and killing eighty islanders. And to cap it all off, a second ship was lost, this time with all hands, in a violent storm off the other Staten Island—the rugged and unpopulated island off the southeastern tip of Tierra del Fuego, the last gasp of the Andean cordillera before it plunges into the sea.*

All told the Ex-Ex was a profoundly disagreeable venture, and when the remaining vessels limped back home into New York harbor nearly four years after leaving Norfolk, the commanding officer, a man named Charles Wilkes (who ostentatiously wore a captain's uniform even though a lieutenant), was cashiered and successfully court-martialed for having punished his men so harshly—especially by having some miscreants *flogged around the fleet*, a peculiarly cruel punishment that allowed bosuns from each ship to have an opportunity to whip the unfortunate man until he was within an inch of his life. There have been subsequent attempts to rehabilitate Wilkes; but his legacy as a captious and rank-obsessed commander, combined with the slapdash manner in which he arranged for the publication of the expedition reports—the last volume emerged thirty-two years after the ships got home—cast a long shadow over what could have been a spectacular American entrance into the world of oceanography.

Yet the expedition's best-known no-show, Matthew Fontaine Maury, was to redress the balance and restore America's oceanic reputation, and in short order.

When Maury was offered the position on the Ex-Ex, he was on half-pay leave from the U.S. Navy and working as superintendent of a failing gold mine in western Virginia, close to home. And soon after he turned it down, and with the six ships of the

---

* In the nineteenth century, Argentina built a tiny lighthouse on the Isla de los Estados to help mariners navigate their way around the treacheries of Cape Horn: Jules Verne wrote a little-known action novel, *Lighthouse at the End of the World*. The light fell into disuse, but its cause was taken up by a wealthy Parisian enthusiast who had a facsimile built, and which still stands, powered by the fitful sun.

Ex-Ex battling their transoceanic way from debacle to debacle, he was involved in a stagecoach accident in which he broke his pelvis and his legs. The accident effectively ended his career as a seagoing sailor, at thirty-three years old. The turn of events might well have dissuaded him from ever thinking of the sea again. But in fact the opposite was true.

After a stagecoach accident, U.S. Navy officer Matthew Fontaine Maury devoted his energies to marine cartography and oceanography. His book The Physical Geography of the Sea and Its Meteorology *is a classic. All American naval charts owe their design and accuracy to his pioneering methods of survey and organization.*

Nine years earlier Maury had been a junior officer aboard the first U.S. naval vessel to circumnavigate the world, the 700-ton sloop *Vincennes*. He had left New York on a brand-new and much larger vessel, the *Brandywine*; he received orders to transfer to *Vincennes* in port in Chile, after he had endured the doubling of Cape Horn and had written extensive notes about the most efficient way to do so. The journey home was nothing short of remarkable for a Virginia farmer's son who had grown up on a hardscrabble estate in Tennessee. The vessel sailed home first by way of Tahiti, Hawaii, Macau, the Philippines, Borneo, and the Dutch East Indies. It next plowed on across the Indian Ocean to Somalia, doubled Cape Agulhas, and rounded the Cape of Good Hope to take on supplies in Table Bay, before finally crossing the Atlantic to St. Helena and following the constant southerly winds—"a Gulf Stream in the air," as an enthralled Maury was to write—back to Sandy Hook. The *Vincennes* dropped anchor in Brooklyn a full four years after the *Brandywine* had carried Maury away.

The voyage left him a changed man, a man with a mission—a mission that no amount of rejection and injury could apparently deter. On that long circumnavigation he had learned the most complicated aspects of mathematics, and had consolidated in his mind what would be a lifelong fascination with maps, charts, currents, tides, and winds. Though nine years afterward he would have his career as a deck officer brought to a sudden stop, he was now so obsessed by the sea and its physical mysteries that he managed to persuade his superior officers to give him a desk-bound job—first as head of the Depot of Charts and Instruments, and from 1844 as head of the newly formed U.S. Naval Observatory. From this position for the next thirty years he would direct America's mapping of the seas and all the remarkable phenomena found within them.

Maury's most enduring triumphs involved the ocean on his

doorstep, the Atlantic. Their best-known manifestation was the great map he published in 1854: *A Bathymetrical Chart of the North Atlantic Basin with Contour Lines drawn in at 1,000, 2,000, 3,000 and 4,000 fathoms.* This document, based on the soundings he had ordered taken by as many naval vessels as he could find—which was actually not very many, rendering the map less accurate and comprehensive than its title suggests—left two important legacies.

The first was the incontrovertible evidence brought in by his survey vessels that the deep water of the ocean shallowed considerably along a line that seemed to run roughly north-south halfway between the European and American coasts. He named this shallowing the Dolphin Rise, after one of his ships: it was the first hint of the existence of what is now recognized as the longest and most dramatic mountain chain in the submarine world—the Mid-Atlantic Ridge.

And second: the revelations in Maury's map whetted the appetite of a millionaire Massachusetts industrialist named Cyrus W. Field, who had made his fortune in the paper business. Field had long nurtured the idea that it might be possible to extend the principle of the electric telegraph across the Atlantic. And when he saw the extent of the mid-ocean plateau on Maury's map, he inquired. The reply was just as he had hoped, for Maury wrote:

From Newfoundland to Ireland, the distance between the nearest points is about sixteen hundred miles; and the bottom of the sea between the two places is a plateau, which seems to have been placed there especially for holding the wires of a submarine telegraph, and of keeping them out of harm's way. It is neither too deep nor too shallow; yet it is so deep that wires once landed will remain for ever beyond the reach of vessels' anchors, icebergs and drifts of any kind, and so shallow, that the wires may be readily lodged upon the bottom.

Little did Maury, or Field, know what the "plateau" was really like—a tortured confusion of peaks and troughs, immense canyons and aiguilles of basalt that would snag and stretch any cable that might be laid. Ignorance of what lay beneath the waves was still profound: the cable layers of those early days—sailors on the USS *Niagara* and the HMS *Agamemnon* first of all, egged on by the eagerness of their equally uninformed investors—behaved like blind men tossing wires from a jet flying across the Himalayas or the Alps. They imagined that their cables would drift like gossamer down onto tracts of endless subsea flatland, but never imagined the sharp summits and ridges and rock-strewn chasms that actually lay below. The very earliest cables, some of them evidently hanging in mid-sea between mountaintops that rose more than two miles high from the abyssal plains, promptly chafed, stretched, and snapped with dismaying frequency. It was not until 1866 that the first permanent link was laid, and for decades afterward cable ships had to scuttle around to repair the breaks that even the well-laid plans set down.

There were other early fears about the cables, too. One shareholder of the Atlantic Telegraph Company wrote to a friend of his assumption that the voices of those who spoke across the line, "subject to such uncommon compression, may only emerge as mouselike squeakings." It was just one misconception among many: Victorian times were abundantly awash with wonderful imaginings about the ocean. One that was widely accepted for far too long held that because the density of water increases with pressure (it essentially doesn't, water being all but incompressible), there were zones in the deep sea beyond which objects could not sink—a wrecked iron ship, for example, would descend into the sea until it reached the level where the water became just too viscously dense for it to pass, and it would remain there, hovering, for eternity.

Gradation in the water's density would reserve different levels for different things, the theory went. Buckets of nails would sink lower than holed rowing boats. Horses would underlie frogs. Dead people would descend to a level determined by their obesity or the thickness of their clothing—with perhaps, as some of the more churchly said, the pressing weight of sins or of a guilty conscience forcing the less virtuous ever lower. In the end the submerged miscellanea filed into weight-related strata—with separate layers of lost cattle, drowned children, unpopular office furniture, sunken oceangoing tugs, executed bandits, hastily dumped six-guns, derailed railway trains, unwanted pets—would be fated to wash eternally around the lower reaches of the sea, a template of the world above, marooned forever in the cold and salty gloom.

It took a while for the lunacy to lift. Those who clung to this belief—and it was surely no great stretch for those who disliked the idea of the earth as an oblate spheroid, but instead thought of it as a flat disc with dangerously vertiginous edges—were additionally skeptical of the reported ocean depths reported by the sounding lines. For how, they argued, could the lead or brass ball-weights on the ends of the much-employed galvanized Birmingham twenty-gauge piano wire possibly penetrate into the viscous zone? Surely all of the soundings merely bounced up against the upper edge of the zone rather than the bottom of the sea?

But then Maury's men created a number of devices to bring back to the surface samples of the seabed, no matter how many thousands of feet below—and when in time masses of sand and gravel and crushed shells and broken shards of coral were retrieved, and the flat-earthers and the skeptics saw them with their own eyes, adherence to this strange belief abated and good sense returned.

Other fantasies came and went. One was also linked to the

viscosity question: down in that stagnant region of intense pressure, low temperatures, and endless dark, there could surely be no life, said some: it was, in the coinage of the time, an *azoic* realm. But soon after the first cables were laid, sections that had broken had to be grapneled to the surface from thousands of feet down, and when laid out on deck the twisted wires were found to be alive with barnacles and worms and other creatures demonstrating the existence of a happy and abundant living universe, even down deep in the darkness.

Another bad idea eventually made good by nineteenth-century oceanographers related to the presence, most especially in the Atlantic, of a large number of very long-lived phantom islands. A map dated 1570 by the great Flemish atlas maker Ortelius showed many of them: the *Isle of Demons* in the mouth of the St. Lawrence River, *Saint Brandan** to the south of Iceland and *Frisland* to its north, *Santana* a little to the northeast of Bermuda, and *Antillia* (or the *Isle of the Seven Cities*) to the southeast—and at the time of the map still playing host, so it was said, to Spanish bishops who had left eight centuries before, ahead of the invading Moors. Ortelius was unable to show on this map either the *Isle of Buss*, which Martin Frobisher claimed to have found during a storm, and had been placed almost six hundred miles west of Rockall, nor was he able to draw Mayda, off southern Ireland, nor the outline of *Hy-Brasil*, which sat with extraordinary persistence in the identical place in scores of earlier and later maps, fifty miles off Connemara.

None of these islands existed; they were as ephemeral and illusory as Atlantis. As was one further and final oceanic peculiarity that gripped the Victorian maritime mind for a short

---

*   Which Charles Kingsley used as home to the water babies and Mrs. Doasyouwouldbedoneby in his 1863 fable.

while: a supposed protoplasmic form of early life, an ur-slime. This was dredged up by the survey frigate HMS *Cyclops* and handed over to an initially not very interested T. H. Huxley, the paleontologist whose eventual coinage of the words *agnostic* and *Darwinism* indicates his strongly rationalist views. But rationalism failed him when, ten years after being handed the samples, he came to look down the microscope at this jelly-like ooze: he became irrationally excited by it, promptly gave it a name (*Bathybius haeckelii*—in honor of the German evolutionist who coined the word *ecology*) and declared it to be a primordial life-form that would surely carpet the seafloors everywhere.

Six years later there came an outbreak of public embarrassment as another biologist performed some very basic chemical tests on the slime and discovered that *Bathybius* was not a life-form at all, but a simple chemical reaction in the test tube between seawater and the preserving alcohol. Perhaps, bleated later supporters of Huxley—who after all, was a great man in his field, a giant of his times—it could have also been caused by a seasonal taint of plankton bloom. But most sided with the facts, and so in very short order the *Bathybius* that never lived was officially killed off. With mordant dignity Huxley renamed it *Blunderibus*, admitted his folly, thus recaptured his reputation in an instant, and promptly went back to naming other creatures, something he was particular good at. He first christened a kind of Mesozoic crocodile with the beautifully sonorous name *Hyperodapedon* and then moved on to a family of fishlike Devonian beasts that he named *Crossopterygians*.

The *Bathybius* mystery having been solved meant that when HMS *Challenger* left the dockside in Portsmouth just before Christmas of 1872, she was less on a mission to discover the undiscoverable and correct the misconceptions of ages, and more on a scientific jamboree the likes of which had never been known, and which has seldom been repeated since.

---

6. TAKING THE MEASURE

HMS *Challenger* was initially a warship, a large, 2,600-ton corvette with three masts and a large funnel to sweep away the exhaust from her 1,200-horsepower engine. All but two of her guns had been removed to make way for laboratories and equipment. Her commander, at least for the long outbound Atlantic sectors of what would be a globe-girdling expedition, was George Nares, an indefatigably correct sailor who would later win fame for his Arctic explorations (though his fame was somewhat tarnished by a later official report that blamed him for a scurvy outbreak on one polar voyage, since he had omitted to stow aboard a sufficient quantity of limes*). The head of science was C. Wyville Thomson, a professor of natural history in Edinburgh and a man who had become intrigued from two earlier surveying voyages by the question of whether life could be sustained at the vast depths of the ocean. He went out of his way, even in the early, shakedown portion of the cruise, to send down dredges and sounding mechanisms—which were always lowered on hemp twine rather than the piano wire favored by most oceanographers because the many miles of line thought necessary to plumb the greatest expected depths would put too much strain on the ship's cranes—to prove his assertion. At first his men found precious little in

---

* His very Victorian rectitude also once prompted him to chide one of his officers for shooting a seal, a prime source of vitamin C, during Sunday service aboard one of his arctic ships, thereby interrupting the colloquy with the Divine. Nonetheless, the Nares name lives on in a harbor in the Admiralty Islands north of New Guinea, two capes in Canada, a mountain chain in Greenland, a peak in Antarctica, the seaway between Greenland and Canada, and a deep in the North Atlantic.

the red clays they hauled up from off the African coast, but then, nearly four miles down off the West Indies, the dredge drew up a pair of miserable-looking annelid worms, proof, for which the rather dour Welshman reported tremendous excitement on deck, that life did in fact flourish without boundaries of depth, and that "animals . . . exist over the whole floor of the ocean."

The great ship shuttled back and forth across the Atlantic, the Canaries to Bermuda, Halifax to Cape Verde, Madeira to Fernando do Noronha, Fernando Póo to the Falkland Islands, all the while taking soundings, recording temperatures, dropping its supply of dredges and epibenthic sledges to the seafloor and having powerful donkey-engines haul them and their dripping contents back to the surface.

On occasion the hauls provided some heady moments: from the six-hundred-fathom-deep shelf off Argentina, the sledge meshes were found to have snared sea cucumbers and sea urchins, starfish in a rainbow of colors, barnacles, corals, squid, slugs, amphipods and isopods, and scores of the very primitive hermaphrodite chordates known properly as tunicates but more familiarly known by sailors as sea-squirts or sea-pork. Generally, though, and in the deeper sea, the routines became a tedious business, even for the scientists, who would come to dread the arrival of yet more grim-looking sludges, especially if the dredge returned at dinnertime. Sixty-one sailors deserted before the journey was done, and a small number died—two going mad, two being drowned, one being poisoned, another having the indignity of his face turning bright red before he dropped dead, and one unfortunate man named Stokes being hit on the head by a flying block-and-tackle, and having to be buried at sea (which prompted his shipmates to ask Captain Nares if the body would indeed float eternally within the viscous zone).

There were as many diversions as the science allowed. At

Christmas there was dancing and whisky and plum pudding, followed by readings and recitations and fiddling contests staged beside an always-refilled punch bowl. Birthdays, of scientists as well as their bluejacket crew, were celebrated with raucous activity. Afternoon tea, then a growing commonplace ashore, was served every day both to allow for interruption in the drear routines of dredging and to serve as a reminder of the civilities of home—even though the Darjeeling was often to be poured into the bone china cups during a lashing hurricane or in some tropical corner of the sea under an unimaginably fierce sun. Someone had brought a harmonium-like instrument called a melodeon on board, and its sounds often wafted up from the 'tween decks during quiet nights, reducing some of the homesick men to tears.

Whenever *Challenger* put in to a foreign port, curious visitors from ashore were invited aboard, especially ladies. To some she seemed like a steam yacht on a world cruise, and her officers were careful not to forget she served as a floating embassy, and her expedition was invariably publicized as an example of British grit and determination, so flocks of the fascinated came to gawp. But the ladies came on to dance and entertain as well, and the ships' fiddlers and the melodeon player were kept busy during the port calls.

There was sport, too: the more middle class of the scientists had brought their shotguns along and went after the more common seabirds with abandon. The traditionalist sailors were initially horrified that, once in the Roaring Forties of the South Atlantic, the sportsmen would shoot specimens of the wandering albatross, a bird long considered taboo; but no serious harm befell the ship, accidents were trivial, and the deaths and other casualties proved to be within the statistical limits for so large a crew on so long a haul.

All told, the ship was away for three and a half years—during which she lightly hit an iceberg (probably thanks to the spirit of the well-shot albatross), was given two Galapagos tortoises that wolfed down all her pineapples, found sea-bottom waters at the equator off Brazil that were almost at freezing point and so deduced the existence of a deep current flowing north from the Antarctic, and discovered, to much zoological celebration, a tiny and excessively pretty squid called *Spirula* and regarded by some as a missing link in the newly revealed Darwinian scheme of species origin. The vessel returned to Portsmouth—encountering on the homestretch off Portugal a massive flotilla of patrolling British warships, one of which had its band play "Home Sweet Home" from its afterdeck. When finally she tied up at the quayside, *Challenger* had logged almost seventy thousand miles, at an average rate of rather more than two miles an hour. Men could walk more quickly.

But, oh! the specimens she brought home: hundreds upon hundreds of crates, of animals, of plants, of bottles of seawater from various depths and places, of test tubes and Kilner jars and Petri dishes of oozes and slimes and gelatinous animals and plants. It took four years for the first volume of the Official Report to be published and another fifteen years—almost the end of the century—before the final one, and the hapless Wyville Thomson went mad and collapsed under the intense and sustained pressure from the publishers.

All told there were to be eighty volumes. It was a formidable intellectual achievement, arguably the most comprehensive study of the ocean ever undertaken, and it remains a landmark to this day. The information assembled and disseminated represented what was at the time the sum total of humankind's knowledge of the sea, and especially the Atlantic Ocean. And once it was done, oceanography was set steadily to become what it is today, a greatly more professional calling. It would not be much longer before the sailors retired to the bridge and the specialists moved in—the chemists and zoologists and submariners and physicists, the mathematical modelers and paleoclimatologists and high-temperature bacteriologists—and changed forever what the science of oceans originally had been.

7. FASHIONING THE CHARTS

Some of the romance then inevitably bled away. With a new fashion for oceanographic progress in the twentieth century and its consequent exponential growth, and with the creation of the great institutions—Scripps in California in 1892, Woods Hole in Massachusetts in 1930, Lamont-Doherty in New York in 1949,[*] and the National Oceanography Centre in Southampton, and smaller European ocean stations in places like Roscoff, Kiel, and Heligoland—the vision of the sea that had impelled the pioneers began to dim somewhat. The routines of the laboratory and the computer started to slowly take over from the rhythms of the old days: the ever-shifting horizons, the knife-sharp winds, the smell of fish and Stockholm tar, the coils of rope, the flap of sails, the keening of gulls, and the thud of marine engines made way for the hum of machines and air-conditioning and the silky sounds of laser printing.

Prince Albert I of Monaco was one of the last of the gifted amateurs to be fully invested in field oceanography, before the calling was enveloped by the realm of technocracy. His inter-

---

[*]   It was scientists from Lamont-Doherty on their steel-hulled research yacht *Vema* (formerly belonging to the banker E. F. Hutton) who in the 1950s established the true nature of the vastly long and profoundly important Mid-Atlantic Ridge, leading to the 1965 theory of plate tectonics.

est came about at a time when nineteenth-century France was developing an acute (though rather short-lived) passion for the sea, and since it was a passion that very much involved an aristocracy that had been somewhat underemployed since the revolution of 1789, it was conducted with great style and élan. The fabulously wealthy Marquis Léopold de Folin was an early entrant. After some years spent searching the floor of the Brittany coast in a comfortably converted trawler, he managed to persuade the French navy to supply him with a full-dress paddle-steamer, the *Travailleur,* and in her he conducted surveys of the seabed in the Bay of Biscay and beyond; they remain classics of scholarship and brio.

Prince Albert followed suit soon after and bought a sleekly elegant yacht of his own, the *Hirondelle.* His subsequent studies of the North Atlantic—especially the Gulf Stream—brought him fame and wide respect: he was evidently not the silk-stockinged dilettante that had been initially supposed.* His work on the great current took him three years and involved sailing *Hirondelle* on several voyages between the Azores and the Grand Banks and dropping into various sections of the stream almost 1,700 floating objects—beer barrels, glass bottles, and spheres of copper—and seeing where they ended up. Beachcombers responded to the scrupulously polite notes inside by writing to say they had found rather more than two hundred of them— discoveries that enabled the young prince (he ascended to the Monegasque throne just as this work was coming to its end, in 1889) to draw highly accurate maps of the direction and

---

* Privilege of rank allowed Albert to win from the Vatican permission to divorce without church sanction his first wife, the daughter of the Duke of Hamilton, and by whom he had a son. This tough Scotswoman, despite being the victim of an annulment, later married a Hungarian noblemen. One of their great-grandchildren would be the fashion designer Egon von Fürstenberg, husband of Diane.

strength of the Gulf Stream, of the North Atlantic Drift that branched from it, and of the clockwise nature of the North Atlantic gyre generally.

He continued his work for most of his years as ruler. He had a 175-foot schooner built as a research vessel, the *Princess Alice*— the first in a long line of vessels built solely for oceanic investigation. His particular interest was in catching and cataloging the fishes and other animals that lived in the halfway-deep seas between the continents and the abyssal plains. His life of leisure and wealth meant that, unlike most salaried scientists or those subsisting on grants, he and his ships could remain on station for weeks at a time, with battalions of stewards, cooks, and valets on hand, and could unwrap the mysteries of oceanic biology as patiently as necessary.

The prince, who died in 1922, left behind three enduring and ocean-related memorials to his thirty-three years of generally congenial rule. Two of these legacies quite deliberately blended the academic approach to the sea with the growing public interest: he had an oceanographic institute of great size and style built in Paris, and another that was similar (only larger) in Monte Carlo, and which had aquariums and exhibits of ships and exploration equipment. (Both of these were financed in large part by profits from the very fashionable casinos for which Monaco was rightly famous.) The third memorial was the one with which this chapter begins: Prince Albert arranged the financing and housing for an entirely new international body, initially called the International Hydrographic Bureau, which would on one hand seek to regulate and standardize all the world's charts and navigational aids, and on the other would seek to define the boundaries of all the world's oceans and seas.

The bureau's famous Special Publication No. S.23, and its

fourth edition,* created by what is now called the International Hydrographic *Organization*, is perhaps the most celebrated and controversial of Prince Albert's bequeathals. Buried within its pages—looking quite prominent in the slender 1928 edition, but rather less obvious when jostled among the enormous collection of brand-new maritime names (the Ceram Sea, etc., as mentioned earlier) in the modern document—was and still is the formal definition of what and where, exactly, the Atlantic Ocean is.

The Atlantic turns out to have grown considerably in the eighty years of its supervision by the admirals of Monaco. To be sure, it has physically widened by about six feet, under the relentless pressure of seafloor spreading, the inch-a-year movement from the Mid-Atlantic Ridge. But that is not what the IHO meant: the newly published enlargement is more metaphorical than actual, and is all a matter of where the ocean's boundaries are deemed appropriate to be drawn. Back in 1928 they were defined in relatively—*relatively*—simple terms.

The 1928 Atlantic was notionally divided into two—North and South—and the boundaries of each sub-ocean were established according to the cardinal point of the compass. The North Atlantic was thus seen to be bordered according to the Monaco-devised formula: on the west it ran to the eastern limits of the Caribbean Sea, the southern limits of the Gulf of Mexico, then from the north coast of Cuba to Key West, and along the American and Canadian coasts up to the southeastern and northeastern limits of the Gulf of St. Lawrence; on the *north* it was limited by the beginning of the Arctic Ocean, then a line from the coast of Labrador to the tip of Greenland, and from there to the Shetland Islands; on the *east* it ended with the northwestern limits of

---

* At the time of this writing it was still unpublished, eight years after its completion. A dispute between Japan and Korea over whether to name the sea that separates them the Sea of Japan or the East Sea remains stubbornly intractable.

the North Sea, then the northern and western limits of the Scottish Seas, the southern limits of the Irish Sea, the western limits of both the Bristol and the English channels, the Bay of Biscay and the Mediterranean Sea; and finally on the south the limit was defined by the latitude line of 4° 25'N that ran between Cape Palmas in Liberia and Cape Orange in Brazil.

The South Atlantic in 1928 was even less complicated. The *northern* limit was the Liberia-Brazil line, above; the *western* limit was the entire coast of South America, but for the estuary of the River Plate; on the *east* the ocean was formally bounded by the coast of sub-Liberian Africa, except for an enormous tract of sea in the continental armpit, otherwise known as the Gulf of Guinea, which was cut off by a straight line between Liberia and Angola; and on the *south* an arbitrary line was drawn by the IHO draftsmen, connecting Cape Agulhas with Cape Horn.

Matters are a great deal more complicated today, and under the new guidelines the Atlantic occupies a far, far greater proportion of the planetary surface than it ever did before. A single example, delineating a part of the northern boundary of North Atlantic, will offer a general idea of the new complexity:

> . . . *thence a line joining Kap Edward Holm southeastwards to Bjartangar, the western extremity of Iceland, thence southeastward along the western and southern coasts of Iceland to Stokksnes on the eastern coast of Iceland, southeastward to the northmost extremity of Fuglöy in the Føroyar, thence along a line joining this extremity to Muckle Flugga, the northernmost point in the Shetland Islands . . .*

Very basically, the expansion has been prompted by the IHO's decision to include as subdivisions of the ocean many seas and embayments that once were considered entirely separate from it. The Gulf of Mexico, for instance, is now regarded as entirely Atlantic (so the 2010 oil-pollution catastrophe, occasioned by the explosion and subsequent collapse of a BP drilling platform off New Orleans, is classified as an Atlantic problem); the Caribbean Sea is part of the Atlantic Ocean; and so is the North Sea, and the English Channel, the Bay of Fundy, most of the St. Lawrence Estuary up to the western tip of the immense and sparsely inhabited Anticosti Island,* the Celtic Sea, the Skaggerak (but not the Kattegat), and the Bay of Biscay. And the idea of the Gulf of Guinea being separate has long been discarded: now the division between the North and South Atlantics is the equator on the Brazilian side and Cape Lopez in the republic of Gabon.

(There happens to be what seems a rather eccentric small kink in this otherwise die-straight southern boundary line. The border takes a lightly angled turn to pass across a tiny palm-covered islet known as Ilhéu das Rôlas, which lies a few yards off the southern tip of the almost equally obscure island of Sao Tomé. There is a cartographic reason: Rôlas is the only Atlantic island on—or essentially on, but for a few feet—the equator. Using it as a mid-sea marker made sense—though it has to be said that in 1928 the delineators of the ocean did not worry their heads about such things. Today, though for no obviously sensible reason, they apparently do.)

This then is the full de facto extent of the Atlantic Ocean—fully 81,705,396 square kilometers (32 million square miles) of seawater, one quarter of the planet's total water area, with the deepest water of well over five miles—8,605 meters—found off Puerto Rico, and a total of 307,923,430 cubic kilometers (74 million cubic miles) of water in its entirety.

---

* Which was once owned by a French chocolatier, was nearly bought by Hitler, and now is home to a tiny community of lighthouse keepers.

8. WIDER STILL, AND WIDER

Only when one includes a human dimension to this story does it present a final but enriching complication. It is when one begins to add up the total numbers of the vast aggregation of humankind who live in some kind of communion with this sea, of those who can rightly be considered belonging to an Atlantic community, or who are—if they are in any communal sense ocean-blessed or ocean-styled or ocean-crossed—to be considered in some regard Atlantic people—that this complication appears.

It is a complication offered up by the great rivers that flow into the Atlantic Ocean.

A very great number do. Many more rivers flow into the Atlantic—especially the more widely defined Atlantic of the fourth edition of S.23—than flow into either the Pacific or the Indian oceans. There are the big rivers of Europe: the Seine and the Loire, the Severn and the Shannon and even, since the North Sea is now properly "Atlantic," the Thames and the Rhine. There is the Niger, the Kunene, the Orange, and the almost impossibly vast network of the Congo flowing in from its headwaters dotted throughout central Africa. There is the Amazon, with its headwaters in Peru, and which brings more water and rain forest mud into the Atlantic Ocean than the next eight largest rivers in the world combined bring into their respective seas. There is the St. Lawrence, originating in the Great Lakes. And there is the Mississippi-Missouri river system, which hauls trillions of gallons of water each day down from the prairies and the Rocky Mountains into that officially sanctioned embayment of the far western Atlantic, the Gulf of Mexico.

So to those who wish to cast the Atlantic's net of influence as far and as wide as is technically possible, consider that it does not stop simply at Cape Race or Heart's Content, at Montauk or the Outer Banks, or at the Argentine beaches of Bahia Blanca or Isla de los Estados or Cape Horn. Nor does it begin on the cliffs of the Faroes or on the Aran Islands, or on Ushant, on Land's End, at Cape Bojador or Robben Island or at the rocks of Cape Agulhas or near the sea caves at Pinnacle Point.

Rather it begins and ends, to be pedantic about it, in the lakes of Zambia (where the Congo rises) and in the Swiss Alps (where a glacier drips to form the tributaries of the Rhine). It also begins in a valley near America's Yellowstone National Park, where a late Victorian explorer named Bruce found the source waters of the Missouri River, and beside which today a Greek farmer, a long way from his old home beside the Mediterranean, lives out his life as an American rancher, raising sheep.

And the ocean also begins and ends beside an eight-thousand-foot mountain in far northern Montana named Triple Divide Peak. This is the hydrologic apex of the North American continent. Rainwaters that fall onto its northern flanks flow into Canada and into the Arctic Ocean. Waters from its western and southwestern sides slip into creeks that eventually take them to Oregon and on the Pacific. Any precipitation that happens to fall on the southeastern slope seeps down eventually into a tiny canyon at the base of which is an even tinier creek—and which makes its way to the north fork of a river that becomes the Marias River. Near the town of Fort Benton, Montana, this Marias stream flows into the Missouri; at St. Louis the Missouri joins the Mississippi; and at New Orleans the Mississippi finally reaches the Gulf of Mexico, from where its waters are connected to the Atlantic Ocean.

With great prescience the explorers of that rugged and ice-bound corner of Montana where Triple Divide Peak rises gave a name to that tiny creek that spills off the summit. They named the very first river that snakes its way downhill, from below the snow line at seven thousand feet to the grassland at five thousand, and its waters coursing swift and pure through a Rocky Mountain canyon. It was almost as though the river knew what the explorers knew—which was where its waters were going. For they called it quite simply Atlantic Creek. They named it for an ocean with which the state of Montana is now ineluctably connected, but which most of its people will seldom if ever see.

Chapter Three

OH! THE BEAUTY AND
THE MIGHT OF IT

———————

And then the lover,
Sighing like furnace, with a woeful ballad
Made to his mistress' eyebrow.

1. THE PLAY OF THE SEA

Although William Shakespeare wrote often and with an easy familiarity about the ocean—about tides in the affairs of men, about fleets majestical, about a thousand fearful wrecks, about fathers lying full-fathom five, about sea changes and sea nymphs and winds sitting sore upon the sails—there remains no firm evidence that he ever boarded a ship, nor ever went to sea, nor that he ever set eyes upon the Atlantic.

But such was the Atlantic's importance to all England in his time, that Shakespeare would surely have known of its existence, and he would have heard of its many dramas. It is of little surprise to find that he deftly wove one of the most celebrated of sixteenth-century Atlantic tales into the centerpiece of his final, and perhaps his most boldly imaginative theatrical work, *The Tempest.* Like a few before him, and like so many after, Shakespeare plucked an image from this ocean of innumerable moods and dispositions and transformed it into art.

The play, which he wrote in 1611, happened to be staged with great imperial pomp in 2009 on the island of Bermuda, as part of the four-hundredth-anniversary celebration of Britain's northernmost Atlantic island colony. It was put on in a theater in the capital, Hamilton, and was staged there for one very overarching reason: most literary scholars like to believe that, unlike any theatrical work created before, *The Tempest* was very much an Atlantic Ocean play, and that it was the original accidental settling of the island of Bermuda, four centuries before, that played a key role in the play's creation.

It might not seem so at first blush. After all, the island to which Prospero and Miranda are found to have been exiled, and where Caliban was found to live, looks upon close reading of the text most likely to be in the Mediterranean. The city of Milan seems from the writing to be unduly close to the scene of the action; and when at the end of the play, the wrecked sailing ship that had brought Antonio and Alonso to the island is repaired and allowed to go home, it was to undertake a quite unexceptional journey merely back to Italy, which probably lay nearby.

But a closer study of Shakespeare's motivations uncovers evidence beyond the text itself that supports an otherwise rather radical idea—that his inspiration for writing *The Tempest* came from a real shipwreck that occurred in 1609, and which happened not in the Mediterranean at all, but right in the middle of the western Atlantic.

There is, moreover, a hint within the text: a reference in passing to the "still-vex'd Bermoothes," which shows that Shakespeare must have known something of the islands and their existence.

The dramatic circumstances of the foundering were well known in the London of Shakespeare's day. They involved a vessel, the *Sea Venture,* which had been chartered in London to the Virginia Company, and which set off from the Plymouth docks to cross the ocean in June of that year. Its captain was a Dorset privateer and adventurer named Sir George Somers. His mission was to resupply the six hundred or so pioneers who a year before had settled in the infant British colonial settlement of King James's Town, sited in one of the estuaries south of the Potomac River.

Cruel chance intervened, however. Somers and his sorely insubstantial vessel were caught in a fierce summer season hurricane. The little ship was dashed onto the reefs of a barely known group of islands—wrecked, though without loss of life, and after the crew saw as augury a spectacularly ominous display of St. Elmo's fire among the masts and spars. The *Sea Venture* was a total loss, left high and dry though perched safely upright, wedged between a pair of rocks at the northeastern end of what is now known as the Bermuda island chain.

News of the shipwreck soon became the talk of the inns of early-seventeenth-century London, and Shakespeare almost certainly heard of it. The story, when told in full, had all the elements of fine drama, and the lurid tales of the strangely dancing illuminations that were seen just prior to the collision must, it is said, have led him to conjure up the notion of Ariel, the island sprite.

The story continued well beyond the wreck itself. There were aristocrats among the survivors, and ladies of some gentility, and all were soon obliged by Somers to work under the direction of his shipwrights to build from Bermuda's abundant cedar trees a pair of replacement vessels for the *Sea Venture.* In these two ships, the *Patience* and the *Deliverance,* almost a year later the party sailed on—only to find the Jamestown settlement nearly completely decimated, with the sixty remaining colonists reduced to near starvation. The rescuers spent some time getting them back on their feet, whereupon Somers returned to Bermuda, an island he liked very much. But in a cruel irony he was to die there soon after his arrival. His body was returned to Lyme Regis, the Dorset village where he was born—but his heart has stayed to this day in Bermuda, in a tomb in what would become one of Britain's earliest Atlantic possessions.

The island remains a British colony. Since 2009 marked the four hundredth anniversary of Somers's unintended but compulsory landing, and since that landing effectively began the island's long relationship with the British Crown, and since Shakespeare probably used elements of the story as the basis for his final play, what more appropriate way is there to celebrate Bermuda's birthday than to have a performance of *The Tempest* on the island where it had all begun?

So the play was staged in the Hamilton Town Hall, a boxy limestone structure that was modeled on its much larger namesake in Sweden's capital, Stockholm. All of the island grandees were there, including in the Royal Box the colonial governor, who arrived in the back of a BMW sedan driven by a uniformed soldier. It would be something of a stretch to say that the performance was memorably great, although Prospero was played by an English actor well-known for his matinee-idol looks, which thus brought out a large contingent of Bermudian middle-aged paying customers, most of them excitable ladies all atwitter.

They had come to see a mystical, magical piece of theater, a play conceived from an Atlantic story by a playwright at the top of his powers who was writing little more than a century after the Atlantic had been crossed by Columbus and then recognized by Amerigo Vespucci as the distinct and separate ocean we all now know it to be.

2. FIRST WORDS

Long, long before the Atlantic was recognized to be an ocean, when it was just an unknowably vast and man-devouring mass of waves and spray and far horizons, the artists were aware, were fully engaged with its awful beauty. The poets were among the earliest to take notice. Classical poets had of course long composed around the sea—but the only sea they truly knew was the Mediterranean, which in terms of its drama is a flat, warm, quite subdued, and almost suburban body of water, rather wanting of an appropriate majesty. The heaving gray waters of the Atlantic were quite another thing, and it was the Irish, when finally they were brave enough or foolhardy enough to launch their curraghs into the boiling surf off their western coasts, who seem first to have employed their literary sensibilities to meditate on their unique maritime environment.

St. Columba's epic voyage north, from Ireland to the west coast of Scotland, in the sixth century A.D. was much written about—and there are stirring images of fleets of curraghs crossing the rough waters between Antrim and Galloway. But the literature surrounding Columba—or Colam Cille, as he is more properly known—is more narrative than contemplative. The poetry associated with the great apostle's missions is regarded as Europe's oldest vernacular verse, but its treatment of the ocean is mere bycatch, and it is another two centuries before the first snippets of imaginative appreciation of the sea start to become apparent.

Rumann son of Colmán was an eighth-century Gaelic poet who is said to have enjoyed a standing among the Irish equal to that of Virgil to the Romans or Homer to the Greeks. His best-known short poem, "Storm at Sea," written around 700 A.D., is rightly regarded as one of mankind's earliest artistic ruminations on the Atlantic. It has eight stanzas and was translated in the 1950s by the great Irish novelist and poet Frank O'Connor:

> *When the wind is from the west*
> *All the waves that cannot rest*
> > *To the east must thunder on*
> > *Where the bright tree of the sun*
> *Is rooted in the ocean's breast.*

The Celts clearly had the sea in their veins, and early Anglo-Saxon writers across ancient England were soon similarly caught up by a mighty vision of the sea, the first writings almost coeval with their Irish neighbors. It is perhaps hardly surprising that so maritime a race as the English produced, early on in their history, powerful poetry about their coastal waters. The best-known of the eighth-century Saxon poems about the ocean is to be found today in a secure loft above the Bishop's House behind Exeter Cathedral, in Devon. Since the year 1072, when the great scholar Leofric died and left his sixty-six-volume manuscript library to the cathedral, one unremarkable-looking volume has stood head and shoulders above the rest in the quality of its contents. It is a codex known simply as the Exeter Book, and it contains unarguably the greatest collection of the poetry of its time in existence.

The precious little volume has had a life as tough as it has been long. The book's original cover is missing, and of its 131 pages, eight have been lost, one was evidently once used as a wine coaster, others were singed by fire, and still others incised with notches suggesting they were used as cutting boards. Yet to the thanks of all, it is a survivor, and the Exeter Book is now recognized to hold about one-sixth of all the Anglo-Saxon poetry ever known to have been written. A single scribe is believed to have copied out all of the poems sometime in the tenth century, using brown ink on vellum, and wielding his quill with an impeccable, monastically steady hand. There is almost no illumination or ornamentation in the book, and just a few small drawings in a number of the margins. It is a priceless work of art: only one other of the four known Anglo-Saxon codices is more famous, and that is the Nowell Codex, which includes the great epic poem *Beowulf.*

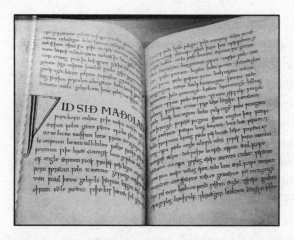

The Exeter Book, a tenth-century anthology of poetry written in minuscule Roman script, is one of the greatest treasures of English literature. It contains the Anglo-Saxon poem The Seafarer, *perhaps the earliest English poem about the Atlantic Ocean.*

But *Beowulf* is mainly about battles and funerals, and it takes place mostly on the land, in Denmark and southern Scandinavia. In the Exeter Book, on the other hand, there is one much shorter poem called *The Seafarer,* and this ranges a great deal further. The poem is dominated, at least in its first half, by a lengthy and mournful meditation on the trials of the sea. It is in truth an elegy to the Atlantic, in the voice of a man—though no one knows his name—who has suffered hard times winning a living from its waters, but yet who, when he is away from it, yearns for the ocean life more than he could ever imagine.

There are many translations of *The Seafarer*: these lines are from one of the better known, made in 1912 by Ezra Pound. It begins with a lament that all worn sailors will recognize:

> *. . . Lest man know not*
> *That he on dry land loveliest liveth,*
> *List how I, care-wretched, on ice-cold sea,*
> *Weathered the winter, wretched outcast*
> *Deprived of my kinsmen;*
> *Hung with hard ice-flakes, where hail-scur flew,*
> *There I heard naught save the harsh sea*
> *And ice-cold wave, at whiles the swan cries,*
> *Did for my games the gannet's clamour,*
> *Sea-fowls' loudness was for me laughter,*
> *The mews singing all my mead-drink.*
> *Storms, on the stone-cliffs beaten, fell on the stern*
> *In icy feathers; full oft the eagle screamed*
> *With spray on his pinion. . . .*

But then, in an instant, even though the summer on shore is fast coming, the mariner's mood changes to one of longing, a mood that all old salts will also know well:

> *Bosque taketh blossom, cometh beauty of berries*
> *Fields to fairness, land fares brisker,*
> *All this admonisheth man eager of mood,*

The heart turns to travel so he then thinks
On flood-ways to be far departing . . .
So that but now my heart burst from my breastlock
My mood 'mid the mere-flood,
Over the whale's acre, would wander wide.

Did our anonymous melancholic see the Atlantic as a thing to love, or as no more than a means of getting away? The small armies of translators who have tried to come to terms, or come to grips, with the poem have worried over the inner meanings for decades. Some concluded that a journey by sea was merely a necessary inconvenience to be endured—Pound was probably counted among these. Others of more romantic mind, however, prefer to suggest that the trials of voyage itself somehow set the voyager aloof from ordinary land-bound folk, making him a superior sort, a man with reason to swagger. (And *aloof*, the term: this is nautical, from the order given to the steersmen to keep *a-luff*, away from a lee shore. So many of the words that gain traction in such early times served as reminders that Britain was steadily coming to be a maritime culture, steeped in the traditions of the sea. *Luff* was thirteenth century.)

But whatever the mysterious mind of its subject, *The Seafarer* established a fashion even as it confirmed a reality. It was on one level an allegory—the idea, to be repeated many times in later poetry, of life as voyage, and which consumes the narrator even as he contemplates returning to the ocean that has treated him so hard. Yet on another, more nationalistic level, it appears to acknowledge that the English of the time had come to understand that they inhabited a place set firmly in the ocean, surrounded by sea and strait and channel. It let it be known, unequivocally, that the accumulating identity of the English people was that of a race of islanders, a people who were in time both certain and obliged to win a living from their borderings of deep waters.

Caedmon and Cynewulf, two of the greatest Old English poets of the time, both probably lived and worked in monasteries by this bordering of sea—Caedmon at Whitby, Cynewulf probably at Lindisfarne—and their works are similarly steeped in maritime motifs. The shadowy figure of Cynewulf, who lived well on into the tenth century, writes of the sea with curiosity and passion, as with his thoughts on *The Nature of the Siren:*

> *Strange things indeed are seen in the sea world:*
> *Men say that mermaids are like to maidens*
> *In breast and body. But not so below:*
> *From the navel netherward nothing looks human*
> *For they are fishes, and furnished with fins.*
> *These prodigies dwell in a perilous passage*
> *Where swirling waters swallow men's vessels . . .*

Two hundred years later came the writers of Norse mythology, and the Icelandic sagas. Most likely—unless missionaries had brought manuscripts, or the evangelizing voyagings of St. Brendan had literary purpose, too—those in Iceland were unaware of the poetry of the Celts and the Saxons. In any case, they would forswear the poetic form for essays, most of them epics of great length and substance. And this new form of writing was in no sense a meditation, but rather entirely narrative in form, stories of heroism and privation, full of action and excitement.

The two essays that most memorably relate the exploits of Iceland's seagoing explorers, the *Greenland Saga* and the *Saga of Erik the Red,* do certainly talk of the power of the North Atlantic, but they do so as part of a much larger story—that of the ocean as a passageway, albeit a very trying one, to discovery. The primary interest of the Nordic sailors was in reaching new territories, to explore and to colonize, as illustrated by a passage in the opening pages of the *Greenland Saga:*

> . . . *they put to sea as soon as they were ready and sailed for three days until land was lost to sight below the horizon. Then the fair wind failed and fog was set in, and for many days they had no idea what their course was. After that they saw the sun again and were able to get their bearing; they hoisted sail and after a day's sailing they sighted land. They discussed among themselves what country this might be. Bjarni said he thought it could not be Greenland . . . "for there are said to be huge glaciers in Greenland."*
>
> *They closed the land quickly and saw that it was flat and wooded. Then the wind failed and the crew all said they thought it advisable to land there, but Bjarni refused . . . "for this country seems to me to be worthless."*

What they had found was almost certainly the coast of Labrador. So from the point of view of a would-be settler, Bjarni's harsh-sounding assessment was probably and shrewdly right.

3. MONSTERS AND MAELSTROMS

The stories of the Norsemen—and one must note that their unimaginably complex mythology is still enormously popular in some circles today—brought with them one further departure from the merely written word, and that is the coming of imagery—sculpted, incised, drawn, or painted, though very little of it remains. Most of the representational art known today—and from which we have become familiar with the look of Odin and Thor and the Valkyries and the others of this vast pantheon—were re-created by nineteenth-century artists who became enraptured by heroic tales that suddenly burst out from a small legion of Scandinavian scholars in Victorian England. Some vague, contemporaneous incised images remain—of ships, for instance, including the giant vessel Skidbladnir (its name is given to cruise ships and fictional space vehicles to this day) and Naglfar, a vessel made entirely out of the finger- and toenails of the dead. There are tapestry representations, too—the medieval Swedish hanging known as the Överhogdal tapestry, found in a church warehouse at the beginning of the last century, show Viking *knarrs*, while the far better-known Bayeux tapestry in northern France shows the eleventh-century invasion fleets hurrying toward England, sailing over a sea populated with fantastic creatures.

There are many images of sea monsters, too—the horrifyingly enormous Midgard Serpent, the *Jörmundgandr,* being one of the better known. And there are real-time pictures of the ever-present maritime dangers, of waterspouts and whirlpools and aqueous myths and legends that have been spun across the entire northern half of the ocean, from Cape Farewell in Greenland to the Scandinavian coast between North Cape and the Skaggerak. Images and tales present the *maelstrom* at the southern tip of the Lofoten islands, for example, and they show Corryvreckan, the terrifying water feature known as the *old hag,* or the *cailleach,* which still booms and thunders with each running tide between the western Scottish islands of Scarba and Jura;* and there are any number of other ferocities and perils whose crudely executed pictures

---

* George Orwell wrote *1984* in a farm at the northern tip of Jura and reputedly nearly drowned himself by going too close to the ferocious streams and tide rips of the Corryvreckan whirlpool.

and vivid descriptions would have served to terrify any would-be North Atlantic oceangoers well into the fifteenth century.

The famous *Carta Marina*, the first map to show and name in detail the Nordic countries, and which was drawn in Rome in the sixteenth century by the Swedish cleric Olaus Magnus, is famous for showing a bull's-eye representation of the *maelstrom* at the very southern tip of the Lofotens; it also has a description of an Atlantic beast, and the translation of this has a poetry all its own:

> *Those who sail up along the coast of Norway to trade or to fish, all tell the remarkable story of how a serpent of fearsome size, 200 feet long and 20 feet wide, resides in rifts and caves outside Bergen. On bright summer nights this serpent leaves the caves to eat calves, lambs and pigs, or it fares out to the sea and feeds on sea nettles, crabs and similar marine animals. It has ell-long hair hanging from its neck, sharp black scales and flaming red eyes. It attacks vessels, grabs and swallows people, as it lifts itself up like a column from the water.*

Until the first of the sixteenth-century crossings, most of the imagery of the Atlantic—as in the cartouches drawn in the margins of maps—was peppered with frightening sea creatures like this, and with dragons and monster fish. Even as late as the relatively more sophisticated seventeenth century there were still engravings published showing giant fish and whales interfering with the passage of ships: St. Brendan, for example, is pictured offering Mass on the back of a whale. In the picture, engraved in a book published in 1621, the immense creature, alarmingly extravagant of teeth, grins with evil intent while letting fly twin spouts of water; yet serene on his back is the priest, gathered before an altar with corporal, chalice, and paten all laid out as neatly as if back in Clonfert, still stoically intoning the liturgy.

The monstrous and the terrifying were not so prominent in the portrayal of the ocean on its western side, however. Such pre-Columbian art as depicts the sea is more accepting, more sympathetic to the ocean caprices of calm and storm. The Incas—not an Atlantic people, true—gave thanks to *Mamacocha*, the goddess of their sea. Those who lived on the Pacific coasts saw her as representing a protective embrace, as the supplier of fish and whales from which they could win sustenance, and generally radiating a mood of benevolence that only altered, albeit with occasionally lethal ferocity, whenever humankind had not been suitably attentive to her needs.

The Mayans, farther north and on the side facing the Atlantic, were perhaps rather less spiritually involved with the ocean. There is precious little art that indicates the sea or anything like it, even though their best-known color, Mayan blue, would seem the ideal candidate for the making of paintings relating to the sea. They were very commercially connected to the ocean, though, constructing great trading canoes to move goods and people from peninsula to peninsula and island to island. The greatest Mayan seaside city, the port of Tulum at the tip of the Yucatán peninsula, is magnificent—but it has little specifically about the sea to its buildings and murals, and its discovered emblems are more concerned with the force of the wind and the beauty of sunrise. If Tulum is more than merely a great port for the decidedly inland Mayan city of Coba, and was constructed as some kind of homage to the ocean—as many of today's other great Atlantic cities clearly are—it is one of the more discreet in its approach.

Similarly the Mayan creation myth, as seen in their art and literature, makes scant mention of the sea. Mountains somehow emerged from it, and then wooden men were created in the jungles on the mountain slopes, and out of them in time, real men. But the sea is not the *fons et origo*, has little of the comfort-

ing power known from Incan legend, is less of a source of sustenance, is more of a carriageway to profit and prosperity.

In Atlantic Africa, however, there is still today much of the
kind of reverence for the ocean that was known among the Inca.
Female water spirits, benevolent and erotic by turn, are enormously important in the tribal cultures of the sub-Saharan
coast—especially among the Yoruba of Nigeria and in the various
voodoo cults of Benin and Ghana, as well as in Liberia, Gabon,
and on the island of Fernando Poo. A popular figure, *Wata-mama*,
or more popularly now, *Mammywater*, has appeared for hundreds
of years in the folk art of West Africa. Since the onset of slavery,
it has also popped up among members of the African diaspora on
the western side of the ocean, especially in Brazil.

Wata-mama is usually depicted as being quite pale-skinned,
blond, and decked out in a singular array of jewelry; it is finned
like a mermaid and usually possesses full breasts between which
nestles, invariably, a sturdy-looking python. Anthropologists believe the origin of the spirit form is the large marine mammal the
West African manatee, also somewhat unfairly named the sea cow.

Men of a certain disposition like to claim that the spirit of
Wata-mama resides in promiscuous city girls, especially pros-
titutes—a belief that leads some of the bolder types to claim to
their wives that visits to brothels are therefore somehow sacra-
mental. Though their belief in the water spirits remains robust,
few African spouses are believed to be entirely sympathetic to
this argument.

4. A MORE COMFORTABLE SEA

The fifteenth-century crossings of the Atlantic coincided with,
or some might argue were in part impelled by, the restless intel-

lectual and commercial urgency of the Renaissance. This was a period when, so far as visual art was concerned, all manner of new concepts were being born—perspective most significantly, but also the incorporation of scientific realism into art, and a yearning to record the newly acquired knowledge about the natural world. It was a change in artistic direction that had a signal effect on the perception of the sea. As the ocean world became increasingly better known, and as it started to be feared less than before, and as its waters and cliffs and creatures began to become steadily more amenable to calm artistic appreciation, so the fantasy world of the unknown started to make way for a more conventional, familiar representational appreciation of a great sea.

Initially the sea is only a backdrop—it appears in some of Dürer's work, for example, as merely a flat expanse of water tucked away as part of the scenery. To be sure, one of Dürer's better-known works, *The Sea Monster*, which he painted in 1498, has at center stage a giant Triton-like figure, with scales and antlers, holding a none-too-incommoded-looking nude woman in his arms while her hysterical friends shout and gesticulate in the background. One might say this was a throwback to the more primal view of the ocean—except that the sea itself in this image is calm, the mirrored ripples suggesting no more than Beaufort wind scale No. 1. And five years later, when Dürer paints his *Lamentation for Christ*, the body of water—though it may well be a large lake—is quite still, mirrorlike, a reminder that whatever the tribulations of morality, the sea continues—to resuscitate Derek Walcott's poetic idea—*to go on.*

As incidental as this was to Dürer, it was front and center for a young Spanish painter named Alejo Fernández: thirty years on and the Atlantic now begins both to insert itself and to assert its claim to a certain standing in art. *The Virgin of the Navigators,* which Fernández is thought to have painted around 1531, is the first known representation of the transatlantic achievements of Christopher Columbus and the implications of the first central European contact with the Americas. In it the Virgin Mary floats on clouds, gazing benevolently down at Spanish explorers and at converted Native Americans alike. Beneath her feet is the ocean, blue and calm and alive with ships of a variety of styles and ages.

The painting was enormous, commissioned to hang in the great Hall of Audiences at the Casa de Contratación, the agency that, from its headquarters in the Alcázar in Seville, directed all official Spanish exploration and imperial expansion. It was there to serve as inspiration for the explorers who would go out in the wake of Columbus and Vespucci; it was to serve as the centerpiece of an altar when divine blessing was sought for another trying voyage out west, or else it was there to offer thanksgiving for a homebound journey successfully accomplished.

The Atlantic Ocean was now seen, by the Spanish at least, as having been placed under the eternal, and eternally maternal, invigilation of the mother of God. It was an ocean divinely designed for the use of man, and such paintings as followed would offer it honor and respect in equal measure. Maps and charts, altarpieces and altar cloths, and hangings for ecclesiastical walls all around Europe and in the far possessions were soon emblazoned with formal imagery of what, with no intended pun, might justly be called the Holy Sea.

And then, all of a sudden, the sea was everywhere. Or rather the ships were, and with the sea beneath, beside, and beyond them, and in an endless variety of moods. It was a swift and sudden upsurge of interest beginning in the middle of the sixteenth century, and it had much to do with national pride. The sight of a cog or a carrack or galleon in full sail, or in later years a ship of the line firing a broadside into a rival mass of broken spars and shredded sails, seemed always to send a shiver of pride through the national mood. The British, the Spanish, and the Portuguese certainly all produced an abundance of paintings in this era; yet it was the Dutch who, from the mid-sixteenth century onward, seemed to hold a brief monopoly on the artistic depiction of the ocean.

If any region can be said to have invented Atlantic art, it is the Netherlands, and an artistic holy trinity of the ship portrait, the port view, and the tempest by the rocks provided the dominant themes of paintings by such men as the Flemish painter and printmaker Pieter Brueghel the Elder;* the father and son who were both named Willem van de Velde and who emigrated to England to improve their craft by winning a slew of royally patronized marine panoramas; and the man who essentially invented the craft, and who was known for creating the finest of all battle scenes, and with painstaking attention to the most gory details, the Haarlem maritime painter genius Hendrick Cornelisz Vroom.

Even now, nearly five centuries on, these are paintings that grab the attention: invariably there is the hungry sea, its waves translucent green and white-capped, the troughs between them deep and dangerous and all providing a savage contrast to the distant comforts of cow-grazed meadows and church steeples. The foreground waters are alive with a bustle of lighters and painters and churning ferryboats—and then, front and center and gleaming white in a single shaft of watery sunlight, are the sails of an enormous Hollander merchantman, leaning over in the wind and bearing away for some faraway destination, the waters beginning to churn under her mighty oak bows as the breeze catches her canvas and she begins to power offstage and out of view.

---

* Brueghel came from the Dutch city of Breda, famous a century after his birth for the treaty that swapped an obscure British fort in the East Indies for a Dutch-run island in America—Manhattan.

There were subtle differences in the manner in which the Europeans treated the sea in their art. The Dutch favored a draftsman-like accuracy in their depiction of the complexity of great ships, with thousands of details crammed into the immense space of an expensively commissioned canvas, and with tautly composed and harmonious settings in estuaries or beneath impressive headlands. The British, less formally, liked to paint their seaports, the more majestic vessels of the Royal Navy, and the triumphantly messy moments of the more complex naval engagements. The French, on the other hand, did rather little with their Atlantic coast, and little is known except for the works of Claude Lorrain (who in any case worked in Italy) and Claude Vernet, who turned out thirteen magnificent paintings of French Atlantic seaports, Boulogne to Biarritz (together with Marseille, on the Mediterranean) for a royal commission from Louis XV.

Canaletto famously concentrated his maritime mind on the canals of Venice, and the Russians (who had no real exposure to the Atlantic, other than nearby White Sea ports of Murmansk and Archangel) did their best to show interest—and though Catherine the Great persuaded a German artist based in Naples to do most of the kind of seascapes that she liked, when he asked for ideas for painting action-packed battle scenes she sent a squadron of her warships down to Leghorn and had one of them blown up so he could get the general idea.

5. THE LYRICAL TRANSITION

It took the poets some time to catch up with the painters, however.

Europe's sixteenth- and seventeenth-century painters may

have come swiftly to terms with the ocean, seeing its expanses more in terms of trade than terror; but the poets were still not entirely convinced. For example, while Dutchmen were busy recording the new glamour of sail and sea and Sir Walter Raleigh was busy exploring the New World (and composing poetry that is curiously almost entirely devoid of marine references), his good friend Edmund Spenser was composing the highly nautical and highly fanciful epic *The Faerie Queen*. Spenser's take on the nature of the ocean in this work's myriad books and cantos was nothing like that depicted in Dutch painting, was anything but glamorous, in the sense that according to Spenser it was filled with

> *Most ugly shapes, and horrible aspects,*
> *Such as Dame Nature selfe mote feare to see,*
> *Or shame, that ever should so fowle defects*
> *From her most cunning hand escaped bee;*
> *All dreadfull pourtraicts of deformitee:*
> *Spring-headed Hydraes, and sea-shouldring Whales,*
> *Great whirlpooles, which all fishes make to flee,*
> *Bright Scolopendraes, arm'd with silver scales,*
> *Mighty Monoceroses, with immeasured tayles.*

Nor was Shakespeare—whose references to the sea are innumerable, though many so wreathed in fantasy as to reinforce the doubt he ever actually saw it—very much more cheerful. The nightmare of ocean drowning still intrudes spectacularly in his plays, as here with the soliloquy of Clarence, imprisoned in the Tower on the order of his brother, the soon-to-be King Richard III.

> *O Lord! methought what pain it was to drown!*
> *What dreadful noise of waters in mine ears!*

What sights of ugly death within mine eyes!
Methoughts I saw a thousand fearful wracks;
A thousand men that fishes gnawed upon;
Wedges of gold, great anchors, heaps of pearl,
Inestimable stones, unvaluèd jewels,
All scatt'red in the bottom of the sea:
Some lay in dead men's skulls, and in the holes
Where eyes did once inhabit, there were crept
(As 'twere in scorn of eyes) reflecting gems,
That wooed the slimy bottom of the deep
And mocked the dead bones that lay scatt'red by.

And John Donne similarly found Atlantic water a horror, as in "The Storm," written in the form of a letter in 1597:

The south and west winds join'd, and, as they blew,
Waves like a rolling trench before them threw.
Sooner than you read this line, did the gale,
Like shot, not fear'd till felt, our sails assail;
And what at first was call'd a gust, the same
Hath now a storm's, anon a tempest's name.
Jonas, I pity thee, and curse those men
Who, when the storm raged most, did wake thee then.
Sleep is pain's easiest salve, and doth fulfil
All offices of death, except to kill.

But then came the beginning of the Enlightenment, and with it the triumph of reason, the age of Descartes, Newton, and very early on, John Milton. He turned out to be one of the first English poets who would at last strike a more sanguine note about the sea generally. In the seventh book of *Paradise Lost*, for instance, he displays his admiration—perhaps not still entirely rational; this was after all only the very beginnings of a new and less superstitious time—for what he saw as the god-created deeps:

> *Over all the face of the earth*
> *Main ocean flow'd, not idle but with warm*
> *Prolific humour soft'ning all her globe*
> *Fermented the great mother to conceive*
> *Satiate with genial moisture, when God said,*
> *Be gather'd now, ye waters under heaven*
> *Into one place. . .*
> *. . . the great receptacle*
> *Of congregated waters He call'd Seas.*

It would be some while yet before the sea became, as it is today, a thing of great romance—the archetype of the sublime, that philosophical quality of natural creations that manages to combine the magnificent and the horrific at the same time. Mountain chains, with their jaggedly vicious peaks and sheer cliffs, and the dangers of rockfalls and avalanches and lashing storms, are classic exemplars of the sublime, presenting an aesthetic that inspires awe and reverence. The sea eventually came to be seen as much the same—a thing possessed of an awesome mightiness, a lethal beauty, of which one might be fearful and respectful and overcome by, all at the same instant. Come the end of the eighteenth century and the sea—and this, to most Europeans, meant the Atlantic Ocean that washed their shores—was no longer a mere inconvenience to be overlooked in life, in art, in literature, in any creative endeavor. It was a thing to be honored and even embraced, though always warily, for the sea could always strike back, and with irresistible force and power.

6. STONES OF THE OCEANSIDE

At the same time as the Dutch were painting the Atlantic Ocean as something of ineluctable grandeur, so the builders of the European empires were beginning to create and expand suitably grand cities around its periphery. And while it would never do to suggest that those who designed the cities or the buildings were purposefully offering any kind of homage to the ocean, today many of them have an architectural legacy that has a certain splendid uniqueness to it. There can be no doubt that because of the history of colonization and the passage of wealth between Europe, Africa, and the Americas, no other of the world's ocean and seas currently possesses such a concentration of urban magnificence: five centuries' worth of creativity in stone have left an indelible stamp on the Atlantic, as valuable a record of man's dealing with her vastness as is the art and literature that she inspired.

The sheer number of oceanside cities could threaten to make any account seem more like a catalog. From Hammerfest to Cape Town on the eastern side, from St. John's to Comodoro Rivadavia on the west, and aside from the great and the obvious like New York and Rotterdam, Liverpool and Rio, there are places such as Esbjerg, Vigo, Takoradi, Walvis Bay, Puerto Madryn, Wilmington, and Halifax—a sampling of the scores of ports and settlements that have come into existence solely because of their closeness to the ocean. Any selection based on the appearance of the homage each of these seems to offer or of a legacy that seems unique is certain to be contentious, forced, and false.

It is, however, possible to offer up a number of the most distinctive cities in unrelated pairs, with the one on the Atlantic's older eastern side matched in a somewhat logical manner to a partner on the younger, newer, western side. Not for purposes of direct comparison, perhaps, and not necessarily because they enjoy any kind of formal historical linkage—as with the cities of Merseyside and the sugar ports of the Caribbean, for example, or the European migrant centers and the wharves of Ellis Island. The pairs offer simply some indication of the range of urban ambition that the Atlantic Ocean has stimulated. Some Atlantic cities are most notable for their antiquity, some for their beauty or drama or their fading magnificence; some are remarkable for their energy, some for their sheer economic or political importance. And happily, to display each of these qualities there appears by sheer happenstance to be a city on each coast. There is also, moreover, at least one city, and maybe two, that are situated not on a continental coast but in the very center of the ocean, and which also have that same quality—that peculiarly Atlantic identity—that is unique and unforgettable.

. . .

Athens is by most reckonings the oldest major city in Europe; Cádiz in Spain is among the oldest of those on the continent's Atlantic coast. There is a claim that Cádiz was founded in 1104 B.C., a date noted in a diary by a prominent Roman historian. But even the proudest of today's citizenry think this improbable, and are content to use the ninth century B.C. as the city's birth date, at a time when the Phoenicians used Cádiz as a trading base for their later journeys to southwest Britain and to northwest Africa.

And while there has never been a Parthenon or an Acropolis found in Cádiz, by chance what would soon afterward come to be recognized as the most ancient of the city's surviving buildings—a Roman ruin—happened to be discovered on the very occasion that I stayed there, on my first-ever visit to Cádiz, in the early 1980s.

I was on a reporting assignment, journeying between the Atlantic Ocean side of Spain to the Mediterranean by walking fifty miles or so along the cliff tops and through the cork forests of southern Andalusia. My starting point was Cádiz, my destination the British outpost of Gibraltar.

Before leaving from London for the walk, I had supposed the high point of this modest expedition would be the stop at Tarifa, Europe's most southerly town, from where I should be able to see the snow-covered cliff tops of the Atlas Mountains in Morocco. It somehow seemed barely conceivable to me—I was in my mid-thirties, and still wide-eyed in my wanderings—that from the quayside of a small southern European town there could possibly be a view of *Africa,* that unimaginably distant and unutterably different continent of lions and giraffes and Moors and bushmen and Mount Kilimanjaro.

But yes, it was all there, large and looming and pink with Moroccan desert dust, and it was quite the spectacle I had imagined, full of symbolism and portent. Yet somehow it did not quite rival the thrill that Cádiz was feeling at the time I left there, just a few days before—because it turned out there had been a curious occurrence: a fire had broken out in an old part of what was already a very old city, some necessary demolitions had been carried out, and on the first blue-washed seaside morning I spent in the city, the maître d' in the Hotel Atlántico could hardly suppress his excitement with that morning's news: *They have found the ruins of a Roman theater!* he whispered to me as he handed me my pair of boiled eggs. *It is maybe the biggest in the world!*

The second largest, as it happens.* But the discovery of a structure built by one of Julius Caesar's lieutenants in the first century B.C. gave concrete form to this otherwise modestly self-effacing city's notion of itself as a place of once great importance

---

* Pompeii's remains the biggest known.

and antiquity. The Romans had used Cádiz as a naval base, and here was proof that they had had the wherewithal to entertain their sailors. The Carthaginians had done much the same, and before them the Phoenicians—who named it Gadir, the *walled place.* It had been a city of substance well before the Atlantic was even known to be an ocean.

The old center of Cádiz is on a slender spit of land between the ocean and the bay. At the seaward tip there is a fort, with thick walls and cannon and barbicans with slit windows from which wardens could once stand sentry over the sea. Immediately inside the walls is a rabbit warren of old structures, most dating from the seventeenth and eighteenth centuries. Beyond there are mansions and palaces and grand plazas, all built from the accumulated wealth of the two centuries when Cádiz was the principal Iberian entrepôt for commerce with the Americas.

I chose as the starting point of my long march east the plaque under the palms in the Plaza del Candelaria that marked the house of Bernardo O'Higgins, the Irish-Chilean who in the nineteenth century freed Chile from Spanish rule. I first strolled beneath the small scattering of towers from which merchants' wives once used to spy for homebound ships, much as New Englanders' spouses in later years gazed down from their widow's walks. I walked past the old tobacco warehouse, past the impeccably preserved cathedral and the convent, and finally out onto the great south road—the Roman theater under its protective tarpaulins to my left, the causeway to the Andalusian mainland and the road to Gibraltar stretching hot and dusty ahead. I managed to get a little lost, and asked an elegant and elderly Spaniard for directions. He was less haughty than he looked, and not at all brusque: *Keep the ocean on your right hand,* he said, *and you can't go far wrong. And mind you keep a lookout for Africa on the way!*

The oceanfront has been dignified for centuries by architecture that reflects an enduring respect for the sea. Cádiz has reminders of traders and explorers dating from Phoenician and Roman times. There is a muscularly commercial aspect to New York and Liverpool, and Jamestown, on the mid-ocean island of St. Helena, has provided for three centuries a Georgian sanctuary in miniature for passing merchants.

Hispaniola's sprawling city of Santo Domingo, three thousand miles away across the ocean to the west, has few such obvious charms, at least initially. Well over two million people are crowded into what is a generally ugly and uninspiring capital of an irredeemably corrupt and venal island (shared with Haiti, and lying between Puerto Rico and Cuba). But on the right bank of the Ozama River there is the old quarter, the reliquary of the city that Bartholomew Columbus, the brother of the explorer, founded in 1496, and which was rebuilt after a devastating hurricane four years later. And it is much more the kind of thing one hoped for.

Such buildings as remain show just how grand, how very much like Cádiz, this city could have come to look. As recently as a century and a half ago the *ciudad colonial*—formally known as Santo Domingo de Guzman, and in fact first named La Isabella in honor of the expedition's sponsoring queen—was still recognizable as an iconically Atlantic city. There was a huge seawall, the waves crashing noisily below. There was a dock, and a lighthouse, and within the walls a barracks, a powder magazine, and a signal tower. A brief orgy of early-sixteenth-century colonial building then brought an enormous and handsomely styled palace of government, a decently proportioned cathedral, a private merchant's mansion or two, a monastery, a hospital, and even more mundane but elegant structures—a warehouse, a slaughterhouse. On the landward side a great gateway pierced the wall with oak doors and two castellated towers, and from which Spanish troops could fan out on expeditions into the Hispaniolan hinterland.

Santo Domingo was in so many ways a classic of the oceanside fortress town: the narrow streets laid out in a perfect grid, all the essentials of expatriate life and imperial expansion crammed together along them within the protective walls of coral limestone blocks carved three feet thick. What little remains is well protected now: the United Nations sees that its unique standing is kept secure for the benefit of all, and that the great buildings—the First Cathedral in America, the First Castle in America, the First Palace in America—are kept beyond the reach of the developers who have so scarred the capital beyond with their skyscrapers and shopping centers. There are cobbled streets and a bustling Plaza de Espana; seagulls from the sea outside the walls swoop and skirl in the breezes.

Anyone walking at evening time high on the walls beside the iron-black cannon can feel very much at one with some stroller who might be in Cádiz, half a world away. This, one might be tempted to whisper across the sea to the other, is the way that the first cities of the Atlantic must have looked and felt and sounded in their first days. The tramp and clang of iron-shod infantry boots on polished limestone sett-stones, the importuning calls of merchants, the creak of ships' timbers and mooring ropes, the cries of the seabirds, the endless grumbling crash of the rollers and the sea beyond, all bathed in the warm seaside light of early morning or late evening, salmon pink on the high coral walls. Cádiz and Santo Domingo might at such a moment be the selfsame city, linked in style and feeling by the men who first built them, and then by the ocean they have risen beside.

· · ·

And then there are mighty Atlantic cities of today, New York incomparable among them. America's "sea-washed sunset gates," as Emma Lazarus has it in her famous inscription for the Statue of Liberty, remains today what it has been for more than a hundred and fifty years: the entranceway to hope and opportunity for millions upon millions of transatlantic peoples. To be sure, it is the great airports that bring in most of the migrants today, and most from far beyond the Atlantic, but the story of today's New York is still in its essentials that of the great funnel opening into which the huddled masses from the storied pomp of old Europe were poured without cease from the middle of the nineteenth century until today.

Even now, the vision of New York as a great harbor city can be powerfully felt. Just below the massive concrete anchor points of the Verrazano Narrows Bridge, in Brooklyn, beside a stretch of truck-thundering concrete known as Leif Ericson Drive, is a sorry patch of worn grassland, and from there one can be

close enough to the passing ships almost, but for an iron rail-
ing, to be able to reach out and touch their hulls. And what an
endless procession they make! Bulk carriers from the African
ports, fully laden and headed for the wharves of Bayonne, New
Jersey. Sleek container ships from Gothenburg, filled no doubt
with cheap Ikea furniture bound for the quay beside the main
store in Elizabeth, New Jersey; blinding white and windowless
car carriers from the assembly plants in Belgium and France
headed for the docks along the Port Newark Channel; oil tankers
heading gingerly up channel for the tank farms south of Kearny;
and maybe even a liner, a still-elegant Cunarder perhaps, or else
a more vulgar-looking and alarmingly top-heavy ship from Car-
nival headed for the piers on the west side of Manhattan, or the
newly furbished terminal at Red Hook, Brooklyn, right by the
tiny factory where some say the best Key lime pies in all America
are made.

The outbound ships thunder slowly past, too, their gigantic
screws thrashing through the waves as they head past Sea Gate
and Breezy Point in New York, and past Sandy Hook and the low
hills known rather generously as the Atlantic Highlands in New
Jersey, past the highly secure navy piers where ocean-bound
American warships are loaded with ammunition, out into the
standing swells of the Atlantic. The smell of the sea margin is
everywhere, and except on the sultriest of summer days there
seems always to be a good breeze, and there are dozens of smaller
craft scurrying about between the mighty ones, like those in-
sects that as children we called waterboatmen. Police and coast
guard tenders are lurking there, too, just in case, with uniformed officers at the helm and engines capable of great speed
idling quietly.

And then behind is the roadway, with its informal spell-
ing of the Norwegian who first crossed this ocean, and on it are

streams of trucks and private cars and yellow taxis, mostly coming back from Kennedy Airport the long way, because of reported traffic jams on the amusingly misnamed Van Wyck Expressway. Few taxis will care to stop, but if one can be persuaded to, then it is a mere five more westbound minutes to pass beneath the Brooklyn Esplanade, then up over the gossamer basket of the Brooklyn Bridge, to where suddenly in front rises the glittering crystal wall of Manhattan, like a curtain in a spectacular theater. I once brought a young Filipina woman here after a long flight from Manila. It was a crisp winter's day, and when she saw the first snow of her life, and then touched it for the first time, she shrieked with a mixture of shock and delight. But when she first saw Manhattan—and it was late afternoon, and the first lights were begin to prick like diamonds in the windows of a thousand buildings—her eyes became as wide as dinner plates, she cried out, and then she burst into floods of tears.

Manhattan hardly remains an architectural temple to the city's maritime history, a celebration of the sea. Its legions of skyscrapers are totems of other fields of commerce and wealth. But down beside the forts and bastions of the Battery, within a few rough-water cables of Ellis Island and Governors Island and the Statue of Liberty in her park on what once was Bedloe's Island, there are still hints of the city's cis-oceanic origins. Most notable of all the suggestions is the magnificent Beaux Arts Custom House, little used today but mercifully spared the fate of other equally noble buildings that were ground into landfill.

Ranged along the front of this great structure are four immense seated statues of figures. They were fashioned by Daniel Chester French, whose fame depends largely on his gargantuan statue of Lincoln in Washington, D.C. The four Custom House figures depict, with more than a nod to the ethnocentric mood of the times, the great seafaring continents.

Asia and Africa, their statues consigned to the two outer corners of the building, are seen to be sleeping and unmoving, little more than merely forgettable and pointlessly pretty. Europe and America, on the other hand, sit across from each other at either side of the flight of steps leading up the main entrance, and they positively burst with noble attitudes, with a frozen-in-marble energy and an apparently boundless capacity for triumph and fortune. If ever two statues can be said to represent the coming together that has created the new Atlantic identity, then this latter pair of marble giants, little seen and largely overlooked deep in the canyons of far lower Manhattan, have few equals. It is a shame that they both are female and that there is no potential for any marble offspring, with all the genetic markers of the new Atlantic spirit.

Shipping newspapers like the *Journal of Commerce* and *Lloyd's List* are still circulated down in these lower streets of Manhattan, to those with an urgent need, and a nearby shop called New York Nautical still sells charts of the *Approaches to Pernambuco* and the *Estrecho de Magellanes*, and has available the *Admiralty Pilot to the West Coast of Scotland* and guides to a hundred other corners of the world's oceans beside. A sailor home from sea can buy the *List of Lights: North Atlantic*, look over a good stock of sextants and brass-bound chronometers, or contemplate Adlard Coles on *Heavy Weather Sailing* and the *Ashley Book of Knots*. An hour spent on Lower Broadway, and then a taxi back to the piers of Red Hook, and one could feel entirely ready to board a ship, weigh anchor, and ease springs and head out beneath the Narrows Bridge, pushing out into the swells off Fire Island and then past Montauk, to where the Nantucket Light Vessel blinks its farewell to shallow sea, and finally set course for one of the classics of the old-world ports, three thousand sea miles ahead. To Bergen, say, in Norway. Or to Antwerp, or Rotterdam, Liverpool, Cherbourg, Vigo, Casablanca, or even, if courageous and well victualed enough to brave a heading far to the southeast, to Cape Town.

Here at the far end of the ocean's longest diagonal is New York City's polar opposite, its intellectual and spiritual antipode. Here, just a matter of miles from the most southerly tip of Africa, is a city truly born of the sea, yet seemingly paying little man-made homage to it, and rather letting Nature do so instead. The spectacle of Manhattan is something entirely enthroned in her buildings, which serve to display the myriad creative capacities of humanity. Her natural landscape is all but irrelevant. The joys of Cape Town, by contrast, lie not the buildings of the city at all, but in the velvet blue mountains that frame it. And all this great scenery serves to display what the sea already knows, and which is the very opposite of New York's self-delusion: not humankind's creative genius at all, but our utter insignificance.

I arrived in Cape Town recently in a small Greek ship. We were coming in on an easterly heading from the island of Tristan da Cunha, 1,800 miles and three days' sailing away. As promised, the Ukrainian steersman had called me to the bridge soon after five in the morning of our arrival: Africa, he said, was now directly ahead and visible, and the sun would soon be rising over the mountains.

It was a perfectly clear morning, cloudless and cool. One low-slung cargo ship, of Chinese registry, was off to the starboard on an otherwise calm and empty sea. Ahead was the glow of coming sunrise, and silhouetted in mauve below was a ragged skein of mountains, ending in a sharp cliff—the Cape of Storms that was, now the Cape of Good Hope. Northward from this cape the land first rose, then fell into a long defile, then rose again in a flattish tilt. It was from behind this that the sun first appeared, changing the color of the land, now twenty miles away, from blue to rock-dust brown and, where there was grass, to green.

Soon we could see a fine stubble of trees on the brows of the mountains, and some of the coastal suburbs—Camps Bay, Sea Point, and Three Anchor Bay—came slowly into view, though as no more than paler stains on the green slopes. Simonstown, the old Royal Navy base at the north end of False Bay, was somehow obscured by a low morning fog. As we growled steadily inshore, the dominant peak itself became ever more familiar, eventually dividing itself into its component parts: Signal Hill and the Lions Head off to the right, and directly ahead now the immense flat-topped pelmet of Table Mountain. As we turned into Table Bay the streetlights of Cape Town could be seen winking out in the distance, roadway by roadway. Traffic could be seen moving steadily along the coast roads. Down beneath its protective rim of hills, the great city was waking herself up for yet another crisp late spring South African morning.

On we pressed into the calm and sheltered expanse of the bay, passing a scattering of anchored ships, some waiting for a berth in the docks, or others rusty and most likely riding out their time in demurrage. To port lay Robben Island, where the colonial rulers had once kept their lepers securely isolated, and where the Afrikaaners did much the same for Nelson Mandela, though with rather less success. There used to be sheep and rabbits on Robben Island, the only ones in the entire continent, it used to be said with pride. Now only the rabbits remain, as pests, and in their thousands.

We were easing very close now, and slowing. A sudden tomtom of hammer blows could be clearly heard and we could see the sudden blue sparkle of welding torches, all from a new stadium being built on the waterfront. The engines stopped briefly and we bobbed alongside a buoy until a fussy little white pilot boat chugged out to meet us, steered by an elderly-looking black man; the pilot himself, young and breezy and in a newly pressed uniform, leapt aboard and was up on the bridge in moments, steering us to our anchorage in the Victoria and Alfred* Dock—the only structure of any note that ever sought to mark Cape Town out as a landmark Atlantic port.

The *Chamarel*, a French-built cable-laying ship, happened to be in dock, busy with preparations to sail, taking on immense coils of fiber-optic wire to string along the West African coast. Some years before the ship had helped lay the immense SAT-3 line, running six thousand miles between Portugal and Cape Town, and since it broke down quite often she was now on near-constant patrol to help keep it in service. Small Atlantic coast countries like Togo and Benin were hooked up to this vastly important cable and relied on it to keep them in touch with the rest of the planet. Engineers were now bent on connecting the half-forgotten Atlantic countries, places like Gabon and Equatorial Guinea, small nations that might well have remained overlooked for decades more, except that lately geologists had found oil in their territorial waters. Raw economics demanded that they now have the luxury of the Internet.

An Antarctic survey ship was also tied up nearby, her hull bright orange and with the gently sloping bow that meant she was able to break through ice floes, "if there are any left," said the lugubrious German skipper, who lived in Colorado and had been reading up on global warming.

A pair of tugs finally nosed us to the innermost of the quays,

---

* Prince Alfred, Victoria's second son, tipped the inaugural truckload of riprap into South Africa's first Atlantic docks in 1860. As the second Duke of Edinburgh he had already given his name to the tiny capital of Tristan da Cunha (the Atlantic's, and the world's, most isolated inhabited island), had survived an assassination attempt while having a picnic in Sydney (his Irish assailant was hanged for having dared try), and had married the daughter of the Russian czar, Marie, who still has a popular biscuit named after her.

beside a pontoon that was covered with fur seals basking in the morning sun. The only buildings of any distinction close by were the Cape Town Passenger Terminal, a somewhat utilitarian PWD-built structure from Edwardian days, and a scattering of Victorian godowns as well as office blocks with gilded ironwork and fretted balconies, most of them now turned into restaurants and hotels.

The oldest structure in Cape Town—said to be the oldest in all of southern Africa—is the old Dutch castle, a five-pointed, star-shaped structure in yellow ocher walls and tucked into a park beside the main railway station. It is almost hidden between the bland office buildings and condominiums: the only colonial architecture with remnant charm are the larger mansions and hotels on the lower slopes of Table Mountain, pretty when the jacarandas are in bloom, in parks secluded from the bustle and the traffic. And in truth, traffic and flyovers and cranes and the ugly architecture of the 1960s does tend to leave the strongest impression: it is only when one gets into the cable car rotunda and whirls up to the top of Table Mountain that the oceanside uniqueness of the place returns to view—and only then does it become easier to remember why this city is where it is, why the Dutch chose it as their rest stop and victualing harbor four centuries ago, and why, though it is so very different from New York, it remains every bit as much an Atlantic city as its counterpart, thousands of miles away at the far end of the long-sea diagonal.

For from the peak the ocean is everything and everywhere. It is only a moment's walk to get away from the grind of the cable car's engines and the jabber of the stall holders and into the windswept peace of the remoter corners of the cliff top, in the company of the eagles, buzzards, and warblers that soar hopefully and without cease in the thermals. The Atlantic lies to the south, where you can glimpse its rough windswept rollers smashing into the continent's most southerly point, Cape Agulhas, and

onto the crags of its most famous, the Cape of Good Hope. It lies
to the north, up along a coastline that, after the peninsulas of
Saldhana and St. Helen's Bay, runs straight all the way to Nam-
aqualand and the sand dunes of Namibia and the Skeleton Coast.
And it lies down across the city below, to the west—a vast, empty,
crawling sheet of hammered ocean steel, with the currents and
tide rips of Table Bay, the furious eddies around Robben Island,
and the faint white traces of great ships leaving for suitably great
and mostly beautiful ports on the far side of the Atlantic world:
Buenos Aires, Montevideo, Rio de Janeiro, Recife, Pernambuco,
Miami, Fort Lauderdale, Wilmington, Charleston, Baltimore,
Philadelphia, Boston, Halifax, and St. John's.

Each one of them charming, each one of them old, most of
them possessed of a stunning beauty, and all settled, like this
very town in the south of Africa, with wharves and port authorities and dry docks and ornately grand buildings and immense
railway terminals, each set down beside an ocean that each
somehow celebrates, in looks and sound and style and smell and
feel. Or that is how I think of these ports as I gaze enviously down
at the vessels churning off through the ocean, westward.

The *Chamarel* is leaving, too: I can see her crawling gingerly
between the harbor moles, white and sleek, with her twin funnels and her curiously bulbous cable-laying bat nose and the
drums of fiber-optic cable lashed astern: she will be heading for
Angola, where there was a radio report of trouble; and maybe she
will look in at the cluster of Atlantic islands tucked into the armpit of West Africa—the Cape Verdes, perhaps, or São Tomé and
Principe, all of them needing electronic connections to a world
that might otherwise pass them by.

And then there is one other ship, a small, stubby blue and
white vessel, by now well beyond the piers, heading to the northwest. She seems to be on a different track from the big cargo ves-

sels, a track that somehow reminds me of the direction that the old Union Castle liners would once take, back when this port served as a destination for the last great passenger liners, which went, as regular as clockwork, up to England, to Southampton. At four o'clock sharp every Thursday, one liner would leave Table Bay while a sister ship would slip southbound out of the Solent. They would pass one another, saying a brief hello, somewhere off the coast of Senegal. "Seventeen Days!" the newspaper advertisements would cry: "Weekly Mail Service to South Africa. Inquire at No. 3 Fenchurch Street, London EC3."

But this below me is no grand passenger vessel—no lavender-hulled *Pendennis Castle*, no *Stirling Castle*, no *Edinburgh Castle*. Besides, the very last of these ships, the *Windsor Castle*, had made the company's final voyage back in 1977, leaving at four o'clock precisely on September 6, getting back to Southampton seventeen days later. She had many owners afterward, mostly Greek; and then she went to be scrapped in India but suffered the ignominy of a steering-gear failure in the Arabian Sea, so had to be towed to a scavengers' feast in the Bombay breakers' yards.

No, this was no Union Castle ship below. When I was finally able to borrow a sufficiently powerful pair of binoculars I was able to identify her, though barely, as she was vanishing into nothing in the afternoon haze. Her name was painted in white on her stern. Her port of registry was Jamestown: she was the 6,000-ton mixed cargo and passenger vessel the RMS *St. Helena*, the only surviving vessel still formally designated a Royal Mail Ship, and as such supposed to be accorded a degree of respect and precedence by all other ships in harbor and under way. She was heading north to dock in Portland, England, eventually; but she was due on her way to stop in a week's time at the island for which she had been named and for whom she was now the only regular lifeline of supply.

The RMS *St. Helena* was this warm autumn afternoon heading out into the ocean on track for her port of registry, the town I still like to think is the most simply beautiful of all the Atlantic settlements. Jamestown, the capital of the crown colony where the British once exiled the defeated Emperor Napoléon, is a place which remains today preserved to perfection by what until very recently has been her almost total isolation. The island, forty-seven square miles of basalt and supporting an almost unvarying population of five thousand, was a good four days' sailing off the coast of Angola, in the middle of a now-trackless wilderness of sea.

Now trackless because the Union Castle liners used to call at Jamestown but long since abandoned the route. The final call was made by the northbound *Windsor Castle* in the autumn of 1977. The service then stopped dead, and when I first went there the journey was a little less easy to organize: no 4 P.M. *every Thursday* anymore.

Long ago I had been sent to the island to write a story about the curious case of a local man—a *Saint*, as they are still known— who had been convicted of a none-too-heinous type of murder (there had been a fight in a pub, so this was by no means premeditated slaughter) and was to be brought back to England to serve out his sentence.

There were seldom crimes of any seriousness on the island, I had been told—in fact, most islanders were so well mutually disposed that there was an alarming abundance of bastard children, known when they turned up at weddings as "spares." I was also told the local policemen had very little to do and were known as "the toys"; and that the Jamestown jail was so small and made so stuffy by the equatorial heat that inmates were let out each afternoon to go swimming in the Atlantic.

This, I decided one gloomy afternoon in London, was just the kind of place that had to be seen: a mid-ocean colonial posses-

sion of great antiquity where life, it seemed from afar, was lived
with less gravity than in most places elsewhere. After much trial
and error, I managed to get myself aboard the 1980s version of
the RMS *St. Helena*—a smaller, stubbier, and bright red predecessor of the blue one I had watched sailed past Robben Island.
After some delays and ditherings we lumbered out from the
Western Approaches and proceeded to steam south at no more
than ten knots past the Canaries and the Cape Verdes, through
to the warming, flying-fish-filled tropical seas.

There was an interlude when we hove to briefly at another
of the Atlantic Ocean's remote colonial outposts, the expired-
volcano island of Ascension, where there is an airstrip and a
lot of expensive communications equipment (some for broad-
casting, some for spying) and a patch of well-watered grass at
the summit where there used to be a herd of cows administered
from London, for complicated reasons, by an obscure department of the BBC. We had called in to collect a detachment of
Saints who worked for the contractors on an island that mostly
looks like a slag heap, or *Hell with the Fire Put Out,* as some of the
more disgruntled have it. But Ascension wages are good wages,
and there is little to spend them on, so back then not a few Saints
were happy to work there on yearlong contracts.

But they were never so happy as when they finally got back
home, which our passengers did after a further two days' sailing. The eventual arrival at Jamestown had a sweetness all its
own, as seaport returns so often do—with the unions of spouses
long missed, children much grown, and then all the news and
the gossip unheard. But here, although the substance of the
afternoon—we dropped anchor a couple of hours before dusk—
was mostly dominated by the delights of reunion, for me it was
a revelation of quite another kind. Jamestown, seen both from
afar and upon close contact, turned out to be a tiny Atlantic-side

city unlike any other, and in scale and style and manner just exquisite. Jamestown truly is a work of art, and Atlantic Ocean art at that.

The town, which generally has around fifteen hundred inhabitants, a third of the island's total, lies in a steep valley on the island's north shore, and like Cape Town is wholly encircled by hills. But there is no dock of useful size, and vessels of any consequence have to lie at anchor in James Bay, with all passengers and cargoes transferred to shore by lighter.* The legendary Atlantic swells, *born in storms as far away as Newfoundland*, the islanders like to tease, can make this tricky, the waits often inconveniently lengthy. But the panorama as the launch chugs in toward the crowded little pierhead seems lifted straight from an eighteenth-century print, with nothing changed or edited. There is a small and perfectly formed white-painted castle on the left, with tiny inner courtyards and cobbled squares; there is a small wooden drawbridge, and a castellated wall a dozen feet thick, built to shield the town beyond from any seaborne hostility, and this is pierced by a gateway with a portcullis and the arms of the Honourable East India Company carved and painted in red, white, and silver above it. There is a microscopic church (the cathedral, St. Paul's, is a ways inland), a town square with a bench for old-timers set in the shade of a pipal tree brought in from India, a minute police station with its aforementioned tiny prison, and then, at the beginning of a single main street that rises gently toward the brown flax-covered hills, there are two rows of

---

* This can on occasion be a risky procedure: when a member of the royal family visited the island in the mid-1980s, the ruling governor, in his best uniform of crisply pressed white duck and a pith helmet topped by swan feathers, stepped carelessly from a pontoon and dropped straight into the ocean, vanishing from sight. Though he survived both the ducking and the ignominy, the Foreign Office hurriedly packed him off to the rather drier post—though still on the Atlantic—of Guyana, where there is (unlike St. Helena) an airport.

Regency houses, each brightly painted and with iron trelliswork and sash windows, which look patiently across at each other as they have for centuries.

It is a town that is, as they say, *all of a piece*. There is the Consulate Hotel, with a gleaming brass plate outside. There is Jacob's Ladder, an iron-railed stairway of 699 stone steps that rises at a vertiginous angle up the valley side, and which was built to help supply a garrison of sentries placed on the cliffs to ensure that any would-be rescuers of Napoléon were seen, and then seen off. There is a public park, with a zigzag path among the jacarandas and bamboo fronds, and reserved, originally, for Ladies. There is a bustling covered market where the floors are wet with seawater and crowded with baskets of dripping fish. And once in a while the crowds part for a Jaguar, flag flying from the hood and a crown instead of a license plate, bearing His Excellency down from his mansion, Plantation House (the gardens of which have giant tortoises that were there at the time of Napoléon), to his offices in the castle.

The Atlantic eases itself into every conversation, every thought. The weather, of course, is made by it—the morning mists, the evening winds, the slapping swells that set the dockside pontoons swaying and creaking. The timetables of the ships are set by it—there is no airfield still, and many islanders suppose never will be, and the RMS in its many incarnations is still the only way to leave and to return. The daily tuna catch comes from the Atlantic; and such economy as survives on the island—for once they grew watercress for the Royal Navy, and they harvested and scutched flax for rope and string, but when the British Post Office decided to tie its parcels with plastic twine, all production ended—now depends almost entirely on the sea. The French *drapeau tricolore* still flies defiantly over Longwood House, where the island's most infamous Atlantic visitor was compelled to live out his final, post-Waterloo years, having been landed there from Plymouth aboard one of the ocean's better-known naval vessels, HMS *Northumberland*.* Even the island's ad-dress—*St. Helena, South Atlantic Ocean* (and with an island postal code, STHL 1ZZ, said to be recognized by the sorting computers in London)—displays a formal and official bond between the is-land and the sea, unique in all the world.

Of course there are nobler architectural confections around the Atlantic; there are many places of greater oceanside charm; and there many places of equal inconsequence ranged up and down the ocean, between Thorshavn in the Faroes up north, to Stanley in the Falklands down south. There is no lighthouse on St. Helena—so no opportunity for one of the great lighthouse builders, like the Stevensons of Edinburgh (Robert Louis among them), who created some of the greatest, most beautiful, and most technically challenging of all Atlantic structures.† That single lapse aside, however, it is tempting to list St. Helena high among the great visual triumphs of the ocean sea.

As a place, pure and simple, the island is perhaps best cor-ralled alongside the odd and the eccentric corners of the Atlan-

---

* There have been eight with the name *Northumberland*, most of which ended their days in the Atlantic—either off Ushant or in Biscay, and with the first having sunk during the Great Storm of 1703. I once came across the newest, a sleek Type 23 destroyer, as she was perform-ing high-speed turns off the island of South Georgia. She nearly came to grief there, too: in a freak accident her sonar equipment was wrenched off, nearly breaching the hull, and she had to limp over to Brazil for temporary repairs and then be brought home to England for a costly refit.

† Even the names of the Stevensons' best-known Atlantic lights have a poetry all of their own: the Bell Rock, Dhu Heartach, Eddystone, Muckle Flugga, Skerryvore! I once visited the light on the Alguada Reef, at the mouth of the Irrawaddy River in Burma. Its plans had been drawn up by Stevensons, it had been built by a Scotsman named Fraser, and the sole keeper kept all the brassware gleaming bright, in case, he said, of a *sudden inspection*. The rulers of today's Burma have committed many crimes: one, I have long thought, was to close down the old Alguada Reef light.

tic, with such places as Puerto Madryn in Argentina, where a number of the locals (descendants of indentured Cardiganshire railway workers) still speak Welsh; or with Axim in Ghana, where there is a magnificent Dutch-built castle; or with Devil's Island off the coast of French Guiana (where Captain Dreyfus was sent from Paris, to be held in solitary confinement). But I have always thought this tiny colonial outpost was deserving of something more. Without being too fanciful, I have long felt that in the architecture that defines both her and her exquisite little capital city, the island of St. Helena somehow represents, somehow stands for, somehow has become the essence of the ocean. That she somehow *is* a period of the Atlantic's human history, unaltered, untinkered with, and preserved for posterity in fine old Regency stucco and well-wrought imported English iron.

7. SOUNDS OF THE WATERS

The ocean is also represented, with equal vigor, in more contemporary writing, in painting, in music. The terror that the great sea stimulated in early times has been long assuaged; the formality with which the sea was represented in her newly crossed years in the sixteenth and seventeenth centuries has been long reversed; in modern times the Atlantic has become an entity to be recorded in all of her moods, in part for the obvious reasons of her drama, her beauty, her spectacular violence. But it also happens that she has come to be on very much more intimate terms with today's humankind—and that, it seems, has much to do with the present state of shore-bound civilization, a condition of existence to which the ocean is now seen as the very antithesis. Many nowadays think of the ocean admiringly, as a place of refuge from the numberless cares and wants of the landlubber. With the trials of modernity the sea has come to be regarded as refuge, as a place without the crowds, the dirt, or the want, without the slums of a vast modern city: as a place well beyond the pullulations of industry, money, and greed.

Of course, the Atlantic is a body of water that still needs to be crossed and navigated through, for reasons of commerce and curiosity and as we shall soon see, for reasons of war. But it is also a body of water that has come to be seen—and if one has to hazard a date, it began to be so from about the beginnings of the nineteenth century—as an entity for more pleasant purposes, of which human recreation, quite literally the re-creation of the human spirit, was one. It was still an illimitably large and powerful body of water, true, but so far as humankind was concerned it was now also something pure, something clean and uncongested, with a certain nobility about it sorely lacking in the slums of industrialized cities.

The ocean—and the ocean best known to nineteenth-century sophisticates was still the Atlantic—was thus something to be envied, an entity well deserving of our respect and admiration. This was a major shift in emphasis—and the art, writing, and music of more recent years has been quick to reflect it, to undergo what can without apology be called a real and very apparent *sea change*.

. . .

In music, it was the growing size of nineteenth-century orchestras that helped this change along, since for the first time it had become possible for the composer to reflect to the full extent the sheer complexity of the sea. Eighteenth-century music had an intellectual rationalism about it, limited by the kind of instruments available and the numbers of players that could be marshaled to employ them. The Romantic movement of Victorian music, on the other hand, vastly expanded both the kinds and the numbers of playable instruments—and so the ocean, with its sudden and sweeping changes of mood and color, suddenly seemed a highly appropriate subject for composers to tackle.

Beethoven, to take an early example, adapted in 1815 two short poems by Goethe to create the little-known cantata *Calm Sea and Prosperous Voyage*, which has as its theme the contrast between the quiet solemnity of a ship becalmed and the furious enthusiasm of the wild winds that would later drive her skipper home to port. Mendelssohn was much influenced by this little work and twenty years later produced a longer orchestral overture with the same title (one of Goethe's poems is devoted to the tedium of calms, the second to wind and prosperity). The overture begins with the quiet of the calm sea, then a flute trills to indicate a sighting of a patch of blue sky and the burning away of the sea mist, after which comes a cascade of the strings swelling with the rising winds, and finally a lone cello, in one of the most languorous and beautiful of all Mendelssohn's melodies, and which celebrates the vessel's arrival home safe and sound. This was not a work that could have been composed or performed a hundred years before, back in those times when the ocean was not fully known: there was neither the orchestra nor perhaps the necessary musical confidence of any composer then living.

Not surprisingly, nineteenth-century Italian composers tended to favor the Mediterranean for their seaborne excursions—as with Verdi, for example, in both *Simon Boccanegra* and *Otello*. Their northern colleagues, on the other hand, were more deeply inspired by the Atlantic: Wagner's *Flying Dutchman*, for instance, dealt with the legend of the spectral ship said to be haunting the tidal races around the Cape of Good Hope, and *Tristan und Isolde* had its doomed protagonists shuttling across a patch of that same sea between Ireland and Cornwall. Gilbert and Sullivan took both the mystique and inanities of British seaborne life as a motif for three of their light operas—*HMS*

Pinafore, *The Pirates of Penzance*, and *Ruddigore*. And more modern composers still—Edward Elgar, Benjamin Britten, William Walton, and Ralph Vaughan Williams—all took turns with the ocean, and in a flurry of marine musicality dealt respectively with its majesty (Elgar's *Sea Pictures*), its tragedies (Britten's own *Peter Grimes* and his adaptation of Herman Melville's *Billy Budd*), the bacchanalian habits of its sailors (Walton's *Portsmouth Point*), and the ocean's endless capacity for elegiac melancholy (as with Vaughan Williams's *Sea Symphony*, a seventy-minute choral epic with Walt Whitman's poetical work, much of it Long Island based and Atlantic related, from *Leaves of Grass*, providing the libretto).

Frederick Delius, who had some knowledge of the ocean from his time spent working on a grapefruit plantation in eastern Florida* and who had lived in Virginia, was also captivated by the Atlantic beaches of Long Island, which he visited in 1903. Like Vaughan Williams, Delius was captivated by *Leaves of Grass*, most especially by the section titled "Sea Drift." From a single poem within this collection, "Out of the Cradle Endlessly Rocking," Delius produced his own *Sea-Drift*, a twenty-five-minute work for baritone and orchestra that remains one of the most poignant of ocean pieces, relating as it does to Whitman's story that begins with the love and loss of a pair of Atlantic seagulls.

Claude Debussy, who employed an equally deft but even more solemn and reflective style, wrote at about the same time three symphonic sketches about the Atlantic—one concerned with the look and feel of the sea between dawn and midday, a second devoted to the complex and subtle play of the waves, and the last

---

* He said later that it had been the sound of his plantation workers in the distance singing their working songs that triggered his interest in the kind of evocative musical compositions for which he later became known. Previously, the older Delius, a Yorkshire wool merchant, had wanted his young Frederick to become either a sheep farmer or a citrus tycoon. Neither took.

to what he termed "the dialogue between the wind and the sea." Collectively Debussy's three works were known simply as *La mer*, and their supreme success in the concert halls of Europe helped attach the word *Impressionism* to a new style of sea-centered music; somehow its sounds managed to leave the audience with a distinct feeling of their having *experienced* the presence of the sea, without any need for the kind of signs and symbols—like Mendelssohn's trilling flute—that were required for the earlier, more direct representations.

8. CATCHING THE LIGHT

Painters had already long seized the concept of Impressionism—the deliberate vagueness, the studied imprecision, or, as one early critic had it, the conveying of "misty sentiment"—and soon found it particularly well suited to the sea. The French were in early: the newly built railways that took Parisian vacationers to the bathing resorts on the Atlantic and Normandy coasts sped painters to the seaside as well: Monet, Signac, and Seurat all famously painted the waters: the rocks, the coast, the summer indolence, the winter fury. The very name *Impressionism* is taken from an Atlantic ocean painting—that by Monet, of sunrise in the harbor of Le Havre, done in 1872. When asked by his dealer in Paris what he would call this quickly executed view of masts and morning mists and scattered sunlight caught from his garret window, he remarked casually that since it could hardly be called a study of Le Havre, it might as well be styled more simply an impression—so write down, he instructed: *Impression, soleil levant*.

John Ruskin once noted that "to paint water in all its perfection is as impossible as to paint the soul." Many have tried. Of all those Victorian and early-twentieth-century visual artists who tackled the ocean—and right up to today's Latvian-American pencil-genius, Vija Celmins, whose drawings severely test Ruskin's assertion, and including the extraordinary Japanese photographer Hiroshi Sugimoto—perhaps none has been more memorably effective than an impeccably transatlantic pair: an American, the Atlantic-born Boston Yankee Winslow Homer; and rather earlier, a Londoner, J. M. W. Turner. Between the two of them they drove a galleon at full tilt through the boom of documentary sea painting and changed the view of the ocean forever.

Few English artists had such a powerful command of the sea as the great nineteenth-century romantic painter J. M. W. Turner. Here, in The Wreck of the Minotaur, *Turner was able to catch the feel and power and impression of the sea in a manner that has survived him.*

Turner, who devoted the first half of his life to the depiction of storms and sunsets and wrecks in oils and watercolors, was well ahead of the age, producing a huge number of paintings in a highly saturated, vividly impressionistic, and immediately recognizable style, decades before the likes of Monet. He would have been well on in years when Winslow Homer was born, dead before Homer created the first of the engravings with which he began his career. He would never see the astonishing power of Homer's *Homeward Bound*, for example, a woodblock that was done for *Harper's* magazine in 1867 and that shows passengers struggling to keep their balance on a sloping deck of a ship in heavy seas, and which makes those who see it feel quite seasick. He would never know of the more famous pictures—*Gulf Stream* or *Breezing Up* or *After the Hurricane, Bahamas*—that depict with a masterful economy, but very much born of Turner's maritime vagueness, the power and majesty of the Atlantic. Homer loved the sea's austerity and integrity; he loved its loneliness; he loved its calms—*Rowing Home* is a perfect example of the still sea in the evening afterglow—and he loved it most of all when it crashed and thundered, in the height of a storm.

I write this just a day after a storm-related tragedy—a score of people swept from a cliff top in Maine and out into the Atlantic, all the fault of a rogue wave from a distant hurricane. A child drowned, her father was saved. They had come up from Manhattan and had been watching the drama of the surf on what should have been a pleasantly dramatic sunny Sunday afternoon.

It was just the sort of day that would have brought Winslow Homer out from his home on Prout's Neck nearby. He would have sat on the cliff top, patiently watching, his walrus mustaches flittering in the gale, struggling to keep his canvas down as he applied the first layers of paint to the gesso. The struggle between humankind and the Atlantic waters was of endless fascination to him—a painting he named *Undertow*, in particular, made in 1886, shows on a day just like this the rescue of two young women, their burly rescuers beside them, all four struggling out from the pull of the sea beyond.

There were to be many paintings like this. Winslow Homer recorded a series of especially heroic images while spending two years in the northeast of England, where the North Sea is especially ferocious, the wrecks legion, the drowning deaths frequent. I worked there as a young newspaper reporter and knew the coast well: how often I would have to drive with a photographer to where the lifeboat had been launched at Cullercoats, or Whitley Bay, or up by the Farne Islands, and then watch as a dripping, blanket-wrapped body was wheeled out of the raging surf to an ambulance that was driven slowly away, with no blue lights. This day's news from Maine would have saddened Homer, as it would sadden any human being: but it would remind him, a connoisseur of the awful power of the sea, that the ocean invariably wins in any contest with humans who dare it, and that is the natural order of things.

9. PEN, PAPER, AND SALT AIR

The modern sea is awash with literary sailors, and an immense treasury of literature has been produced over the years. Dickens, Trollope, and Poe have all tried their hands; Melville, Thoreau, Emerson, Virginia Woolf, Belloc, Eliot—at times one wonders if there is anything left to say, whether any marine scenario has been left undissected, undescribed. Writers have dealt with oceans known or unknown, crossed or uncrossed, from craft propelled by sail or by steam, over waters friendly or hostile, ice cold or drenched in steamy heat, and in ports gigantic or minute and with cargoes of any kind, volume, and worth. (With one exception: little contemporaneous literature of lasting quality appears to have been born directly from the slave ships of the Middle Passage. Much was written later, but little enough at the time—perhaps not too surprisingly, given the terrible exigencies of the experience.)

It is possible to distill from the raw mash of sea writing those pieces that concern the sea itself rather than using it mainly as a backdrop to some other yarn. And in this regard I have come to think that American sea writing has an energy about it that somehow overtops the writing found elsewhere—even though overall (and certainly when dealing with the North and South Atlantic) the experience of any ocean-bound writer is much the same, the sea is in a general sense similar, and the wrecks and dangers and storms and calms are none too different, no matter which port you leave from or in which direction you choose to sail.

But there is a discernible difference in the approach taken by writers in English on the two sides of the Atlantic. Some say it is a difference that stems from the fact that America possesses a continent of a magnitude about equal in scale to its neighboring seas, and a continent that is a body of land that, with its impenetrable forests and deserts and mountain ranges, is just as able to challenge and force endurance, of loneliness and privation, as is the sea itself. The British, on their side of the sea, inhabit a small and crowded set of islands, and their attitude to the ocean is not the same at all—for though Britain's surrounding seas may be vast and cold and dangerous, they are for a romantic about the only means of escape, mountaintops aside, from the busy nuisances of land. So although the British regard the sea as something that is always there, close to hand, it also somehow manages to be precious and unique, something of a refuge. To Americans, on the other hand, the sea may be notionally very much more distant and foreign, but it has a stature that is of some equivalence to their continent, and so it is to be viewed with a greater degree of understanding and a more casual acceptance.

Thus the Briton more often than not ventures onto the sea as an act of great daring and comes home with a story of great moment. But when a Richard Henry Dana or a Joshua Slocum sails down from New York and into the wastes of the Sargasso or through the tidal rips off Cape Horn, he does so with the same happy fascination and wide-eyed innocence that he might have exploring the badlands of South Dakota or the deserts of Death Valley. As narrator he seems to get in the way less often; the sea is forefront, and it is addressed the more directly.

Joshua Slocum is my particular hero. I have long felt a connection: my first North American summer—the summer that followed my arrival in Montreal aboard the *Empress of Britain*, in 1963—was spent in a cottage on the Bay of Fundy, in Nova Scotia, and in these parts Joshua Slocum was a local-born hero, even though most of his later years were lived either at sea or down in Massachusetts, a state that today regards him as a favorite son, and is the state where I live now. It was in the Massachusetts waterfront town of Fairhaven—just across the Acushnet River from the great whaling town of New Bedford—that in 1892 Slocum rebuilt from scratch a worn-out thirty-six-foot sloop, the *Spray*, until the hard-nosed local whalers, to a man, pronounced her "A-1" and forecast that she had been built so well, of pasture oak and Georgia pine and with a mast of New Hampshire spruce, that she "would smash through ice."

She rode at anchor "like a swan," said Slocum, when first he floated her off the hard—and it was with his beloved *Spray* that he then proceeded to circumnavigate the globe, quite alone, and wrote a book, *Sailing Alone Around the World*, that remains perhaps the finest example of modern sea literature. There is a laconic quiet about the writing that is almost hypnotic in its evocation of the ocean. Here he is well out of Boston, heading across the North Atlantic:

I put in double reefs, and at 8.30 am turned out all reefs. At 9.40 pm I raised the sheen only of the light on the west end of Sable Island, which may also be called the Island of Tragedies. The fog, which till this moment had held off, now lowered over the sea like a pall. I was in a world of fog, shut off from the universe. I did not see any more of the light. By the lead, which I cast often, I found that a little after midnight I was passing the east point of the island, and should be clear of dangers of land and shoals. The wind was holding free, though it was from the foggy point, south-south-west. It is said that within a few years Sable Island has been reduced from forty miles in length to twenty, and that of three lighthouses built on it since 1880, two have been washed away and the third will soon be engulfed.

On the evening of July 5 the Spray, *having steered all day over a lumpy sea, took it in her head to go without the helmsman's aid. I had been steering southeast by south, but the wind hauling forward a bit, she dropped into a smooth lane, heading southeast, and making about eight knots her very best work. I crowded on sail to cross the track of liners without loss of time, and to reach as soon as possible the friendly Gulf Stream. The fog lifting before night, I was afforded a look at the sun just as it was touching the sea. I watched it go down and out of sight. Then I turned my face eastward and there, apparently at the very end of the bowsprit, was the smiling full moon rising out of the sea. Neptune himself coming over the bows could not have startled me more. "Good evening, sir," I cried; "I'm glad to see you." Many a long talk since then have I had with the man in the moon; he had my confidence on the voyage.*

Slocum's is a plainsong kind of writing, honest and four-square, and with a good New England Shaker simplicity to it. He was perhaps a little mad, but the affliction was a gentle one; and his writing demonstrates the author's deep knowledge for the sea, his respect for its moods, and his fond hope for its fair treatment of his little boat. And it was a hope amply borne out: for three years after he had set out, almost to the day, Captain Slocum sailed little *Spray* into the harbor at Newport, Rhode Island—to rather little excitement, for the Spanish-American War had the headlines—and to the beginnings of his writing career and his brief flirtation with fortune. A decade later, when his funds had run low, he took off once again—only this time he vanished, somewhere in the West Indies, presumably taken by the sea in circumstances that were specifically unknown but generally familiar. But his literature remains: *any child not interested in Slocum's book*, wrote Arthur Ransome in a review, *should be drowned at once.*

Sailing alone around the world has since become almost a commonplace: there was Francis Chichester, and Robin Knox-Johnston, and the sadly mysterious affair of Donald Crowhurst (who cheated, went slowly mad, and drowned himself, all within the boundaries of the Atlantic), and since then about a hundred others. At the moment I am writing this—and hard on the heels of the news about the sad occurrence on the coast of Maine—comes the announcement that a boy of barely seventeen, by coincidence from the same small English town where I grew up, has sailed alone around the world as well. The Royal Navy sent a warship to greet him as he crossed the imaginary line between Ushant and the Lizard Point, from where such efforts are now timed and measured. That Joshua Slocum's achievement aboard *Spray*—without a chronometer, and certainly without any kind of GPS—has evolved into a mere high-technology stunt, and one in which children can compete, seems, though perhaps only to the churlish of mind, some kind of a diminishment.

Economy of writing, like Slocum's, is all too rare. This is hardly surprising, given the effort that any modern writer must now feel is essential to say something about the sea that has not already been said. But Rachel Carson—of whom the Blue Ocean Institute's Carl Safina once wrote, *her very name evokes the beatific luminosity of the canonized*—recognized its occasional presence, and in an unlikely source. In a chapter of her classic work *The Sea Around Us* that was devoted to foul weather and furious waters, she quotes from one of the British Admiralty Pilots, the blue-backed volumes of coastal description that line the chart-room bulkheads of every ship that ever made passage to foreign shores. She writes:

> . . . it seems unlikely that any coast is visited more wrathfully by the sea's waves than the Shetlands and the Orkneys, in the path of cyclonic storms that pass eastward between Iceland and the British Isles. All the feeling and fury of such a storm, couched almost in Conradian prose, are contained in the usually prosaic British Islands Pilot:
>
> > In the terrific gales which usually occur four or five times in every year all distinction between air and water is lost, the nearest objects are obscured by spray, and everything seems enveloped in a thick smoke; upon the open coast the sea rises at once, and striking upon the rocky shores rises in foam for several hundred feet and spreads over the whole country.
> >
> > The sea, however, is not so heavy in the violent gales of short continuance as when an ordinary gale has been blowing for many days; the whole force of the Atlantic is then beating against the shores of the Orkneys, rocks of many tons

in weight are lifted from their beds, and the roar of the surge
may be heard for twenty miles; the breakers rise to the height
of 60 feet, and the broken sea on the North Shoal, which lies
12 miles northwestward of Costa head, is visible at Skail and
Birsay.

Joseph Conrad wrote of the stormy seas, too, and in *Typhoon* (where the sea was the Pacific) just memorably; Richard Hughes wrote unforgettably, in *In Hazard,* about a storm in the Atlantic. Charles Tomlinson devoted an entire short poem to the analysis of one spectacular Atlantic wave:

Launched into an opposing wind, hangs
Grappled beneath the onrush,
And there, lifts, curling in spume,
Unlocks, drops from that hold
Over and shoreward. The beach receives it,
A whitening line, collapsing . . .

But as envoi to this chapter—which is, after all, about the romantic love for the ocean—I will offer up the words of one of the most remarkable transoceanic solitary sailors, the Frenchman Bernard Moitessier. The decision that lifted him into a different maritime realm from all of those others who have circumnavigated the world was one that he took in the far south Atlantic, during the race in 1968 that was won by Robin Knox-Johnston and in which Donald Crowhurst so tragically died.

Moitessier was spotted passing the Falkland Islands, heading northbound, and fast. Fast enough, in fact, for it to be assumed that he would win. But then suddenly, and for no apparent reason connected with the race, he decided he would not continue north at all, but would turn due east, would pass out of the Atlan-

tic Ocean altogether, and would head into the Indian Ocean for
the second time. In due course he explained himself, in a letter
squeezed into a can that he fired from a slingshot toward a passing merchantman:

> *My intention is to continue the voyage, still nonstop, toward*
> *the Pacific Islands, where there is plenty of sun and more peace*
> *than in Europe. Please do not think I am trying to break a*
> *record. "Record" is a very stupid word at sea. I am continuing*
> *nonstop because I am happy at sea, and perhaps because I*
> *want to save my soul.*

He was later to write his testament, an ode to the sea as the
center of his happiness. Within there is a paragraph that goes to
the heart of his beliefs, and which are held by most who love the
Atlantic Ocean, and all the other seas besides:

> *I am a citizen of the most beautiful nation on earth. A nation*
> *whose laws are harsh yet simple, a nation that never cheats,*
> *which is immense and without borders, where life is lived in the*
> *present. In this limitless nation, this nation of wind, light, and*
> *peace, there is no other ruler besides the sea.*

Chapter Four

HERE THE SEA OF PITY LIES

Then a soldier,
Full of strange oaths, and bearded like the pard,
Jealous in honour, sudden and quick in quarrel,
Seeking the bubble reputation
Even in the cannon's mouth.

1. MOURNING HAS BROKEN

The missile hit home shortly after lunchtime on a cool mid-winter's day in early May 1982. The weather was overcast, with a steady westerly wind and the kind of hefty swell that is typical for the far south Atlantic. Almost no one saw the rocket coming. It was a French weapon, small and sleek and inexpensive, and it had been dropped from an Argentine fighter aircraft ten miles away. It hit square in the vessel's midsection, just above the waterline. Sailors aboard remember only a surprisingly small explosion—shoddy bomb-aiming meant the missile was fired too close, did not have time to arm itself, and hit the ship without initially exploding—yet within moments the rocket's remaining propellant caught ablaze and set off a volcano of fire within the ship, with torrents of black belching smoke. The aircraft that had dropped the device zoomed overhead to confirm the strike had been lethal.

Indeed it had. Within just a few hours, HMS *Sheffield*, a gleaming and nearly new destroyer, a pride of the Royal Navy and on station in the South Atlantic to guard the great aircraft carriers and other warships converging for the beginning of the Falklands War, had been reduced to a burned-out hulk and was abandoned and drifting. Six days later, while she was being towed toward home, she sank. The lonely site in the deep ocean where she and more than a score of her incinerated and suffocated seamen now lie was formally declared a war grave, with the official request that the site be respected by all.

Sheffield was the first royal naval vessel to be destroyed by enemy action since the Second World War. She would not be the last to founder during the ferocity of the brief Falklands conflict: eight other vessels, five of them from the Royal Navy and one a giant Argentine cruiser bought from the Americans, still lie at the bottom of the Atlantic Ocean. Each still leaks a fine filigree of engine oil that floats to the surface and colors its gray waters with Newton's rings, the stricken vessels' only remaining visible memorial.

Being the first British ship to go, the *Sheffield* has become the one most poignantly remembered. Most shocked Britons remember with vivid exactness just where they were and what they were doing when the sinking was announced in the manner of so many recent tragedies. I had good reason to remember especially well, too, because at the time I was locked up on espionage charges in a prison cell not too far away in the grim sub-Andean town of Ushuaia, in southern Tierra del Fuego.

It was a particularly bitter cold evening. I remember a sudden commotion in the prison, and an officer from the Argentine navy running to my cell. He was jubilant and breathless, roaring in Spanish, like a football announcer. He came to the cell, gripped the bars, and with evidently undiluted glee yelled at the three of us being held there: "We have sunk one of your ships! We Argentines have sunk a Royal Navy ship! You are going to lose this war!"

But Britain, as it happened, did not lose that war; and the Falkland Islands remain today as British as they have been for almost two centuries. The war to confirm and secure this curious colonial status—a war that was akin, in the memorable phrase of the great Argentine writer Jorge Luis Borges, to "two bald men fighting over a comb"—was brief, bitter, and exceptionally bloody. Hundreds died on both sides; the Falkland Islands remain littered today with graves and land mines and memorials, and there are battalions of ever-vigilant British soldiers posted there still to make sure no invasion force ever tries its luck again. Beyond the South Atlantic, though, the conflict is thought of little more than as a rather ridiculous skirmish; it has faded from the collective memory, and few besides those directly involved care to speak much of it today.

Except for one subsequent event, I might well have half forgotten, too. But a great many years after the fighting was over, that same Argentine naval officer who had delivered the dismaying news of the *Sheffield*'s loss on that terrible midwinter evening somehow managed to track me down in Hong Kong, where I was living at the time. He wanted to meet, he said. He had something to say. And so after some complicated arrangements, the two of us did manage to meet, largely by virtue of my flying back once again to the by now very much larger and more prosperous Patagonian city of Ushuaia.

Outwardly I could see he was a much changed man. No uniform, for a start: he was now a civilian, grizzled and careworn in appearance, and the gruff air of machismo that was so much of his naval personality in 1982 had plainly evaporated. He told me, with evident sorrow, that he had left the navy many years before, had for puzzling political reasons been put in prison himself—in my old cell, in fact—and then had sold powdered soap door-to-door in Buenos Aires just to keep his family financially afloat. Then he reinvented himself: he managed to go to university, took a degree in history, and was now teaching at a small campus of the national Patagonian university.

He took me to dinner—he wanted me to try *centolla*, the giant crab for which the waters off Cape Horn are famous, and a soufflé of *calafate* berries, which Patagonians insist have the magic to lure anyone who eats them back to this strange and eternally gale-torn part of the world. And then, after pouring liberally from a second bottle of Malbec, he said that he wanted to explain himself.

He cleared his throat and looked rather nervous. He would preface his remarks, he said, by reminding me that in his view *las Islas Malvinas*—he couldn't bear to give the islands their British name, the Falklands*—should still be recognized as sovereign Argentine territory. The argument with Britain would go on until this was agreed, he said. But on the other hand, back in 1982 the dispute should have been settled by negotiation, he said. The war had been wrong; the imprisonment and ill-treatment of the three of us—we were reporters sent to cover the war, and we had been arrested on patently trumped-up charges—had been wrong. But most of all, he said, and this had weighed on him for many years, was the exultation he had expressed that night over the sinking of the *Sheffield*. That, he said, was terribly wrong.

---

*   Though he saw no inconsistency in using the name Malvinas, which was given by early French settlers who came from the Breton port of St. Malo.

For, he said, it had betrayed his principles as a navy man. Even though the British at the time were his enemies, he said, no sailor should ever take the kind of delight that he had taken on that cold May night in the foundering of another ship. No one should so ardently wish a vessel of any navy, or indeed any ship, ever to be sunk in the ocean. For it was his certain belief that to die alone at sea, in the emptiness of a wilderness of cold water, was just a terrible, terrible thing. "I am a good sailor," he kept saying. He stared sightlessly into his glass, his eyes brimming. "I am a good sailor," he repeated. "There is no pleasure to be taken over a thing like this. There is a brotherhood of the sea."

2. LITTLE LOCAL DIFFICULTIES

Brotherhood or not, the Atlantic seabed is littered with the wrecks of many thousands of ships and the long-decayed skeletons of many millions of men. War has been a constant feature of the ocean's experience, and wars have been fought on its surface ever since there has been iron with which to fight them. Discounting any undocumented coastal skirmishes among the seagoing Caribs, or the Beothuk of Newfoundland, or the Aztecs or the Mayas, the first recorded use of naval vessels in Atlantic conflict was probably made by the Romans two thousand years ago, when they sent wooden troopships to ferry their land armies across the sea into Britain for a century-long slew of invasions.

Biremes and triremes, with a mainsail each and two or three stories of oarsmen-slaves providing motive power, set out from ports either in northern France—Boulogne most probably—or from the river Rhine, and then wallowed slowly and dangerously across the English Channel. Eighty of these ships took part in Caesar's famous first invasion in 55 B.C., very many more when

Claudius made his much more successful landings almost a century later.

But the fights that then followed, pitched battles that would eventually bring England under the formal rule of Rome for the next three hundred years, were land battles: any oceanic component to the Romans' ambitions was severely limited. So the first real conflicts associated with the Atlantic Ocean were not Roman invasions at all. Rather they were the many centuries of seaborne maraudings that would become the scourge of all northern Christendom, and they were fought principally by another people entirely, the Vikings.

For the years when the Vikings were so occupied—for the most part in coastal waters, on the eastern fringes of the ocean—their invasions offer a textbook illustration of one of the basic reasons why humans engage in this extraordinary form of activity in the first place.

They were a highly mobile people—the maritime equivalent of the so-called wagon folk, those who in early civilization favored wandering as opposed to settlement, who were determined and nomadic pastoralists rather than fence-building and wall-constructing agriculturalists. The clash between those who built fortresses and those who drove wagons or sailed ships was a central part of early human life—from the time in the second millennium B.C. when the highly mobile Indo-European hordes swept down from the Caspian grasslands and crossed the Danube to begin populating central and southern Europe. That series of events marked the beginning of European warfare: what the Vikings unleashed when three of their marauding longships drew up on the beach beside Portland Bill on the English Channel in 789, and when four years later they murdered a group of monks at Lindisfarne, the great English Christian monastery on the North Sea's Holy Island, marked the true beginning of Atlantic warfare.

Historians argue over why the Vikings began their rovings and rampages. Those who believe it was a need of more agricultural land to feed a growing population are countered by those who wonder why they didn't just push backward into their northern forests and make agriculture there. Others suggest it was a decline in the trade to which the Vikings had long been accustomed—with the expansion of Islam in the Mediterranean having an unanticipated impact on the old trade routes and prompting the Vikings to try to open new ones. Some say climate may have also played a role: the period between 800 and 1300 A.D. coincided with a period of warming in the Northern Hemisphere, which increased sea temperatures by a degree or more and would have caused the ice in many of the fjords used by the Vikings to melt, furthering their ability to sail away more frequently. Still others—pointing to Viking grave sites dotted along the Atlantic shores, which show lots of dead Viking men buried alongside a scattering of clearly local women—insist that the seafarers had gone abroad looking for wives, to better their genetic stock.

Whatever the reason, these initial raids were followed by three centuries of Viking expansion, which rendered the eastern and northern Atlantic a zone of unpredictable and ceaseless unpleasantness. Longships thereafter left the Atlantic settlements in droves and plowed through the seas to places as far away as Archangel in the north of Russia, the various ports of the Baltic, the islands off the west coast of Ireland, and the coasts of France and Spain, and through the Mediterranean and past today's Istanbul and Anatolia into the Black Sea and to the cities of the southern Ukraine.

Moreover, longboats were sufficiently shallow that they could easily sail from the sea up into the estuaries of European rivers. Paris fell to Viking attacks after Ragnar Lothbrok took 120 longships and five thousand men up the Seine, declared he had never seen a land so fertile nor a people so cowardly, and refused to leave until King Charles the Bald paid out three tons of gold and silver. Soon afterward Dublin became a Viking *longphort* when longships sped up the Liffey, and a Norse base was established high up along the Loire, allowing the Vikings there to attack towns in northern Spain. Seville came under the harsh impress of Viking malevolence, as in later years did such places as Nantes, Utrecht, Hamburg, and Bordeaux. Add to all this the fact that Norsemen were in Iceland, Greenland, Labrador, and Newfoundland, and it can be fairly said that in their heyday the Vikings ruled the North Atlantic with much the same degree of hegemonistic influence and power that the U.S. Navy wields today.

But as with all imperial ventures, the Vikings' influence eventually waned. Their apogee, at least in England, came during the rule of the famous King Cnut, who had not only secured the English throne but united it with Denmark's, thus briefly welding together the two countries under a common Viking rule. But by 1066, a mere thirty years after King Cnut's death, the Viking rule in Britain was essentially at an end. The Normans—from the part of northern France that had not long before been under Viking control—then invaded England, speeding across the English Channel to defeat King Harold, who just weeks before had driven the last Vikings out of the north of England.

The sea must have been an exceptionally cruel place in that autumn of 1066. First an invasion fleet arrived from Norway that had to be defeated in the north of England, and then a second invasion fleet sped in from France in the south. King Harold had managed to inflict a savage defeat on the Vikings at the Battle of Stamford Bridge: of the three hundred longships that been had sent over from Norway, in an invasion that must count as the last of the Viking hurrahs, only a tenth of that number were needed to take back home the paltry number of the survivors and the wounded. But this victory had exhausted and depleted the king, and so when the Norman fleet arrived a month later, Harold was no match. England fell to the invaders, Harold was killed with a longbow's arrow through the eye, and the Norman conquest got formally under way, with consequences political, cultural, and linguistic that have remained to this day.

3. THE THEATER EXPANDS

For the four centuries following the Norman invasions, the maritime activities of the western nations were principally concentrated in the Mediterranean. Thanks mainly to the Crusades, the inland sea became one of the many battlefields on which, in essence, the European Christians found themselves pitted against the increasingly powerful forces of Middle Eastern Islam. Yet, ironically, the growing power of the Muslim world—and most particularly, the eventual stubborn intransigence of the Ottoman Turks—would provide the spur that led to a wholesale change in the status of the Atlantic Ocean. As a direct consequence of Islamic behavior in the Mediterranean, the Atlantic became the main highway for Christian war and conquest and imperial ambition for centuries to come.

It all had to do with Spain, which in the early part of 1492 engineered the final defeat of the Moors and the departure of the Islamic leadership from Granada and the Alhambra. Spain, suddenly and after an interlude of some seven centuries, was a united Christian kingdom again, ready to assume her place among the great nations of Europe. She also became—very rapidly—profoundly authoritarian in her attitudes and ambitions (demanding the expulsion or the conversion of the Jews, for example). She was transmuted into a Christian kingdom poised at the very beginnings of an imperial moment.

There was one other factor, more geographical than philosophical. In the fifteenth century Spain was poised, this time quite literally, between the two seas that, at this remove, can be seen to have shifted suddenly in their relationship and their relative importance. To the east was the Mediterranean—which was now blockaded at each end by Muslims, the Moors at one, and the Turks at the other. To the west was the Atlantic—a body of water that was largely free of the predatory and hostile Islamists, and into which Spanish vessels could sail unchallenged and unmolested. So the Spaniards must have seen the Atlantic as a means of furthering Spain's imperial ambitions and as a way of forgoing and forgetting the now suddenly uncongenial Mediterranean.

Portuguese explorers had already blazed a trail to Asia, had found the spices, ivory, gold, and other delights of the Indies and Japan and Java and Sumatra. But ever since the fall of the previously Christian stronghold of Byzantium to the Ottoman Turks in 1453, the land trade routes between the Christian West and these rich and exotic and possibly Christian (and certainly not Muslim) countries in the East had become severely frustrated by the Ottomans in between. If only Asia could be approached from the other direction, then the Turks and their allies blocking the passes between the Bosporus and the Khyber could be circumvented.

The geographers of the time thought it was an easy enough thing to do. They believed that the distance from Spain to Asia, going westbound by sea, was quite minimal. According to the calculations of their cartographers, Japan was a little more than three thousand miles west of the Canary Islands, and the Chinese coast stood just about where Oregon lies today. So if the sea that lay off the west coast of Spain could be easily crossed, and if Christian ships could navigate with ease all the way to Japan and China and then perhaps beyond to India, and if these friendly and much-prized states could all be reached, in essence, *from the other side*, then the commercial and political benefits were obvious. Just a few months after the expulsion of the Moors, Columbus's fleet was officially contracted to proceed west out of Spain to head for Japan and the Spice Islands of the Indies. But in the late autumn of 1492, the island of Hispaniola was found to be placed inconveniently in his way. "My intention in this navigation," Columbus wrote later to the Spanish throne, "was to reach Cathay and the extreme east of Asia, not expecting to find such an obstacle of new land as I found." Hence his later voyages around the Caribbean, and hence the eventual discovery by others—Vespucci, most famously—of the actual American continent that lay there unsuspected, and that the body of water between America and Europe was not simply an easily crossable minor sea but was in fact, and as we have seen, a brand-new ocean, the Atlantic.

This newly defined Atlantic Ocean was soon to become the main roadway along which the Spanish warships would journey to mount their attacks along on the newfound continent's margins; it was to be the main supply route for their subsequent conquest of the continent itself; and it was to be the sole highway home for all the plunder and treasure that poured without cease from the vaults and the mines of the Spanish Main.

All of this, the beginnings of what might be called the new American enterprise, coincided with the dawn of the Age of Discovery. This was to be a worldwide phenomenon, beginning in the fifteenth century and lasting for the following four hundred years, and involving European explorers and merchants who fanned out across the world in search of treasure, trade, and knowledge. And with these two burgeoning trends, the old world of the Mediterranean—"that little inland sea where there had been so much scuffling and struggling among European peoples for centuries," as the historian Fernand Braudel described it—suddenly and precipitously collapsed. The New World, washed by this huge new ocean, began dramatically to flourish, as it continues to do today. This was the true beginning of the Atlantic's primacy; it was a hinge point in world history and was accompanied by the traditional handmaidens of commerce, plunder, and war.

Spain's much-feared conquistadors were to be the first engineers of the many transatlantic colonizing missions that then followed. The template for their ruthless behavior was first struck in 1502, when a pious Castilian soldier-administrator named Nicolas de Ovando was appointed *Governor and Captain-General of the Indies, Islands and Firm-Land of the Ocean Sea,* as his orders had it, and brought 2,500 colonists, in thirty ships, to settle the island of Hispaniola. Over the next seven years he suppressed the locals with a massive display of force and violence, supposedly in the process reducing the native population drastically, from half a million to some sixty thousand. He imported scores of Spanish-speaking slaves and used them and such willing remaining locals as he could find to build the rudiments of the first cities. He planted sugarcanes he had imported from the Canaries, he opened gold and copper mines in the hills, he ordered large galleons to speed the crops and the metals back to Spain, and then he sent legates to other West Indian islands nearby to spread the benefits of Castilian rule as widely and as quickly as possible.

The one person Ovando was unable to bring with him on that first voyage was a relation of his wife, a minor noble from the southwestern Spanish town of Medellín, Hernán Cortés. The excuse, perhaps apocryphal, was that on the evening before the ship's departure, the then eighteen-year-old had injured himself escaping from the bedroom of a local married woman—the kind of story Cortés, who would go on to become the archetype of the swaggering, bombastic, and savage conquistador, would very much welcome. Cortés did eventually reach the West Indies, and like so many of his kind—daring warlord-adventurers with private means and good connections to the Spanish court—he used the islands as a springboard to reach the American mainland. Once ashore, he began his famously ruthless campaign of suppression and cruelty that resulted in the defeat of the Aztec Empire and the establishment of a permanent Spanish viceroy in the capital of New Spain, Mexico City.

Like all the solemn, full-bearded, and greedily determined conquistadors, Cortés came by ship, with thousands of Spanish fighting men, with limitless armories of sophisticated European cannonry and well-tempered steel swords, and, most crucially, with horses and specially trained and armor-clad war-dogs. He employed all of these assets against the bewildered Aztecs, without hesitation—although the extent of his cruelties is said by many of his defenders, and by the defenders of Spanish colonial policy generally, to have been widely exaggerated. But by the end of 1520, after a siege-and-destroy march from the coast and the clever forging of alliances among the other native peoples—which notionally was said to spread both the rule of Spain and the benevolent balms of Christianity—the Aztec lake-capital of Tenochtitlán had been utterly destroyed by Cortés and his armies. By the beginning of 1521, the Aztec Empire, which in some senses had been as sophisticated and advanced as any of the European civilizations, had been entirely destroyed.

. . .

The tragedy of the Aztecs and of their melancholy leader Moctezuma (who was taken hostage by Cortés and died mysteriously soon afterward—some say at his own people's hands, others by Cortés pouring molten gold down his throat) was a story that would happen again and again—to the Mayas, to the Incas, to the various Native American tribes in North America—until the Viceroyalty of New Spain became a vast imperial possession reaching from the fogs of Northern California to the fogs of Lima, and from Panama and Darien to the city of Santa Fe and the peninsula of Florida. Great tracts of the western coast of the Atlantic were under Castilian rule by the close of the sixteenth century, thanks in large measure to the organized dispatch of so many superfast westbound ships and the hosts of well-armed soldiers aboard them.

In due course the Portuguese, the French, the Dutch, and the English would sail their colonizing ships across the Atlantic, too, and though with generally rather more moderate displays of violence, would subdue the indigenous peoples they encountered and establish settlements themselves. The stories of these coastal colonies, the creation and extinction of some, the remarkable survival of others, have long passed into the legend of America's making: the stories of Walter Raleigh and Francis Drake, of John Smith and Pocahontas, of the Pilgrims and the Puritans, and of Peter Stuyvesant, are all familiar—and in almost all of them, the role of the sea is paramount. But their sea was not a sea of pity: it was a barrier to be bridged, a source of wealth to be plundered, and eventually a passageway for the New World goods—tobacco, lumber, rice, indigo, furs, gold—that could be sent back home to Europe.

As the sixteenth century became the seventeenth, and as the settlements in the Americas started to coalesce into permanence, two new seaborne phenomena began to develop, a direct result of the swift European colonization of America. And then, coincident with these two, and to a degree also a consequence of each, there came a vastly important third. And pity was a common component of them all.

First of all, a new generation of pirates began to operate in Atlantic waters, which suddenly were being increasingly traveled by treasure-laden vessels, heavy with New World bounty. The sea, especially in the close waters of the West Indies, became a maelstrom of unpredictable violence, with the masters of eastbound galleons nervously on watch for the sudden appearance of black-flagged attackers, with consequences quite likely to be as lethal as they were surely financially ruinous.

Secondly, slaves were being brought across the sea and put to work for those settlers who ran the large plantation estates of the American South: the seventeenth-century Atlantic became the superhighway of the so-called Middle Passage, the triangular journey that took vessels from England—largely—down south to West Africa, where they were laden with forcibly taken Africans, who were then transported in the most atrocious conditions to the slave ports of the Americas, after which the ships were swabbed down and laden with trade goods to be taken—provided there were no pirate attacks—back to the home ports of England.

The third development, and to a degree a consequence of both of these, was entirely military. It came about in part because the crushing of piracy and the abolition of slavery both eventually became matters of state policy in the European countries that had first nurtured them. One might see this as ironic—and yet it was no more than a result of the Enlightenment, for as times and mores became increasingly enlightened, both activities were seen as they are today: as wicked and nakedly criminal. The change in heart, especially in London, brought about an increase in harsh oceanic activity designed to bring this criminality to an end. During this period, navies, the seaborne state forces that were used to root out the pirates and chase away the slavers, became steadily better organized and equipped and tactically more sophisticated.

But these naval forces were not employed simply to put down maritime misbehavior. At the same time as these new sailing fighting ships were being designed, built, deployed, and improved, and as techniques of admiralty were being honed, so there developed a slew of disagreements between various of the oceanic states themselves. Conflicts arose between England and Spain, for example, or France and Holland, or England and France—all countries that now had well-developed navies. As a result, an entirely new kind of fighting was born. The naval forces that had been created could now fight one another, at sea.

To be sure, boats had fought one another before. But the battling boats in the early Mediterranean—propelled by oars in the early days—employed techniques of ramming or forcible boarding, with one vessel attempting to sink or overwhelm another. In this new sixteenth- and seventeenth-century world, in this world where pirates had to be sent packing and slavers dissuaded from their calling, there existed a new generation of sailing craft that were swift and nimble, and, most crucially, were mounted with powerful metal guns. This led to a whole new school of naval warfare—the birth of the naval engagement, in which one of these ships, or in time a whole fleet of them, might mount attacks upon another, with guns and fireballs and chain-shot, all of the battling conducted at sea, until the fight was concluded by capture, by rout, or by wreck.

Piracy, slavery, and sea battles: all three phenomena connected in a frenzy of military activity. The first two became inadvertent godfathers to the third: and the great later naval battles—Trafalgar, Jutland, the Battle of the Atlantic, even to a degree the very much earlier British defeat of Spanish Armada—owed much in

their conduct and their tactics to lessons learned in the fight to
cleanse the seas of pirate and Middle Passage villains.

4. THE SCAVENGERS OF THE SEA

Pirates—those who, as the law has it, *take a ship on the high seas
from the possession or control of those lawfully entitled to it*—have created havoc in the world's seas for as long as mankind has been
sailing in them. Long enough to have passed firmly into folklore:
the Jolly Roger, the eye patch, the parrot perched on the shoulder, the disfiguring scar, or perhaps a wooden leg, or a hook for
a hand—and cruelly appropriate punishments like walking the
plank—all these ingredients have created a fictional confection
of pirates as somewhat capital fellows with a liking for bellying
up to the bar. Only when one knows that a far more common pi-
ratical punishment was to gouge open a living captive's stomach,
drag out his entrails, and nail them to the ship's mast, then force
him to dance backward along the deck, running his guts out like
a clothesline—does the romance begin to fade.

To be attacked by a pirate ship was a terrifying experience.
The scenario had a certain routine to it: under the steady press
of the westerlies the cargo vessel, laden with treasure or trade
goods, would be lumbering heavily east through steady seas of
warm aquamarine, minding her own business—when suddenly
a suite of sails would appear on the horizon, and a small sloop
would sweep swiftly into sight. At a distance it might be flying
the flag of a friendly nation; when within sight or hailing distance, it would unfurl the plain black flag, or one adorned with
skull and crossed bones (or cutlasses), that was the widely recognized pirate flag. The sloop would then would come alongside, its
crewmen firing warning shots across the bows or into the sails

and so ripping them to shreds, and would then tack wildly so that its own sails would begin to flap madly from the mast. The victim, slowed by its loss of sail power, would then be forced to lower her own ruined canvas and come to a dead stop. Grappling hooks would then be thrown, hawsers drawn taut, and as soon as bulwark smashed against bulwark, scores of heavily armed, wild-eyed young men would swarm over the rails.

The reality of seventeenth-century Atlantic piracy was often colored by the fanciful imaginings of artists, such as the creator of this nineteenth-century wood engraving. Most pirates were cruel beyond belief, took little pity on their victims, and enjoyed boisterous celebrations at sea.

They would be brandishing cutlasses and sabers and light axes that they would slash at anyone showing the slightest resistance or disapproval. Some of the pirates would round up the crew, begin interrogating them, beating them, stabbing, all too often eviscerating or strangling them—in one famous case nailing a sailor's feet to the deck, whipping him with rattan canes, and then slicing his limbs off before throwing his carcass to the sharks. Others would rummage through the ship's holds and through the cabins, searching for anything of value or of interest. There might be gold aboard; there would certainly be guns and powder; and maybe skilled crewmen who could be forced or persuaded to join the pirate ship. And then, perhaps mounting a final violent assault on the passengers by way of Parthian shot, they would all swarm back onto their own ship, detach the ropes, and slip rapidly away, soon passing over the horizon and leaving whoever remained of the crew and the surviving passengers to limp away for refuge and repairs.

The golden age for the pirates of the Atlantic—a term that in this context includes both the *buccaneers* of the Caribbean and the *privateers*, the fleets of state-sponsored brigands who attacked enemy ships on behalf of nations whose own ships were too busy elsewhere—lasted for no more than seventy-five years, from about 1650 to 1725. Thanks to writers like Robert Louis Stevenson and Daniel Defoe, the exploits of the most notorious found their way into the popular prints: men like *Blackbeard*—or Edward Teach—who conducted his business in the shallow waters off the Carolinas; or Captain Kidd and *Calico Jack* of the West Indies; or Bartholomew Roberts, *Black Bart*, whose beat was off West Africa; or Edward Morgan, who was pardoned of his early buccaneering and, as British privateering naval tactician of legendary skill and prescience, went on to be appointed a governor of Jamaica—all became celebrated, familiar figures. Writers had a heyday, too, with the small number of female pirates, most infamously Mary Read and Anne Bonny, who dressed as men and by chance encountered one another while serving on the same pirate ship—learning to their mutual dismay that each, of heterosexual inclination, was a woman.

Mary Read and Anne Bonny escaped capital punishment by declaring they were pregnant. Men had no such luxury: and as the naval patrols in the Atlantic and in the West Indies swept up more and more of their like, and as the world began to weary of their exploits, and as the scourge of piracy began to wear itself out, more and more men were brought home to England, many to suffer an especially appropriate execution.

Arrested pirates were tried in London in the Admiralty courts; and if found guilty, as most were, they were hanged on a special gibbet set up in the Thames at Wapping, on the muddy foreshore between the low- and high-tide marks. Captain Kidd was hanged in 1701 at this point, the so-called Execution Dock; the sentence handed down to him read, as was the custom, that his body must be left in the noose until three tides had passed over it and "you are dead, dead, dead." Afterward the body was taken down, covered in tar to deflect the attention of seabirds, and hanged in chains at the mouth of the Thames at Tilbury. It was an advertisement, a warning to other mariners of the terrible sanctions that would be mounted against anyone planning to sail on a vessel that might unfurl the Jolly Roger.

The sanctions took their time—after all, there was so much money out there in the uninterrupted sea-lanes of the ocean. By the turn of the eighteenth century, however, a combination of policing by the Royal Navy and the rigid determination of the Admiralty courts conspired to begin to break the pirates' grip. By 1725 the menace was ebbing away, and though it was not until 1830 that the very last pirates were hanged at the Execution Dock, the story of piracy in the Atlantic in the later eighteenth century steadily became more fanciful and romantic, and the reality of life on the ocean became more a matter of discipline, regulation, and the rule of law.

The British in particular enjoyed an early edge in suppressing the activity. But there was another evil that was very much more insidiously dreadful than piracy. By chance one of the more famous British piracy trials, and one not conducted at the Admiralty in London but in a corner of West Africa, shed some long-needed light on it. It was a curse of the high seas that was eventually to be among the most severely policed as well, such that in time it was finally abolished. Yet it was an extraordinarily long-lived maritime cargo-carrying phenomenon, the memory of which now scars and shames the world: the unseemly business of the transatlantic slave trade.

The Trial of Black Bart's Men, as it came to be known, took place in 1722, in the dauntingly magnificent-looking, pure white cliff top building that still stands well to the west of the capital of Ghana: the famous Cape Coast Castle. It was adventurous Swedes who first built a wooden structure here, near a coastal village named Oguaa, as a center for gold, ivory, and lumber trading; it next passed into the hands of another unlikely Scandinavian colonizing power, the Danes; and then in 1664 it was captured by the British, who had an enduring colonial interest in West Africa and held on to the Gold Coast—as Ghana was then called—for the next three hundred years. At the beginning—and at the time of the piracy trial—the Castle became the regional headquarters of the Royal African Company of England, the private British company that was given "for a thousand years" a British government monopoly to trade in slaves over the entire 2,500-mile Atlantic coastline from the Sahara to Cape Town.

Though the monopoly ended in 1750, slavery endured for another sixty years and British colonial rule for another two hundred. The British turned the Castle into the imposing structure that remains today—and it has become sufficiently well known and well restored that it attracts large number of visitors, including many African-Americans who naturally have a particular interest in its story. The American president Barack Obama visited with his family in 2009, to see and experience what remains one of the world's most poignant physical illustrations of the evils of slavery. The dire reputation of the place is reinforced by its appearance: though Cape Coast Castle is the smallest of the three surviving slaving forts on the Bight of Benin,* it was designed to be by far the most austere and forbidding. It also has the infamous "door of no return" through which tens of thousands of hapless African men, women, and children were led in chains and shackles onto the ships that then crossed the Atlantic's infamous Middle Passage, eventually bringing those who survived the rigors of the journey to the overcrowded barracoons of eastern America and the Caribbean.

The trial, in which piracy and slavery overlapped in a way that intrigued the faraway British public, involved one of the Atlantic's more notorious and commercially successful brigands, Bartholomew Roberts, a Welshman who was better known after his death as Black Bart. He had worked as third mate on a slave ship, the *Princess*, and in 1719 was lying off the Ghanaian coast when his vessel was attacked by two pirate

---

* The Danish fort that still stands in the Ghanaian capital, Accra, is both named and modeled after Christiansborg Castle in Copenhagen, where the Danish royal family lives to this day, in storied splendor; and the old fort at Elmina, built by the Portuguese, had lots of decorative crests and a big sundial. Cape Coast Castle, on the other hand, is almost entirely unadorned, has dungeons with walls fourteen feet thick, four enormous bastions, seventy seaward-facing cannons, and gardens for the resident officers—but until 1820 it did not even have a chapel and gave the appearance of being a place of an overwhelmingly gimcrack creation, offering the outbound slaves only the most wretched venue for their final African farewells.

sloops, captained by Welshmen also. A connection was duly made; Roberts joined one of the pirate crews and over the next three years captured and sacked no fewer than 470 merchant vessels—making him one of the most successful pirates in Atlantic history, and grudgingly admired even by his most implacable enemies.

Cape Coast Castle was seized by the British and used as a central export hub for its West African slave trade. Like other former slave castles, with their infamous dungeons and doors of no return, Cape Coast Castle has become a place of pilgrimage for visiting statesmen, including President Barack Obama in 2009.

His luck ran out while he was careening his ships after a successful raid on a slaving convoy, once again off the Ghanaian coast. A Royal Navy antipiracy patrol, led by HMS *Swallow*, duped him into battle, and Roberts was fatally wounded in the neck by grapeshot. The 268 men on the three pirate sloops were taken away by the *Swallow* and her attendant vessels, and sent to the dungeons in Cape Coast Castle to await their sensational trial.

5. HUMANS, OFFERED WHOLESALE

Back in England, the men's fate drew the most excited comments because among the captives were 187 white men, all alleged pirates, and seventy-seven black Africans, who had all been taken as booty from the captured slave ships. Of the white men, nineteen died of their battle wounds before the trial, fifty-four others were found guilty of piracy and were hanged from the cannons on the castle walls, twenty were sentenced to long prison terms in colonial African jails, and the remaining seventeen were sent back to London, to be detained in prisons there.

The seventy-seven black African slaves, innocent victims of all this mayhem, were not treated with any great leniency. They were returned to the Castle dungeons, were forced to walk once more in shackles and chains back through the door of no return, and were put on yet another slave ship and sent back across the Atlantic for a second time. This time they encountered no pirates and were delivered to the slave markets in the coastal cities, and became fully a part of the still-growing slave population of colonial America. A poetic injustice, if ever there was one.

And though many thinkers at the time recognized this, and though a tide of common opinion was beginning to turn, at the beginning of the eighteenth century there was still enormous official and intellectual support for the trade, in England and elsewhere. The better read of the slave traders were content to note that two thousand years before no less a figure than Aristotle had written of mankind that "from the hour of their birth, some are marked out for subjection, others for rule." And even though some critics pointed out that the trade required that "one treat men of one's own tribe as no more than animals," still both the Church and the state accepted slavery as part and parcel of human behavior, part of the natural order of things. As an example: John Newton, an eighteenth-century clergymen of considerable piety—and talent: he composed, among other well-known pieces, the hymn "Amazing Grace"—was a slaver of some prominence and found no difficulty coming to terms with the fact that, as the *Dictionary of National Biography* has it, he was "praying above deck while his human cargo was in abject misery below." Thus cleansed of any moral ambiguity, slaving could be an exceptionally profitable business.

Beatings and torture were familiar practice on the Middle Passage slave ships. Even though the teenage girl shown being hanged and beaten in this famous cartoon subsequently died of her injuries, John Kimber, the notorious Bristol-based slaver charged with her murder, escaped conviction and mounted a lifelong campaign against his accuser, the abolitionist William Wilberforce.

Eleven million Africans were carried westward across the Atlantic between the middle of the fifteenth century and the end of the nineteenth. Three million of them were carried in British ships, owned by slave traders based in Liverpool, Bristol, Lon-

don, and such smaller west coast ports as Lancaster and White-
haven. (The comparable French slaving ports were Honfleur,
Le Havre, and the biggest, Nantes.) The entire British establishment—from the royal family to the Church of England—won
dividends from the business. And even beyond the rarefied
world of an aristocracy who risked money to back the slavers,
everyone else in Britain who used such mundane products as
sugar, tobacco, or rum benefited from the slave trade as well. It
was not just a singular evil: it was a singularly pervasive evil.

The so-called triangular trade was arranged so that goods
were taken from Britain to the African ports or slave castles, like
that at Cape Coast;* slaves from these ports were then shipped
across the infamous Middle Passage to the American slave de-
pots; and then, once the vessel had been emptied and cleansed,
New World cargoes went from there back to Britain.

And so, in small vessels called *snows*, in barques and brigs or
in three-masted square-rigged vessels that, somewhat oddly to
modern ears, were formally known as *ships*, the slave-ship captains set off fully laden, from England. Their orders were quite
simply to proceed to West African ports and, using the cargoes
they had shipped with them from England as barter, according
to their standing orders "to procure as many good merchant-
able slaves as you can." With most of their crewmen pressed into
service through the work of crimping gangs who found drunken
and persuadable young seamen in the shoreside inns, the ships
set sail filled to the gunwales with Africa-bound trade goods.
They took such marketable items as muskets, felt hats, iron
knives, brass casks, gunpowder, cotton, and gun flints, and on
one ship, the *Pilgrim*, which left Bristol in 1790, the somewhat

---

* Scores of such buildings litter the African coast from the Sahara to the Cape—sixty of
them in Ghana alone, so close packed that many lie within sight of one another.

more bizarre inventory of "1 trunk East India goods, 4 chests
bugles, 12 cases calicoes, 2 puncheons rum and 15 dozen bottles
wine." Hugh Crow, a successful (albeit one-eyed) slaver from the
Isle of Man, always made a point of first calling in at Rotterdam
and Jersey to buy extra spirits (more cheaply than he could in
England) to use as trade goods for the African slave merchants,
who liked nothing better than a drink.

Most vessels took what the French called *la petite route* south,
sailing via the Canary and the Cape Verde islands before turning inland along the now-east-trending African coast. They first
bartered their goods, usually for some rather prosaic set of objects—iron bars, brass bars, swatches of cloth—that had become
a crude currency for the buying of slaves. The prices in this currency—the iron bars looked rather like stair rods—remained
fairly constant for years: a male slave bought on the Senegal
River in the mid-eighteenth century fetched seventy bars; a
woman, somewhat more costly despite being offered "with a
bad mouth," went for sixty-three bars, another for "the exces-
sive price of 86 bars," according to the famous journal kept by
the Reverend Newton. (For purpose of comparison, a two-pound
bag of gunpowder went for one bar.)

And then, armed with wagonloads of such bars or swatches,
the British captains went either to the slave castles, which were
run by the Royal African Company, and bought officially sanctioned and price-regulated slaves, or else threw aside convention
and visited the more competitive (and in later years, more commercially successful) upriver slave markets, where they bought
a clutch of such black-skinned humans, either men, women, or
young boys, who seemed most suitable for work on the far side of
the ocean.

Whether these unfortunates came from the rivers, or through
the doors of no return found in the Gold Coast castles and the

other slave factories, they were first marshaled roughly onto the waiting boat. Next they were branded—often with the initials "DY" for the Duke of York—and shackled into pairs, the left wrist and ankle of one to the right wrist and ankle of the next. They were then taken belowdecks in the storage areas, where, it was hoped, they would survive the crossing—a hope born not of compassion, but of commerce.

Normally slave merchants were allowed—for there was regulation—to carry some two slaves for every ton of the ship's burthen, later raised slightly to five slaves for every three tons up to 207 tons, and one slave per ton thereafter. A 500-ton ship was permitted to carry more than 360 slaves—and for reasons of commercial efficiency these beings were stacked like so much tightly packed lumber, lying on shelves with no more than thirty inches of headroom. Even in calm water and cool days, the conditions were intolerable; when it was hot and the waters rough—which was common on the eight-week voyage—they were insufferable. The sanitary conditions were execrable. Privacy was nonexistent. Security was everything: the men were closely watched and guarded, and any attempt at insurrection or mutiny was put down with terrible force. The slaves were fed two meals a day—yams, rice, barley, corn, and ship's biscuit boiled up together into an unattractive mess—and to guard against scurvy (for the contracts with the American and Caribbean slave importers specified that the slaves be delivered in good physical condition) they were made to wash their mouths with lime juice or vinegar. They were also made to "dance"—being brought up on deck to be exercised, jumping rhythmically on the deck to the extent that their shackles allowed, crew members armed with whips standing nearby to make sure everyone moved with equal energy and kept their muscles in tone.

The ferocity of the slave masters is legendary—men were bru-

talized, women sexually assaulted, sick slaves thrown overboard
(as long as they were covered by the ship's insurance policy).
One passage will serve to illustrate the piteous conditions under which the human freight had to live, and on all too many occasions, die. It comes from evidence given to a British House of
Commons committee by a crewmen named Isaac Parker of the
Liverpool slave brig *Black Joke*, and of his commander, a certain
Captain Thomas Marshall. There were ninety slaves aboard
this fifty-six-ton ship, all collected from a castle in Gambia and
headed for South Carolina.

> *What were the circumstances of this child's ill-treatment? The*
> *child took sulk and would not eat . . . the captain took the child*
> *up in his hand, and flogged it with the cat. Do you remember*
> *anything more about this child? Yes; the child had swelled feet;*
> *the captain desired the cook to put on some water to heat to see*
> *if he could abate the swelling, and it was done. He then ordered*
> *the child's feet to be put into the water, and the cook putting*
> *his finger into the water said, "Sir, it is too hot." The captain*
> *said, "Damn it, never mind it, put the feet in," and so doing the*
> *skin and nails came off, and he got some sweet oil and cloths*
> *and wrapped them round the feet in order to take the fire out*
> *of them; and I myself bathed the feet with oil, and wrapped*
> *cloths around; and laying the child on the quarter deck in*
> *the afternoon at mess time, I gave the child some victuals,*
> *but it would not eat; the captain took the child up again, and*
> *flogged it, and said, "Damn you, I will make you eat," and*
> *so he continued in that way for four or five days at mess time,*
> *when the child would not eat, and flogged it, and he tied a log*
> *of mango, eighteen or twenty inches long, and about twelve or*
> *thirteen pound weight, to the child by a string around its neck.*
> *The last time he took the child up and flogged it, and let it drop*

out of his hands, "Damn you (says he) I will make you eat, or I
will be the death of you;" and in three quarters of an hour after
that the child died. He would not suffer any of the people that
were on the quarter deck to heave the child overboard, but he
called the mother of the child to heave it overboard. She was not
willing to do so, and I think he flogged her; but I am sure that he
beat her in some way for refusing to throw the child overboard;
at last he made her take the child up, and she took it in her
hand, and went to the ship's side, holding her head on one side,
because she would not see the child go out of her hand, and she
dropped the child overboard. She seemed to be very sorry, and
cried for several hours.

Whether or not Parker was telling the entire truth we shall
never know. All that is certain is that this account is to be found
in official British parliamentary papers for the year 1790,* and
that the child in question was said by Parker to have been inde-
cently young, little more than an infant.

Some fifty days after leaving West Africa, the American coast
came into view, and the second leg of the triangular voyage of
what the French called *le trafic Négrier* was over. Most of the slaves
had already been spoken for under contract, and the master's
orders had him head for a seasoning camp at certain island dis-
tribution centers—in Barbados, say, or Jamaica—or to one of the
mainland slave ports, such Norfolk or Charleston. Perhaps the
master would be fortunate and the American slave factor would
manage both to clear the holds of their human cargo—to buy the
merchandise in bulk and wholesale, and sell them individually
at retail prices at a market later on—and then to arrange for other

---

* *Accounts & Paper/Session papers, Minutes of the Evidence taken before a Committee of the Whole*
House on Regulation of Slave Trade, 1790, xxx (699), 122–24, 127.

freight to be carried home on the empty ship. Maybe there would be an auction, held either aboard the vessel or on the quayside below.

Or perhaps the slaves would be subject to the final indignity of their passage—the so-called *slave scramble*. The waiting merchants would have been told that each of the Africans aboard could be had for a certain price; and at a given signal, usually the stroke of a drum, they would all rush board the ship, and like the crazed mob at a department store sale would feverishly make their selections from among the terrified men and women who, still in the shackles, had been herded up onto the quarterdeck. Families would inevitably be broken up, with one merchant demanding the man, another the female partner, still others the children.

And then the ship would be off again, its decks fully cleansed with vinegar and lye, the shelves on which close-packed black humanity had been crammed for the previous weeks now jammed solid with tobacco or furs or the manufactured products of the settlements. Some weeks later the Head of Kinsale would be sighted off the port bow, and a day or so beyond that, the lighthouses off the Mersey, or the Avon, and the long trick would at last be over. There were wives and children to be seen once again, lanes to be walked and churches to attend, and the matter of the black cargo—morally vexing to some, but merely routinely unpleasant to others—could be safely shelved in the very back of the mind, until the next journey.

Slave traders remained cunningly determined for many years—most notably by buying shares in Portuguese slaving boats, since Lisbon kept slavery legal in its African colonies until 1869 and continued to supply Brazil with slaves from Angola until Brazil banned the trade in 1831. But over the years the West Africa Squadron of the Royal Navy did gain the upper hand;

and though service in its enormous Portsmouth-based fleet was wildly unpopular—mainly because of the deeply unpleasant tropical diseases that killed so many seamen—by the middle of the nineteenth century the men of the so-called Preventative Squadron had captured some 1,600 slave ships and freed 150,000 slaves. The final slave ships to cross the ocean were American, the *Wanderer* and the *Clotilde*, and they managed to get through the various cordons and blockades in 1858 and 1859 respectively. The last surviving slave from the last arriving slaver died in 1935, in a suburb of Mobile, Alabama. And with the death of this dignified old man from Benin, a ninety-four-year-old named Cudjoe Lewis, so was severed history's final living link to the transatlantic slave trade, which had begun with the French in Florida and the English in Virginia in the beginning of the sixteenth century and had endured for more than four hundred years.

As coda, though, there is one further account worth relating—that of a white American who after crossing the Atlantic became a slave in coastal Africa, and thus provided history with the mirror image of a trade that was otherwise overwhelmingly conducted in the opposite direction.

He was James Riley, a Connecticut farmer's son who was master of the American trading brig the *Commerce*, which set out from Hartford, Connecticut, in 1815 to trade in North Africa. That August, while trying to make the Cape Verde Islands, he was blown off course inside the Canaries, got himself lost in a fog, and rode up onto the rocks near Cape Bojador—the cape that Gil Eannes had managed so famously to double nearly four centuries before. He and his crew were captured and used as slaves by Sahara nomads, were forced to trek for weeks through the desert, half starved and compelled to drink the urine of camels.

Eventually, using all his of resourcefulness and cunning, but only by the greatest stroke of luck, Riley managed to get a note to the British consul in Essaouira, William Willshire, telling him of their plight. After a tortuous northbound journey through the sands of the southern Sahara, he and his principal Arab owner made it to the coastal city, and once the consul paid his captors off, with $920 and two double-barreled shotguns, he was freed and rescued. Willshire also procured the freedom of four crewmen who had traveled with Captain Riley—describing the five when he met them as "skeletons of men, with bones that appeared white and transparent through their thin and grisly covering."

Once he recovered his strength—his ordeal had caused his weight to drop from seventeen stone to six—Riley was sent home to Connecticut, his wife and five children, and promptly wrote a book about his experiences—*An Authentic Narrative of the Loss of the American Brig Commerce*. It was published in 1817 and sold more than a million copies—and because it presented for the first time the perfect inverse to the story of African slavery with which all Americans were familiar, it became an influential book as well, remaining in print until 1859 and going through at least twenty-three editions.* No less a figure than the young Abraham Lincoln read it: he later said that except for the Bible and *Pilgrim's Progress*, no other book had influenced him more. And Riley himself campaigned vigorously both for the abolition of slavery and for the settling of freed slaves in the newly created Liberia, which altruistic America colonists would establish a few years later on the African Atlantic coast close to where he had first been shipwrecked.†

---

* Under the title *Sufferings in Africa*, Riley's famous book is back in print today.

† James Monroe, U.S. president at the time of Liberia's creation, was to become memorialized in the name of its capital city, Monrovia. William Willshire was also to win his own memorial: a small town on the Indiana-Ohio border, with a population of fewer than five hundred, was laid out by Riley in his later days: Willshire, Ohio, a place once famous for its cheese and built close to a bog known as the Black Swamp.

6. THE RULES EVOLVE

The war against the slavers and the ceaseless campaigns against the pirates did indeed help influence naval tactics by offering instruction to professional sailors in two very basic areas of oceanic fighting. These sailors became more adept at using the seaborne gun, which was in any case changing fast in its design and lethality; and it also had an impact on just where in the sea the fights with this type of weapon would take place.

Traditionally, all early naval engagements took place within sight of land, or very close to it—in part because early mariners had such difficulty knowing precisely where they were, once they were in the gray and heaving sea that was entirely without landmarks. But as the techniques of determining both latitude and, more crucially, *longitude*, improved, ships' masters were able to tell more or less exactly where they were. Then they were able to determine where their enemies were on the high seas, which made it possible to fight them there. Once that happened, the expression "command of the sea" started to become a reality: in the early days, fighting navies who might claim command of the sea in reality had command only of the coastal waters in which they operated; post-longitude, they could extend this command into the deep oceans. And command of the sea was becoming of paramount importance in the new age of commerce and trade: the secret at the heart of imperial ambition held that winning control of the sea was becoming much more important than winning control of land.

Whoever exercised the most influence over the Atlantic—over the ocean *sea-lanes* that were just beginning to make themselves apparent—would enjoy an enormous commercial advantage. The European nations grouped around the eastern Atlantic shores—

and as time wore on, the American powers on the western side—would each dispute who had the ultimate sovereignty of the sea. Most often, such disputes were settled with the application of common sense. But from time to time fighting would develop—and rather than any need for the deployments of armies to settle such fight on foreign fields, this kind of fighting would and could be settled by confrontations between navies, and by battles that would be staged out in the neutral wilderness of the open ocean.

To conduct these fights there had to be an new set of tactics, and in tandem with these the sensible and efficient use of the new death-sport of naval gunnery. The first such confrontation came at a battle known only by its date—the Action of the 18th September 1639—and it took place in the English Channel, between the navies of Holland and Spain. Up until this point all naval confrontations were highly chaotic affairs,* spray-filled donnybrooks with the sailing vessels ponderously wheeling and turning this way and that in a furious melee, colliding with one another, firing at each other from guns mounted in the bows, not infrequently committing friendly-fire errors, sending flag signals to one another that could not be seen through the smoke, with each master taking his own chance to fight through the fracas as he saw fit. But in the 1639 battle, the Dutch commander decided on the simple idea of standing all his vessels in a line, such that their sides all faced the enemy fleet—and opened fire

---

* Prime among these oceanic shambles was the decimation of the Spanish Armada off the British coasts in 1588. In the context of this account the battles fought and the fireships launched are of less interest than the terrible navigation error made by the Spanish commanders as their defeated fleet rounded the northern coast of Scotland. Not knowing exactly where they were, and discounting the effect of the Gulf Stream, they turned south far too early and were set by the westerly storms onto rocky coasts that became a lee shore. Far more ships of the would-be invasion fleet were lost as wrecks on the Irish and Scottish coasts than in the earlier naval engagements; five thousand men died; only half the fleet managed to limp home to Spain.

with broadside after broadside, sending a withering cannonade of shot directly at any Spanish ship within range.

This technique, from then on called a line-of-battle arrangement, was to remain paramount in naval actions until the invention of steam-powered ships at the beginning of the Victorian era. Since it required stronger and stronger vessels to keep station in the middle of a line of battle—more particularly so when enemy forces did the same thing, and battles became fantastic exchanges of fire between two long lines of opposing vessels— the very best and most suitably strong and well-armed vessels came to be known as ships of the line of battle, a phrase that with the elisions of time became the *battleship*.

The action in the channel—which led to an even greater battle in a roadstead known as the Downs, off the coast of Kent, and which resulted in a rout of the Spaniards, the death of six thousand of their men, and the loss of forty-three Spanish ships— was still a confrontation that took place within the sight of land. The first battle to be joined out in the deep ocean took place more than a century and a half later: this was the battle of 1794 that has come to be known as the Glorious First of June.* It was fought, and using much the same adopted tactics, between twenty-five ships of the line from the British and twenty-six from the French navies, and nowhere near the coast but in deep Atlantic waters some four hundred miles to the west of the French island of Ushant. It was seemingly won decisively by the British, and it made a hero of the fleet's astute and brave commanding admiral, the then sixty-eight-year-old Richard Howe. In fact the aim of

---

* Naval battles traditionally have taken their name for the point of land closest to the engagement—engagements farther away being given as name the date on which they were fought. Calendrical ambiguities abound—and in this case the French call this battle (not that they often refer to it, since they lost) the *Bataille du 13 prairial an 2*, using the Napoleonic systems of months, of which only *thermidor* remains in use, as the name of a lobster dish.

the French navy was to secure passage for a convoy of American grain ships bound for the relief of a starving France—and all of them got through. So the outcome of the first truly oceanic fight was somewhat ambiguous—a tactical victory for England, but a strategic success for France. More important, though, it was an action that presaged fights over convoys that would be much deadlier, less than a century and a half later.

During the remaining age of sail, there would be many Atlantic battles that would quite deservedly pass into the history books, either because of their textbook elegance as naval engagements or because of their profound significance in signaling or triggering some great shift in the placement of the world's political chess pieces. The defeat of the Spanish Armada by Queen Elizabeth's navy in 1588 was an action that led essentially to the creation of the British Empire and the reduction to mere decadence of its Spanish predecessor. The defeat of Napoléon's navy (along with more Spaniards) at the classic battle of Trafalgar in 1805 is best remembered for the death of Nelson—a man still regarded with great reverence by all in Britain, and by all sailors everywhere. (His uniform, with the bloodied hole left by the musket shot from the *Redoubtable*, remains the most prized possession of the Maritime Museum in Greenwich, England; Trafalgar Square in London, with Nelson's Column as its centerpiece, has long overtaken Piccadilly Circus as the most iconically British of London's gathering places; his massive flagship, the 2,600-ton HMS *Victory*, remains in fine fettle in Portsmouth;* and a captain in the French navy is to this day called *capitaine*, not *mon capitaine*; Napoléon stripped away the honorific because of what he perceived as his sailors' unheroic failure.)

---

* HMS *Victory* is the oldest commissioned ship in the world; but the USS *Constitution*, though launched thirty-two years later, in 1797, remains the oldest commissioned *floating* ship. *Victory* has been in dry dock since 1922.

By disposing of the French maritime threat during the battle off Cape Trafalgar, Britain was now able to enjoy total mastery of the Atlantic Ocean and could throw her imperial weight around with almost total impunity there and in seas still farther away. As with all naval battles, there is no memorial—the two square miles of ocean, more or less, some forty miles west of the Strait of Gibraltar, swallowed up all the victims, and the place where twenty-seven British ships fought against a combined Franco-Spanish fleet of thirty-three—2,100 guns against 2,500, and 17,000 British sailors against 30,000 Frenchmen and Spaniards—is just waves and swell. But Nelson's famous flag signal, *England Expects That Every Man Shall Do His Duty*, still flies above his ship now pinned in that Portsmouth dry dock,* and his famous prayer, asking his God for *a great and glorious victory*, is still memorized by many English schoolchildren to this day.

Moreover, Nelson's grand and unorthodox tactic, that of sending his two parallel but well-separated lines of battle directly into the sidewall of the enemy fleet, piercing both the enemy's heart and his lower limbs rather than sailing alongside and hoping to cannonade him into submission—is still taught as an example of bravery and naval chutzpah; and the tragedy of the day, with the admiral lying bleeding to death on the deck, wounded by a sniper's luck, cradled in the arms of his doctors and his trusted captains, all the while warning his fleet to take shelter from a coming storm and with his last words, allowing how humbled he was to have been able to do his duty, remains etched with acid on the British public mind.

The *Pax Britannica* was in essence conceived at Trafalgar:

---

* Nelson actually asked his signaler to send "England confides that every man will do his duty," but the young lieutenant asked to substitute the word "expects," since it already had a purpose-made flag in the signal vocabulary, while "confides" would have had to be spelled out. As would, very oddly, the word "duty," which was not at the time in the naval vocabulary.

and since the British Empire was au fond an oceanic empire—
dependent on the navy to secure it, on islands to coal and sustain
it, and on fertile oceanside countries to victual it and bring it
fortune—and as one might argue further that it was an Atlantic Empire, too, so the siting of its inaugural battle in the heaving gray seas forty miles off the coast of Spain could hardly have
been more apposite.

*Nelson was England's greatest naval hero, and his greatest triumph was
to be his last: the defeat of the Franco-Spanish fleet in the Atlantic off
Cape Trafalgar, on the Spanish coast, in October 1805.*

The romance of the great battle lingers to this day. On Trafalgar Day in October 2009, more than two centuries after the
encounter, one of the last remaining battle ensigns from the
day, the union flag that flew from the jackstaff of HMS *Spartiate*,
one of Nelson's most prized attack vessels, was sold at auction in
London for some 384,000 pounds—more than twenty times its
estimated worth. Perhaps the extraordinary price was a mark
of affection for the Royal Navy's ship, which had been captured
from the French at the Battle of the Nile; perhaps it signified a
more general affection for the battle itself; or perhaps one has

to suspect that in truth it was a formal recognition of the story of the family that had owned it—descendants of the notably courageous first lieutenant aboard the *Spartiate*, a thirty-seven-year-old Scotsman named James Clephan. This young man, uneducated and low-born, had joined the merchant navy out of need, when his job as a weaver was swept away by the industrial revolution. He had then been press-ganged into the Royal Navy, but had then climbed steadily through the complicated and class-bound ranks of the senior service until he became an officer, and an evidently very capable one. The giant flag, eleven feet long and seven feet high, had been hand-sewn by his crew as a gift—a mark of respect and admiration, it was said, for one of the very few—sixteen out of three hundred thousand, naval lore has it—who rose from the press-gang to become officer. Clephan indeed went on to become a commander in the navy, dying a much-honored man in 1851.

7. WALLS OF WOOD, CASTLES OF STEEL

Great sailing-ship battles would take place in the Atlantic theater for many years to come. The War of 1812, the endlessly stalemated conflict between Britain and the United States, which arose as a sideshow to Britain's ongoing war with Napoléon, saw many memorable naval encounters: despite the entire U.S. Navy being a quarter the size of the Royal Navy's force assigned to blockade duties—a mere twenty-two American ships to the eighty-five British—the courage and good seamanship of the crew of the USS *Constitution* thrills to this day: not only did she soundly defeat the thirty-eight-gun frigate HMS *Guerriere* off Cape Cod, but she then took off south to Brazil, where she forced another British capital ship, HMS *Java*, to surrender and scuttle herself. The first battle was all neatly done in half an hour, but the second endured for three hours—a long exchange of shot and shell that gave the *Constitution*—which still floats in Boston Harbor—her current nickname: *Old Ironsides*.

But then, and all too swiftly for some, the age of sail, with all its honors and rituals and romance, came to an end, and in its place there came the ruder replacements of coal and steel and steam, and Winston Churchill's sardonic remark suggesting that British naval tradition was henceforth to be based on *rum, sodomy, prayers, and the lash.* Vessels that had been made of great walls of teak, pine, and oak were soon to give way to ships that resembled nothing more than immense castles of iron. The last British wooden warship to be built was the *Howe*, a three-decker with 121 guns and a full suite of sails, but with a thousand-horsepower steam engine and a screw for good measure, launched in 1860. She swept off for her duties just as the keel of the first British ironclad, HMS *Warrior*, was laid down—a vessel fully intended "to overtake and overwhelm any other warship in existence." The new shipyards on the Clyde and the Tyne and the Wear, equipped with furnaces and foundries, welding torches and rivet guns, would then promptly set to, clanging and fizzing for decades to come, to produce many thousands of successors. They were all wooden ironclads first, then eventually ships entirely made of steel, with their production continuing into the twenty-first century.

The first ironclad ships to enter into battle with one another did so in the Americas during the Civil War. They went at each other with hammer and tongs—and by doing so in the New World, also offered an early indication, unrecognized at the time, of the torch of technological advance being passed westward across the Atlantic.

The first involved a British sidewheel steamer, the *Banshee*, which managed to break through a fiercely imposed Union blockade and sneak into South Carolina waters no fewer than seven times, with much-needed cargoes for the Confederate forces. After more than a year running between Britain, Bermuda, and various ports on the secessionists' coasts, her luck eventually ran out, and she was captured in a battle in Chesapeake Bay. In a delicious example of the cruel irony of fate, a judge in New York ordered this Liverpool-built ship to be transmuted into a gunboat and commissioned into the Union Navy—as the USS *Banshee*. Moreover, she would join the very same North Atlantic Blockading Squadron with which the federal government was then trying to seal off the Confederacy from supplies and outside sympathy—a classic case of poacher turned gamekeeper, even if accomplished by force of arms.

The somewhat better-known early battle that involved metal-encased ships—and here two of them, for the *Banshee*'s captors had been made of wood—also involved an enforced turncoat: in this case a former Union forces' steam frigate, the USS *Merrimack*, which cunning Confederates had plated with iron and festooned with guns and had renamed the CSS *Virginia*.*

On the morning of March 8, 1862, this strange-looking but evidently formidable weapon of war steamed slowly out of Hampton Roads, Virginia, intending to join battle with the local units of the Blockading Squadron. To the *Virginia*'s delight, dawn delivered her a magnificent potential prize: a federal twenty-four-gun wooden sailing frigate, the USS *Cumberland*, was riding in the shallows, at anchor. She and a sister vessel, the USS *Congress*,

---

* The southern commanders had done more than simply weld steel plates around this six-year-old, Boston-built pride of the Union fleet. They had found her abandoned, burned to the waterline, and scuttled in the Norfolk Navy Yard—but were so desperate for ships that they raised her, pumped her dry, rebuilt her without sails, and only then welded on her armor and gave her a new name.

clearly stood no chance: though both ships, along with various other sister ships hastily summoned, rained shot and shell down on the *Virginia*, everything bounced off her flanks without causing injury. When finally the *Virginia* opened up her guns from short range, the USS *Cumberland* and the USS *Congress* were sunk in a matter of hours. Almost three hundred Union sailors were burned to death as the ships went down.

The *Virginia*'s dominance of the waters would be a short-lived affair. Overnight, while she and her crew were resting, the Union admirals were planning. The White House was frantic with the belief that this extraordinary new vessel might well next turn her attentions to the Potomac River, sail into the estuary, and begin shelling the seat of the Union government within a day or two. She had to be stopped by all means possible.

As it happened, the timing was perfect. The union forces' brand-new and purpose-built ironclad, the USS *Monitor*, was that very night battering her way down through the Atlantic rollers on her way from her builders in Brooklyn. She reached the shelter of Hampton Roads just in time to hear the last of the gunfire from the *Virginia*—and despite her crew being dog-weary from the storms en route, took up station immediately alongside the *Minnesota*, her enormous revolving gun offering formidable protection. When the sun came up the following morning, and the *Virginia* steamed out from shelter, an historic battle was immediately joined.

For three hours the two clumsy and heavily armored warships traded fire, salvo after salvo ricocheting from the iron plates, an air of smoke-filled and noisy bewilderment everywhere, crowds onshore watching in horrified amazement, and in the end, after a whole day of fighting, with neither commander inflicting fatal damage on the other. Both ships withdrew, each corps of bridge officers supposing they had won the fight, but neither side having achieved their intended goal. The *Virginia* was scuttled inside the Roads some weeks later, and toward the end of the year *Monitor*, under tow out at sea, took on water and sank off Cape Hatteras. Yet despite the individual fortunes of the vessels involved, the Battle of Hampton Roads, fought at the very site of what has now become the greatest naval base in the world, changed the face of Atlantic warfare—and in time, the face of maritime warfare generally—forever.

From the moment of that battle—news of which spread around the world with surprising speed, considering that a reliable transatlantic telegraph cable had yet to be completed—no major Western navy would build an important sailing warship out of wood again. Iron, steam, engines, coal, oil, trunnions, swivels— these were the new vocabulary of late-nineteenth-century naval warfare. Topgallants and Turk's heads, powder monkeys and marlinspikes and mainsails were words and notions that faded swiftly into memory.

Inventions that had eluded the mind of man through much of the age of sail now started to seep into common use: less than forty years after Hampton Roads came the marvels of wireless, which allowed ships to talk to one another and to their owners or their directors; forty years later still, there was radar, which allowed ships to see one another, or the land which they wanted either to avoid or to reach; then there was sonar, which permitted a mariner to know how far the ocean's bottom was beneath him; and the making of submarines, which changed every rule of maritime warfare. These and a thousand other pieces of wizardry turned the oceans, and the Atlantic in particular, into a very different arena for the conduct of war. Ships that in the sailing age could find and engage one another only with frustrating infrequency now could arrange to rendezvous—whether for reasons peaceful or belligerent scarcely matters—and with accuracy,

regularity, and reliability. Warfare that had become more tactically organized now became more geographically directed; and
when these developments were supplemented by the creation of
weapons of great power and by a new generation of ships of great
strength, and with vessels ordered to ranges of unimaginable
scale and at speeds hitherto unthinkable, so the stain of warfare
spread, cable by cable, fathom by fathom, until it encompassed
the entire ocean.

And stain of warfare it was: Trafalgar had been a bloodbath, a
massacre of wanton ship-killing and man-killing, and no battle that followed would be much less brutal. Decorum was at an
end. Naval warfare was henceforward to be a truly horrible business, and though all the evidence of death sunk into the ocean,
it was every bit as foul and fierce as the great land battles that
were so notorious for their ghastliness. If Trafalgar was the last
great Atlantic battle of the wooden ships, the Battle of Jutland,
which was fought over two days in the early summer of 1916, was
truly the first great Atlantic battle of vessels forged from steel.
It was also the first Atlantic battle that employed guns designed
to hurl explosive projectile shells—not merely the sail-slicing,
spar-smashing balls of iron fired from the muzzle-loading black
cannon that navies had used for centuries. Wooden-ship commanders had in past years come to some kind of unspoken ac-
cord neither to fuse nor to use exploding shells (since both were
likely to set wooden ships ablaze, your own as often as your foes');
but post-*Merrimack* sailors, fighting aboard ships made of non-
flammable metal, could do as they wished with high-explosive
devices, could tinker with them on deck, could use immense rifled artillery pieces to lob these terrifying fast-spinning devices
three miles across the water or more, to scourge and savage an
enemy.

Naval visionaries soon realized that steel ships would at last offer floating platforms to the same kind of artillerymen who had for years been using rifled shell-firing guns on land. At a stroke the world's new navies could become every bit as modern as the world's land armies—but with just one difference: the ships, which were obliged to carry their own highly explosive ammunition with them in their magazines, had to be absolutely sure to protect them against hostile gunfire—for one well-placed shell in a magazine could destroy a ship in seconds, ripping her apart and sending her to the bottom. Armor, and lots of it—a belt of twelve-inch-thick steel plates weighing a quarter of a ton for each square foot enwrapped a battleship's midsection—had to be applied; and vastly powerful new steam turbine engines had to be created to move this ponderous metal edifice swiftly across the seas.

All of this modernization was the brainchild of the then First Sea Lord, the remarkably ugly, autocratic, dance-obsessed,* and much-loved Ceylon-born martinet Admiral Jacky Fisher—a man who first entered a navy of elegant wooden-walled sailing vessels, and who left behind him the biggest and most modern fleet of steam-powered iron ships then ever assembled. By the time of the outbreak of the Great War, Fisher's new navy was a fighting force created in and for the Atlantic, and it gave Britain for the next half century near-total mastery of all the world's seas.

Enormous bases, with quays and piers and cranes, graving docks and fuel bunkers, ammunition and stores, were constructed all around the British coasts and on the fringes of the world's oceans. Though the Indian Ocean was nominally supervised from Trincomalee and the Pacific from Hong Kong and

---

* Fisher would stage whirling dance jamborees on the poop decks of his battleships, insisting that all his officers attend and docking the leave of any who remained behind the wardroom. His decision to fuel all his ships with oil rather than coal led to the founding of the oil company that became BP—an irony, considering the pollution caused by the great BP accident of 2010.

Sydney, the Atlantic was deemed most vital, and it was accordingly policed by squadrons of capital ships and their escorting flotillas based at naval headquarters in Bermuda, Jamaica, and Trinidad in the west, the Falkland Islands in the south, and Freetown, Simonstown, and Gibraltar in the east. Britain, from which the affairs of the North Atlantic were policed, was herself draped in an immense chain-mail curtain of naval protection: destroyers patrolled the western approaches, battleships cruised the North Sea and the deep waters off Ireland, enormous guns were forever trained over the narrow choke point of the Channel. Under Admiral Fisher's explicit instruction, what was called the Grand Fleet was moved north, close to where the ever-expanding German navy might one day try to venture from her Baltic and North Sea bases. The ships were to be based at a sheltered lagoon—Scapa Flow—in the midst of the Orkneys, a roadstead protected from the Atlantic gales and the sub-Arctic blizzards by furze-covered sandstone islands, with waters shallow enough to provide a secure anchorage, and of an area large enough to accommodate the mammoth assembly of hardware— almost forty modern capital ships, which together with flotillas of destroyers and frigates made up the biggest and strongest military force then known in the world.

This fleet was untested, however. Napoléon's defeat and death (on the mid-Atlantic island of St. Helena) was followed by a century of near peace in which scarcely any warships ever fired a shot in anger, nor did any British admirals stage any kind of major battle at sea. The first true test of these men and of their *dreadnoughts*, as the most gigantic of Fisher's huge vessels came to be known—named so simply, for what could possibly cause so mighty a craft ever to be afraid—came in the cold early summer waters of the North Sea, eighty miles off the western entrance to the Baltic between Norway and Denmark, the Skaggerak.

The Battle of Jutland, the greatest confrontation ever between two fleets of steel battleships—more than 250 enormous ships were engaged—started on May 31, 1916, and brought together the German High Seas Fleet and the Royal Navy's Grand Fleet. Twenty-five ships were lost and eight thousand men died, but the results, in strategic terms, were indecisive, with the Royal Navy still able to claim mastery of the eastern Atlantic.

The two fleets—the British Grand Fleet dispatched eastbound from Scapa Flow and the German High Seas Fleet steaming north from Wilhelmshaven, and each with battle-cruiser advance squadrons steaming before them—inflicted the most terrible punishment on one another, with ship after ship pummeled into submission by exploding shells, many sunk or exploded, and thousands of men killed in the most horrific circumstances. Two hundred and fifty steel ships fought with each other—twenty-eight British battleships, sixteen on the German side, and a huge number of auxiliaries. Both sides were astonished to lose capital ships that had been regarded—like the White Star liner *Titanic* four years before—as unsinkable and undefeatable. In the first hours of the battle, the British lost the *Queen Mary* and the *In-*

defatigable, and later the *Invincible*, blown to smithereens when the flash from a German shell enveloped her magazine: all were immense battle cruisers. The Germans lost 62,000 tons of shipping during the two days of the conflict, and the British almost twice as much, 115,000 tons. Six thousand British sailors died, two thousand on the German side. Numerically it looked very much as though the Kaiser's navy had won.

And all of this was in spite of the Royal Navy's adroit success in crossing the German "T"—a performance of that classic naval maneuver that suddenly had German admirals seeing across their bows the entire Grand Fleet, with the British twelve- and fifteen-inch guns all trained to shoot broadsides that could decimate the Germans at will.

Yet the German fleet was not to be broken—a catalog of errors, signaling mistakes, poor gunnery, and bad ship design prevented the British from landing the knockout blow their commanders wanted. And despite all the carnage and loss, when the two fleets broke off the Jutland engagement and made it back home,[*] the cool reckoning suggested only one thing: that submarines, torpedoes, and aircraft would be the dominant ocean instruments of war for the remaining thirty months of fighting. Big-ticket naval engagements, in which admirals of the old school tried to impose Trafalgar-like battle tactics on the new world of high-technology navies, would be short-lived indeed. The great naval encounters of the next war, the Second World War, would largely be fought by aircraft from carriers. Two years later the entire German High Seas Fleet surrendered—thanks not to any direct consequence of the encounter at Jutland, but in essence because the war came to

---

[*] By chance the last survivor of the battle, Henry Allingham, died at age 113 in July 2009, while this chapter was being written. He had helped launch one of the aircraft used as spotters during the fight. The only surviving Jutland ship, the light cruiser HMS *Caroline*, is still in use as a training vessel in Northern Ireland.

an end both as the result of an allied naval blockade of the German ports, which brought the Kaiser's economy to its knees, and because the German army collapsed on the Western Front. The Kaiser's ships were all interned in the Orkneys, behind the same booms of Scapa Flow from which the British Grand Fleet had sailed for Jutland. Seventy-four ships were imprisoned there after the 1918 armistice, with bored and humiliated German skeleton crews manning them, all their guns spiked, all their ammunition confiscated. All of them awaited the outcome of the slow-moving peace negotiations in Versailles.

But then on June 21, 1919, a prearranged secret radio signal, which in essence said simply "Paragraph Eleven—Confirm," was flashed to all of the waiting German ships—and, obeying long-established emergency orders that would be implemented on receipt of this cryptic note, the captains of the anchored warships immediately scuttled every last one of their ships, by opening seacocks, smashing pipes, gouging holes in the hulls, and allowing fifty-two of the ships to slip slowly down into the shallow waters of the lagoon before the British—most of whose ships were away at sea on exercise—could stop them.

The British fumed—they had wanted to divide up the surrendered fleet among other navies—and did what they could to punish the miscreant German officers. But in the end the provisions of Versailles allowed the Germans to go home; and in time some of the larger vessels were raised and sold for scrap, the money going to the British Treasury. (Many of the hulls remain—and high-quality German steel salvaged from some of the remaining wrecks is occasionally brought to the surface today, much prized for certain very delicate scientific experiments, since it was fashioned, forged, and cut long before the spread of radioactive contamination, which has affected most metals that have been made since Hiroshima.)

Naval commanders may have learned as many tactical lessons from Jutland as their predecessors did from Trafalgar more than a century before, but these lessons paled when set against one reality that was little understood at the time but is all too well realized today. And that is that all naval fleets were from the late nineteenth century onward to be made almost entirely of steel— and Britain, despite the vastness of her imperial possessions and the industriousness of her people and the sophistication of her factories and foundries, had less steel than the Germans, and in just a few years the Americans would have much more than the Germans. In the future, whoever had the greatest access to high-quality steel would eventually have the wherewithal to build the greatest navy in the world—which the United States soon and most definitely managed to do. That, and the eventual deployment of different and infinitely more potent kinds of naval weaponry—and kinds of weaponry that were no longer exclusively wedded to traveling on the surface of the sea, but could weave their way beneath the water, or fly thousands of feet above it—was what Jutland has come to signal to the admirals of today.

It would perhaps be invidious to remark that just as the Great War had ended with a famous episode of German naval scuttling, so the Second World War began with another—a scuttling that also involved a German capital ship, and which also occurred in the Atlantic, although in this case the South Atlantic. The ship was the pocket battleship *Graf Spee*, and the event occurred outside the port of Montevideo in Uruguay, at the widest part of the mouth of the River Plate.

The ship, a sleekly villainous-looking Nazi surface raider, was part of Hitler's plan to restore Germany's navy to its former glory—but it had been built as a cruiser, since the Versailles Treaty prevented the Germans from constructing anything larger. She was very fast, and had been armed with an arsenal more suited to a battleship—including eleven-inch guns. She left Wilhelmshaven in August 1939. Her commander, a Captain Langsdorff, had sealed orders to attack all Allied-flagged civilian vessels in the Atlantic once war had been declared.

When the British prime minister made that formal declaration on September 3, *Graf Spee* had already broken out into the North Atlantic, steamed north of the Faroes, turned sharply south, and was well into the calms of the Sargasso Sea, a thousand miles west of the Cape Verde Islands. As soon as Germany was formally at war, Langsdorff ordered his guns unsheathed and began a rigorous program of commerce raiding, attacking every merchantman he came across.

Grain carriers, frozen meat carriers, fuel tankers, it made no difference—the *Graf Spee* went after everything she found in the South Atlantic, successfully notching up a kill every three or four days, causing major consternation up in London and paying no heed at all to the American zone of neutrality that President Franklin Roosevelt had established to protect Allied merchant ships sailing within a thousand miles of the shores of either North or South America.

But then, early in December, the deadly little battleship encountered three smaller Royal Navy vessels that had been ordered to scour the seas in a frantic hunt for her. They were the cruisers *Ajax*, *Exeter*, and *Achilles*, and when they met, and despite being massively outgunned and outranged, they joined battle with the German ship with the madcap enthusiasm and imprudent tenacity of terriers. It did not take long for *Exeter* to be so badly damaged that she was forced to retire, and although *Ajax* and *Achilles* suffered major damage, too, a lucky shell strike from *Exeter*'s eight-inch guns on the *Graf Spee*'s midsection hugely reduced her fuel-processing system, leaving her with limited fuel and (though no one except Langsdorff knew this at the time) a consequently doomed future. The stricken German ship slowly limped into the neutral safety of Uruguayan territorial waters, and up to an anchorage in Montevideo port—her officers knowing only too well that under the neutrality terms of the Hague Convention she had only seventy-two hours to effect repairs.

Huge public interest was aroused by the ship's steadily ticking fate, particularly while British naval reinforcements were gathering—or were believed to be gathering: many clever ruses were being played—in the ocean beyond. It was gripping stuff. In London, Harold Nicolson, the politician and diarist, wrote as follows for his entry of the December 17:

> *After dinner we listen to the news. It is dramatic. The* Graf Spee *must either be interned or leave Montevideo by 9.30. The news is at 9. At about 9.10pm they put in a stop-press message to the effect that the* Graf Spee *is weighing anchor and has landed some 250 of her crew at Montevideo. As I type these words she may be steaming to destruction (for out there, it is 6.30, and still light). She may creep through territorial waters until darkness comes and make a dash. She may assault her waiting enemies. She may sink some of our ships. . . .*

The *Graf Spee* did leave port just before the deadline—but she did none of things Nicolson imagined. She steamed slowly out across the territorial limit, trailed by a small tug. Then, four miles from shore and still within sight of the vast crowds on the Montevideo waterfront, her crews exploded three almighty demolition charges inside her. These set her furiously ablaze and, at risk of widespread public ignominy in Germany and Hitler's private fury, they slowly and painfully sank her, in full view of the astonished mobs and to her equally astonished and relieved enemy. Captain Langsdorff, one of the more honorable of German naval officers at the time, was eventually taken off the burning ship, put in to port in Argentina, and two days later killed himself with a single shot to the head.

Burned and listing, the German commerce raider Admiral Graf Spee *sinks outside Montevideo harbor on December 17, 1939, after being scuttled by her crew. Neutral Uruguay demanded she leave port before she was fully seaworthy—prompting her captain to set demolition charges and send the pride of the Nazi fleet to the bottom.*

For many years the wreck's canted mast could still be seen poking above the muddy estuarine waters at low tide. One of the ship's 150mm guns has been salvaged and is on display in a Montevideo museum, an anchor and a rangefinder are mounted on the foreshore, and the *Graf Spee*'s eagle crest was pulled from the water in 2006, its swastika covered with canvas to lessen any likely offense. Two cemeteries house the graves of those who died in the battle. But otherwise the burned and torn wreck of the ship remains untouched, marked on South Atlantic charts merely as a *hazard to shipping*—though somewhat less lethal a hazard today than she briefly had been in the southern spring of that first year of the war.

8. ENEMIES BELOW

Submarines were by far the greater hazard in the twentieth-century Atlantic, and during both of the conflicts. However, not at first: though they had been invented before the outbreak of the Great War—the world's first was built in England in the seventeenth century; the first German submarine was made in 1850; the first German naval submarine in 1905—and though it was fairly obvious how these sinister boats could best be employed, as the unseen snipers of the ocean—the manner in which they were first used offered an almost courtly regard for the old-fashioned, gentlemanly values of seaborne warfare.

There had never been any doubt Germany would deploy its small but growing fleet of submarines as commerce raiders, using their torpedoes to sink as many of Britain's supply ships crossing the Atlantic as possible. As an island nation, Britain could be supplied only by sea, and Germany's actions were meant to wreck the British economy, starving her people and forcing her to subjection and surrender. But initially there were rules of engagement that had been laid down in treaties signed in Paris in 1856 and then again in The Hague in 1899 and 1907, and they related to what was known as Prize Warfare—the seizing or destruction of merchant ships on the high seas. These agreements all held, for example, that passenger ships should never be attacked; that the crews of merchant ships should be placed out of harm's way before their vessel was plundered and sunk (and lifeboats were considered to be out of harm's way only if they were within sight of shore—if out of sight of land the crews had to be taken aboard the attacking ship); and that formal warnings had to be given before an attack.

These rules were made, however, for the benefit of belligerent surface vessels—sailing vessels, in fact—and not for submarines. But of course, so far as submariners were concerned, the rules were absurd. The first to point out that a diesel submarine could hardly behave in the same way as a sail-powered ship was Jacky Fisher, the British admiral. A submarine had neither the manpower nor the room to deal with the crew of a merchant ship: "there is nothing a submarine *can* do," said Fisher, "except sink her capture."

Churchill objected to this notion in an uncharacteristically Blimpish way; he thought that turning one's back on the rules of naval gallantry was arrant nonsense: no civilized power, he harrumphed, could, should, or ever would do such a thing. And for the first few months of the war he seemed to be right: commanders of ships on both sides—submarines included—behaved in a manner he thought proper. Although German U-boats would torpedo any British warship they found (and for warships there was no warning), each time they came upon a merchantman they would surface, demand the evacuation of its crew, sink her with gunfire, and then submerge again. In purely military terms it was a fairly pitiful exercise (not least because it rendered the floating submarine vulnerable to attack herself), and the attempt to retain chivalric codes in submarine warfare resulted in the loss of only a few British ships and made hardly any economic impact on Britain at all.

And then came May 7, 1915, and a sudden and horrific reversal. This was the day when the German U-20 sank the passenger liner RMS *Lusitania*, entirely without immediate warning, a few miles off County Cork, Ireland. The *Lusitania* had left New York six days before despite a formal notice from the German government that she would be entering a war zone and was liable to be attacked. The submarine that, more by luck than by judgment, managed to do so fired a single torpedo—the only one she had left, after sinking three small merchantmen a few days before. It hit the liner's starboard side just below her bridge and triggered a massive explosion amidships—maybe (according to some survivors) even two. The *Lusitania* promptly keeled over in a steep list, took on water at the bows, and went down to the bottom in just eighteen minutes, within sight of the cliffs of Ireland.

The death toll was staggering—almost as many died in this Atlantic tragedy as had three years earlier when the *Titanic* struck her infamous iceberg. More than eleven hundred *Lusitania* passengers, many of them Americans, drowned in the foggy waters off Ireland in what was long regarded as one of the most hateful episodes of the Atlantic war. The controversy surrounding her sinking has never fully abated, not least because her owner, Cunard, was found to have illicitly allowed the loading into the ship's hold of large quantities of ammunition and other matériel, and which would have provided the German navy with considerable justification for attacking her. The story still fascinates many, not least because the Royal Navy, as recently as the 1950s, was thought to have bombed and depth-charged the wreck in an effort to prevent divers and other explorers from finding out exactly what the vessel had been carrying, and to keep the matter entirely secret.

The thought that the Germans might introduce unrestricted submarine warfare to the Atlantic—that they might effectively throw away the rule book and deal with merchant vessels as harshly and uncompromisingly as with warships—peaked at the time of *Lusitania*. For the rest of 1915 it slowly ebbed away as the Germans, clearly dismayed by the worldwide hostile reaction to their sinking of an unarmed passenger liner, made some effort to rein in their more aggressive submarine commanders. But after the great Jutland battle—which, though it ended in a draw, effectively kept the German surface fleet in port, for fear of ever meeting the full might of the British Grand Fleet again—everything changed. Almost as soon as their surface ships returned to Wilhelmshaven, the German High Command announced that its submarine flotillas, by then based in Ostend, Belgium, would be allowed to roam the Atlantic at will and to sink and kill whatever Allied ship each might find. It was a decision that resulted in relentless Atlantic countersubmarine battles being fought in both wars—from the summer of 1916 onward in the Great War, and for all of the six years of the Second, during which the conflict was of such ferocity and duration that it became known, officially and now historically, as the Battle of the Atlantic.

During the First World War, the German submarine threat was dire and many Allied ships were sent to the bottom of the ocean. But in the end the threat proved to be manageable—and Germany's *unrestricted* submarine warfare helped to draw the Americans into the war. U-boats sank an enormous number of Allied ships in 1917, but in time the various responses of the Allies, which included the introduction of convoys and the employment of newly invented depth charges and other explosive devices, began to take effect, and the threat steadily wound down.

The same manageability was not possible during the early years of the second war, because by then the strategies of the German naval planners, the range and armament of the submarines, and the production rates of the German factories had evolved to an extraordinary level of sophistication. For many years the Allied navies faced an impossible task of suppressing the Nazi submarine attacks. In March 1940, Winston Churchill proclaimed the long fight between the Royal Navy's surface fleet and Admiral Karl Dönitz's German submarine armada as the new "Battle of the Atlantic," and in later years—especially after the climacterics of 1916 and 1943, when Britain's future appeared truly to be on the knife-edge of a balance—he had no doubts of its importance: "The Battle of the Atlantic was the dominating factor all through the war," he said. "Never for one moment could we forget that everything happening on land, sea and air depended ultimately upon its outcome, and amid all other causes we viewed its changing fortunes day by day with hope and apprehension."

Germany's strategy was to wage a tonnage war in the Atlantic, and its arithmetic was brutally simple. More and more submarines were ordered—Dönitz had command of 57 U-boats in 1939, but this had risen to 382 in 1942—and more and more wolf packs began to operate. A noose started to close around the Atlantic approaches to Britain—and as one by one, night after night, huge explosions and eruptions of oily fire marked where yet another lumbering merchant ship and her vital cargoes had been destroyed by a torpedo, the prospect of Britain's maritime asphyxiation seemed all too real. The German fleet, with an exquisite sense of the grotesque, called this period "the happy time."

But then emerged the convoy system—with vast assemblies of ships grouping into bands, first in the shallow waters off Halifax, Nova Scotia,* then making their way like herded cats under the protective supervision of increasingly powerful, watchful, and technically sophisticated naval escorts—and slowly, very slowly, the threat began to recede. Other transoceanic convoy routes were soon established: New York to Gibraltar, Port of

---

* Halifax, until then little connected with twentieth-century conflict, had on December 6, 1917, been the scene of one of the Great War's greatest catastrophes: in a congested part of the city harbor an inbound ammunition ship was struck by an outbound vessel, the MV *Imo*, taking relief cargo to Belgium. There was a fire, and the immense quantity of ammunition packed into the holds of the MV *Mont Blanc* exploded, leveling most of central Halifax and Dartmouth, killing more than two thousand people, leaving nine thousand homeless. Such was the size of the blast that Robert Oppenheimer later studied it as a model of what might happen with the first atom bomb.

Spain to Freetown, Natal (in Brazil) to Gibraltar, Freetown to the Clyde; and though the tales from so many of the individual convoys were all too often the stuff of heroic legend in the face of terrible tragedy—especially those slow convoys designated with the letters "SC," and of dreadful vulnerability—by May 1943 the Battle of the Atlantic had reached its turning point.

This was the moment when Allied aircraft were finally present in sufficient numbers—operating either from bases on land, or from aircraft carriers in mid-ocean—to offer a secure umbrella to the ships passing slowly underneath. The sinkings and the killings continued until the final day of the war, May 7, 1945—a little Canadian steamer the *Avondale Park* and a Norwegian vessel, the *Sneland 1*, were the last to fall victim to U-boat torpedoes on that very day, when they were but a few cruel miles from their destination in Scotland. But the U-boats never managed to bring Britain to her knees, did not prevent the assemblage of matériel that would be vital for the Normandy landings in 1944, and did not bring about surrender. The battle lasted all six years of the war: 3,500 Allied merchant ships and almost two hundred warships were sunk by submarines, and nearly eight hundred German U-boats were sunk in retaliation. The remains of sixty thousand young seamen now lay at the bottom of the Atlantic Ocean. More men had died there in the five years of the Second World War than in all of the conflicts in the ocean since the first Romans had set out on their invading expeditions nearly two thousand years before.

. . .

As a battleground, the Atlantic is a different place today. No longer does a ship make war on another ship; no longer are broadsides fired at walls of steel across miles of empty sea; no longer are vessels built to collide with one another, nor do commanders require adherence to ancient codes of conduct ensuring gentlemanly behavior at sea, something once thought so necessary when all sides were fighting on the playground of an even mightier enemy, as the sea was widely regarded. High technology has usurped the courtly customs of the ocean; war is made today in a more businesslike fashion; senior sailors take a managerial approach to their navies; romance is all but gone.

Maybe the last Atlantic Ocean conflict to sport echoes of Trafalgar and Jutland and the Glorious First of June was the British war fought in 1982 over Argentina's snatched sovereignty of the Falkland Islands; and since this was a war tied up in history, and bound to the security of the distant colony of an ancient island realm, it had some Nelsonian romance and derring-do about it. The fact that a British naval force felt obliged to sail a third of the way around the world from its docks and arsenals, and to come down to the winter storms of the South Atlantic, and to operate there with supply lines eight thousand miles long while the enemy, coming from bases on the Argentine coast nearby, had fresh stores and munitions and men not three hundred miles away—the fact that such a classic imbalance of advantage could be overcome by courage and cleverness and good planning remains remarkable still.

The stated reasons for going to war over the Falkland Islands may never be fully accepted by many, and it may well be that the weapons employed in the fight, and the manner in which the fights unfolded, bore little resemblance to the fights of old; but the heroism, the romance, and the poignancy of many of the events of the war's three months will still stir old sailors for many years to come. Not the least of these is the tragic sinking of HMS *Sheffield*, with which this chapter began.

The sinking of the Argentine cruiser the *General Belgrano* in early May 1982 remains to date the last lethal use of torpedoes in the Atlantic. The cruiser, formerly a Brooklyn-built stalwart of America's Pacific war, had been sold to Argentina some years before; at the time of her sinking she and two escort destroyers were steaming back toward their home port of Ushuaia, in Tierra del Fuego, after patrolling just south of the Falklands. The flotilla was spotted by a British nuclear submarine, HMS *Conqueror*, which sent two elderly torpedoes into the cruiser's port flank. One blew off her bow; the other hit amidships, knocked out her electrical systems, caused floods and fires, and killed scores of men. The great ship, listing heavily to port, was abandoned within twenty minutes and sank shortly thereafter. More than three hundred Argentine sailors died in the attack, which excited much controversy over the legitimacy of the British naval action.

HMS *Conqueror* returned to her home base in Scotland some weeks later. Considering that so much of modern naval battle tactics originated in the seventeenth-century fights with Atlantic pirates, it might be regarded as something of an irony that as the submarine rose to the surface and sped home along the sea loch, she was flying from her sail the skull and crossbones, the Jolly Roger, the dreaded black pirates' flag—the device by which the modern Royal Navy still displays to sister ships and in friendly ports any notable success in doing battle with an enemy at sea.

9. WHAT NO SUBMARINER SUPPOSED

The consequences of the Atlantic wars were many; among the least expected is that which links the ocean, albeit in a tenuous and speculative way, with the foundation of a state very far away from its shores, after a series of events that started to unfold in the autumn of 1915. This was when the Royal Navy began to have particular difficulty in repulsing a relentless series of German U-boat attacks—a problem that arose not from a lack of warships or poor training or any lack of political will, but as a simple matter of chemistry: the Royal Navy's gunners did not have sufficient quantities of the smokeless explosive known as cordite to be able to attack the surfaced submarines.

Cordite is made from a mixture of nitroglycerine and guncotton, acetone and petroleum jelly; and it was in short supply in 1915 because Britain was unable to produce sufficient quantities of one of its key components—acetone.

In the early summer of 1916, the editor of the *Manchester Guardian*, C. P. Scott,[*] happened to have lunch with a middle-aged and avuncular White Russian émigré and science professor at the University of Manchester, a man named Chaim Weizmann. Over coffee after lunch, Weizmann mentioned to Scott that he had developed a new bacterial method for producing acetone in large quantities. The following week, and also at lunch, Scott—who knew about the navy's problems—told all of this to his friend David Lloyd George, the politician (soon to be prime minister) who was then heading the Ministry of Munitions. Weizmann was in consequence hastily summoned down to London, given research space at a big London laboratory, and finally handed the keys to the disused Nicholson's Gin Distillery in east London, where he could employ his new techniques to create the much-needed chemical. All he needed for his process to work, he declared, was a goodly supply of cellulose—something that could be found aplenty in maize, or even, he added, in *chestnuts*.

That autumn, schoolchildren all over England were asked to collect horse chestnuts, which they normally gathered for their

---

[*] A legendary editor perhaps best remembered for his cautionary remark to his journalists that *comment is free but facts are sacred.*

ritual games of "conkers," and thousands of tons of these soft nuts were brought to the gin factory and thrown into the hoppers and vats and stills. Within days pure acetone began to drip, then stream, then cascade, and finally gush into the carboys. Long tanker trains would take the acetone down to the Royal Navy's top-secret cordite factory on the Dorset coast, and before long, boxes of the sticky high explosive of which it was so critical a component would be delivered to the naval dockyards, the ships' guns would start firing once more, and the tide of the Great War's Battle of the Atlantic would very slowly but surely begin to turn in Britain's favor.

Storytelling and mischief-making, handmaidens to much in history, have since made a series of intriguing connections from the bare bones of this story. An oft-repeated yarn begins with British government circles deciding that Chaim Weizmann should be given an official honor for his role in so profoundly changing the direction of the Atlantic war. Lloyd George, by then prime minister, demanded that his foreign secretary, Arthur Balfour, be asked to suggest the honor to Weizmann, who after all was not a Briton but a White Russian. Crucially, he also happened to be the leader of the British Zionist league, and a prominent figure in the worldwide movement to create a state for the world's stateless Jews.

Weizmann was said to be delighted by the successful outcome of his chemical experiments, but desired no official British recognition. The Israeli Foreign Ministry, in its official history, then picks up the story of what followed:

> Weizmann's [achievements] opened doors for him in British
> government circles, where he continued to serve as an eloquent
> spokesman for Zionism. . . . Lord Balfour commented dryly that
> "Dr. Weizmann could charm a bird off a tree." . . .

When Lloyd George, then minister of munitions, was appointed prime minister and Arthur Balfour became foreign secretary, years of persistent persuasion and "sensitization" to Zionism played a decisive role in the decision of Great Britain to issue the Balfour Declaration. A rare constellation of British and Jewish strategic interests, together with personal empathy for Dr. Weizmann and his cause—the fruit of eight years of what today would be considered "networking"—culminated in this document, approved by the British cabinet on November 2, 1917, that proclaimed the sympathy of the British government for Zionist aims in Palestine. . . .

Informing Weizmann of the decision, Lord Mark Sykes, secretary of the war cabinet, declared: "Dr. Weizmann—It's a boy." Indeed, the landmark document . . . was a crucial step towards the birth of a Jewish State, and is considered Chaim Weizmann's most outstanding achievement.

Whether there is a true and direct link between C. P. Scott, Chaim Weizmann, his freshly minted acetone, and the eventual formation of the state of Israel is a matter still open to scholarly debate. But if it truly is the case, then it is not too mischievous to say that present-day Israel is a state that was conceived amid the trials of the Atlantic Ocean. Such a connection, entirely unanticipated, offers a further clear reminder of the central role that the ocean has played and still plays in the workings, great and small, natural and manmade, of the intricate machinery of the planet.

Chapter Five

THEY THAT OCCUPY THEIR BUSINESS ON GREAT WATERS

———

And then the justice,
In fair round belly with good capon lined,
With eyes severe and beard of formal cut,
Full of wise saws and modern instances;
And so he plays his part.

1. LAWS AND ORDER

The far north Atlantic is where parliaments began. The first law-making assemblies were founded there in the tenth century, and soon thereafter some kind of justice and order began to settle, not just on the lands where assemblies convened and laws were first made, but also on the seas between.

The first true parliament is reckoned by most to have assembled in Iceland—and somewhat symbolically, in the curiously fashioned valley in the west of the country known as Thingvellir, where the world's American and Eurasian plates are still tugging apart from one another and new ocean floor is being created.

There is a large basalt slab protruding upward from the western wall of the valley, and it was here beneath it that more than a thousand years ago farmers and peasants and priests and merchants passing through the valley would agree to stop and camp and meet each year to hammer out in some fashion the manner in which they thought their island nation should be run. The assembly was eventually called the *Althing*, and once it had a formal structure—the date generally agreed for this formation being 930 A.D.—it became the sole body charged with fashioning Iceland's laws. The rock, from which the Icelandic flag flies still, day and night, is today without doubt the most revered monument in the Atlantic north: the Rock of Laws, which set the patterns for the governance of much of the rest of the world.

Soon afterward the processes and customs of the Icelandic *thing*—and yes, it is the very word that today signifies an object or a concept—were mimicked by men who made laws in the Faroe Islands nearby, and later on in Norway and Sweden and Denmark, too. It was also mimicked on that British half-possession the Isle of Man, where the assembly was and still is known as the Tynwald. It first met in 979 A.D., and since it has gathered without interruption in all the years since (unlike the Icelandic Althing, which was suspended for many years when the country dissolved itself into anarchy), it lays claim to being the oldest continuously and regularly meeting democratic institution in the world.

There are many other competing contenders for primacy among the various parliamentary assemblies dotted around the Nordic world, and there is little value in delving into their arguments. But accepting that the idea born in Iceland did spread, rapidly and over large distances, one overarching truth appears: that in a large quadrant of the world's northern nations—and all of them nations that happen to have been intimately involved

with the Atlantic Ocean—there was from the tenth century onward both a popularly established means of creating codes of
laws of sorts, and popularly elected or otherwise assembled
bodies that were established and designed to promulgate and
administer those laws.

No such institutions were created this early in Russia, say,
nor in China, nor even in Greece, despite the ancient Athenian
origins of a rather different kind of popular governance. Parliamentary democracy, as it is understood in today's world, was
very much an Atlantic creation—a further reminder, if one were
needed, that while the Mediterranean Sea was clearly central to
the makings of the classical world, the North Atlantic and many
of the countries bordering it were witness to the construction of
many of the foundations, ties, and crossbeams of what we now
know as the modern world.

2. THE RULES OF TRADE

It is entirely axiomatic that any outward-looking society that
agrees to run itself according to a body of homegrown laws will
eventually make contact with neighbor nations whose legal customs may be quite different. This is nowhere more obvious than
in the matter of trade. If Icelandic merchants do business with
their kin in Norway, what law applies? The laws of the Thingvellir
Althing, or the laws of the parliament in Oslo? Such differences
in law may never have been immense—but for greater efficiency
and ease of doing business, it was realized by merchants early
on, these legal systems needed to be somehow synchronized,
made to mesh neatly, one system with another. And so during the
eleventh and twelfth centuries the ships, the trades, the explorations of the neighbor nations, and the relations they enjoyed

with one another and with countries beyond slowly and steadily began to be organized and regulated by a catalog of über-laws, by sets of agreements that once might have been rooted in the domestic laws of each trading country, but which—in the matter of the management of ships and the seas across which they sailed—were somehow transformed into something greater, more all-encompassing.

The ocean that lay next door to these newly democratic nations of Scandinavia and the Baltic Sea steadily became in consequence something of a regulated entity. An Atlantic that had once been no more than a confused and terrifying body of water, full of storm and monsters and mystery, started to succumb to order and discipline. To the ultimate advantage of all, the ocean became, first in its northeast, and then slowly across a broader area as more and more of it was found and mapped and settled, a titanic expanse of customs, habits, regulations, timetables, and tariffs—and rules.

. . .

Exploration, settlement, war, pilgrimage, fishing, and trade had always been the main impulses behind maritime adventuring; and though exploration started to diminish when all had been found, and though migration slowed down when faraway places became full, though wars would end with treaties signed and fealty agreed, and though pilgrims ceased to travel when sufficient of the convertible had become converts—two oceanic things survived above all: the seas as a source of food and the seas as a passageway for trade. In all history neither of these two ever diminished, ever would, and ever will.

From the times when the Phoenicians traded purple murex dye between Mogador and Tyre, to today's container ships that rumble between the Chesapeake and the Mersey, trade across and by way of the Atlantic has been ceaseless, the fortunes involved in the process almost beyond imagining.

Initially maritime trade was run purely as freelance operations—the concept of true international trade, now a nearly essential component of all modern economies, was virtually unknown. Sporadic commercial expeditions were born as the mood took or the opportunity was presented: a group of wharfside financiers would give their backing to a ship and its master, would order him to proceed with a cargo of trinkets or bullion to some distant possible source of wealth, would trust that any foul weather, pirates, and local resistance might be minimal, would wait for a safe return and then divide any spoils the vessel managed to procure. The risks of such ventures were huge, the competition unpredictable, the profits uncertain: a means to fortune for some, but a source of ruin to most.

It eventually fell to a group of thirteenth-century merchants in northern Germany to construct a better and more organized way of conducting oceanic business—and they did so initially to protect their trade in salt fish, which were harvested mostly in the Baltic and the North Sea.

The city of Lübeck, in northern Schleswig-Holstein close to the mouth of the Baltic, south of Copenhagen, is generally supposed to be where the merchants, forming themselves into what they called a Hanse—after the Latin term *hansa*[*] for a military troop or company—decided around 1241 to create an association of like-minded (and initially, nearly always German) merchants in other nearby cities, and to organize maritime trading among themselves. The Hanseatic League was thus formed, and for the next four hundred years it evolved into a heavily protected—and on occasion, heavily armed—monopoly of traders who directed almost all the seaborne trade between Bergen and London in the west, and Danzig, Riga, and the Rus-

---

[*]   The word survives still in the name of the German national airline, Lufthansa.

sian city of Novgorod in the east. The Hansa was a truly formidable organization, and its influences—cultural, architectural, even linguistic (Scandinavian and German words were spread into the lexicons of England and as far away as Spain and Portugal)—are still felt to this day.

Generally speaking the Hanseatic merchants set up a two-way trade along an essentially east-west axis. The Hansa ships—in the early days of the League small flat-bottomed vessels called *cogs*, but often with armed protective escorts following discreetly in their wake—brought raw materials—furs, wax, grain, wood, pitch, tar, flax, and beer—in from the rural east of the Hansa zone. Then, in the various ports the merchants constructed specifically for the purpose—in towns like Rostock, Stettin,* Riga, Königsberg—they traded these for the manufactured or more rare or sophisticated goods—wool and linen cloth, processed furs and skins, wines, salt, knives, swords, and cooking utensils—that their vessels had brought in from Western Europe, where there were also Hanseatic offices.

London was one such western outpost of the League—with a countinghouse or *Kontor*, with its own warehouses and houses for the merchants. The Britons who did business with the Hansa found them trustworthy and reliable. According to many lexical authorities, the word that Londoners used for traders from the Hanseatic eastern cities—*easterlings*—became shortened and incorporated into the English language as the word *sterling*, with its implied meaning of solid reliability. The city of Bruges was another major Hansa city; and as the need for dried or smoked fish—cheap, health-giving, and transportable—grew almost exponentially with the rising population and prosperity of Europe,

---

* Made briefly famous by Winston Churchill in his 1945 speech in Fulton, Missouri, when he spoke of the new "iron curtain" then descending between "Stettin in the Baltic and Trieste in the Adriatic." This former Hanseatic port is now Szczecin, and is in Poland.

the merchants expanded their influence farther and farther north, eventually creating a Hansa outpost as far as the port of Bergen in Norway.

The Bryggen were the wharfside warehouses of the Hanseatic League outpost on the Atlantic coast of Norway. From the thirteenth to the seventeenth centuries, furs, timber, ore, cod, and herring were dispatched south to other league members, while cloth and manufactured goods came up from England and Germany in return.

The Bergen outpost remains today, with a creaking warehouse or two, a rabbit warren of laneways still scented with the smell of tar and wet hemp, and down below, beyond the slippery cobbles, the sea lapping ice-cold against the massive granite walls of the jetties. Cruise ships tie up there now, and small shops and cafés cluster where the portly Hansa merchants would make their deals on a handshake and over a shared pipe. Seen from the hills above the railway station, the little German ghetto still stands apart, visibly different from the rest of this bustling little seaport, just as it was when Baltic order was brought to bear on the legendary waywardness of Viking seafaring, and the Atlantic started to settle down.

The men of the Hansa oversaw many practical improvements to the conduct of North Atlantic maritime trade: they made sure that channels into their ports were properly dredged, they constructed lighthouses to warn of shoals and reefs, they mounted campaigns against piracy, they became powerful enough to stand up to the despotism of the occasional monarch. But for all that, the Hansa were solely concerned with what mariners still call the short sea— the coastal trade, the crossing of bights, the transit of estuaries, the quick passage between neighbors, and with most voyages conducted in the visibly comforting presence of nearby land.

3. SEAFOOD, CANDLES, AND CORSETS

It would be some centuries before the ocean was fully crossed, east to west—notwithstanding the eleventh-century Vikings, who visited Labrador and settled in Newfoundland—and longsea trade began. Until then, major voyages into the ocean were performed not in pursuit of trade but so men of great courage and daring could exploit the one resource with which all the world's seas, and most especially the North Atlantic, were once replete: fish.

It was the Hanseatic League that established a proper footing for commercial fishing in the North Atlantic. The popularity of the highly nutritious and economical cold-water fish prompted the Hansa merchants to order the construction of two fleets of vessels to exploit the massive shoals of fish in two quite distinct Atlantic fishing grounds: the so-called Scania waters off southern Sweden, where there were plenty of herring; and the Lofoten

Islands, above the Arctic Circle in northern Norway, where there were unimaginably large stocks of *Gadus morhua*, the Atlantic cod.

The importance of this remarkable white-fleshed, protein-rich, almost fat-free fish in the Atlantic's history can hardly be gainsaid. It dominated the trade of the Hansa; it stimulated the transoceanic adventuring of the Basques; it provided hundreds of thousands of Britons with work and tens of millions of Britons with food; and for decades it formed the central plank of the economics of all of maritime Canada and the coastal states of New England.

Cod is a demersal fish, meaning that it likes to swim close to the seabed in shallowish waters—a fondness it shares with flat-fish like sole, flounder, plaice, and halibut and with other fellow five-finned gadiforms like haddock, pollock, hake, and whiting. (The second broad division into which oceanic fish are divided is that of the pelagic types, which swim in the surface waters or the middle depths: the herring is a pelagic fish, as is the sardine, the anchovy, the mackerel, the infamous South African snoek,* and the currently endangered bluefin tuna.) Cod was also once very numerous (Alexandre Dumas joked that the female was so fertile that if all cod eggs survived and hatched, within three years one could walk clear across the Atlantic by standing on the fishes' backs), and until recent times most of the adult fish caught were large and muscular with dozens of pounds of white, motherly, nourishing flesh.

---

* Though popular in southern Africa, few Britons still care for it, as a consequence of the importation of millions of tons of canned Atlantic snoek during the Second World War and a highly ineffectual campaign by the then Ministry of Food to persuade people to eat it. It was found to be oily, bony, and bad, and despite entreaties for cooks to prepare such dishes as *snoek piquante* (when it was clearly *piquante* enough already, once the can was opened), most remained unsold. In the 1950s the sudden appearance on shelves of similarly sized containers of cat food suggested its eventual fate.

Cod is also very easily preserved, and its goodness—its protein, essentially—remains intact. This was one of the secrets of the Vikings' success as long-distance mariners—for they would simply open up the fish, hang it by strings or on frames in the cold Arctic air, and let it dry out until it had lost 80 percent of its weight and became stiff as a board, like a piece of plywood. As needed, the Viking captain simply added water and lo!, the dried fish would swell back to their original size and shape, and the flakiness, rich taste, and nutritiousness were all restored, as if by magic.

If the Vikings had the good sense to air-dry their cod, the Basques of northern Iberia went one better: their knowledge of the ancient ways of the Mediterranean fishermen taught them how to use as a preservative one of the major mineral components of seawater: salt. Northern peoples had little access to crystalline salt, mainly because their climate seldom offered the heat necessary to evaporate it from the sea. Mediterranean peoples, however, were blessed in this regard; and the Basques, being a seafaring people with ready access to an ocean rich with cod, and, thanks to an accident of geography, having ready access to salt, combined the two—and at a stroke invented a preserving technique unknown in the Atlantic Ocean before. They split open the fish, salted it liberally, and only then hung it out to dry: the resulting salt-fish survived for much longer than previously known by those salt-starved others (like the French) who knew only how to "wet-cure" their fish and then watched helplessly as it eventually turned green with age. The new technique allowed the Basques uniquely to make ever-longer sea journeys, even for many months, since they knew they always had supplies.

The fish they caught and preserved also turned out to taste much better—meaning that the Basques could readily conduct an energetic trade in it. Thus they had found the perfect combination: a magnificent, protein-rich, fat-free, highly attractive, abundant cold-water Atlantic fish, and an impeccable means of preserving it for their own use or for selling on to others. Armed with this, the Basques promptly left their ports on the Galician coast and began a period of long-distance sailing across the North Atlantic that has left its commercial imprint to this day.

Their fishermen particularly favored the ocean close to the headlands of America, off Newfoundland. In those hundreds of square miles where the sea shallows dramatically—the Grand Banks and the Flemish Cap—and where the warm waters of the Gulf Stream and the cold waters of the Labrador Current sidle past one another and kick up clouds of nitrates upon which feed the phytoplankton, the zooplankton, and the krill, cod thrives in fantastic, churning abundance. Just when they found these breeding grounds is a matter of much argument: some insist that John Cabot first found and named New-found-land on behalf of the British in 1497[*] and lured the Basques to sail northwestward; others believe, with little evidence, that the Basques discovered the cod-fishing grounds on their own, before Cabot, but opted to tell no one.

Certainly by the time the Breton explorer Jacques Cartier arrived nearly forty years later and planted his famously huge cross—with *Vive le Roi de France* inscribed—on a cliff on the Gaspé, and named the surroundings *Canada* and claimed it all for France, many hundreds of Basque fishing vessels were already working away with energy and zeal, though making no imperial gestures or claims. Moreover, the name *Gaspé* is widely

---

[*] John Cabot, not as British as he sounds, was in fact a Venetian named Zuan Chabotto (or more generally, Giovani Caboto), who sailed from Bristol westward on commission from England's King Henry VII. His eventual landings in Newfoundland and on the Labrador coast make him in all likelihood the first post-Viking European to reach North America—an achievement that was, of course, denied to Christopher Columbus.

assumed to derive from the word for shelter, *gerizpe*, in Basque—
which admirers of the Basques say further adds to their claim
that they were pursing North American cod, and settling in
North American harbors, well before any other Europeans, Vikings aside.

The precise timing of the Basque arrival is of less interest,
however, than the simple fact that with it—just as with Christopher Columbus's arrival on San Salvador; or with John Cabot's
discovery of Newfoundland; and most especially Vespucci's realization that the Americas was a separate and discrete continent
and the Atlantic was a separate and discrete ocean—an entirely
new phenomenon could finally be allowed to develop: from now
on, voyages—whether taken for curiosity or commerce, for God
or war or a host of other reasons—could be made *transatlantic*.
Sea journeys could at last be undertaken between coasts that
provably existed on opposite sides of the sea. Voyages no longer
needed to be limited merely to coastwise wanderings and con-
fined and limited to certain quadrants of the sea.

Basque fishing vessels, for example, now no longer needed to
venture westward into a fog-shrouded, unknown sea, with their
purpose only to catch cod, with their success questionable, their
safe return more a matter of chance than certainty. No, they
could now and for the first time travel to a destination. The captains of Basque fishing vessels were now aware, as they pushed
off from home into the confused waters of the Bay of Biscay, that
there was a *far side* to their journey, with harbors and victuals
and shelter and repair—and in time, settlements of their fellow
countrymen—that were all available to them. But this was true
for others as well; in short order Spanish galleons and Portuguese carracks and English ships of the line also understood
that there was a far side to their journeying—and by the early decades of the sixteenth century, transoceanic passages were be-

ing made, trades were being conducted, and the bounty of the sea was being exploited.

And while the Europeans involved themselves in this voyaging, the new Americans did so as well. Whether they were sailing as settlers or as colonials or whether, as after 1776, they did so as citizens of a newly independent nation, Americans were particularly quick off the mark in exploiting all manner of transatlantic venturings.

They first got their sea legs by chasing whales.

Yet once again it was the Basques who showed the way—since they, for the previous six hundred years, had been hunting for these warm-blooded oceanic mammals with the same determination and ruthlessness they had displayed toward the much smaller and nonmammalian codfish. Instead of the unsophisticated methods used by others in the past, of waiting for the whales to land near the beach, the Basques took their boats out into the deep Atlantic, hunting the whales well offshore just as they might hunt for any other sea creatures.

Their principal target, first in Biscay and then in the waters south of Iceland and beyond, was the great baleen type[*] known as the Atlantic right whale—a mammal eventually so named by its American pursuers because it was self-evidently the "right" whale to hunt. The Atlantic right—an all-black creature, weighing about a hundred tons, and with a fatal liking for swimming in leisurely fashion dangerously close to shore—and its rather bigger Arctic cousin the bowhead were animals all too easy to hunt down. The Basque technique for snaring them was so dev-

---

[*] Baleen whales lack conventional teeth but have a series of filters in their (often enormous) mouths. The other division of cetaceans, the toothed whales, includes the sperm whales, the belugas, narwhals, porpoises, and dolphins—only a few of which, most notably the sperm whales, managed to excite the same degree of commercial interest that attended the baleens.

ilishly simple it soon became the world technique: it involved tying floating drogues to a harpoon line, such that a harpooned whale found it nearly impossible to dive and so in time would weary of swimming endlessly on the surface; it would slow and allow his pursuers to kill him.

Right whales generally float when dead, and they could be towed home or to a nearby island base, there to be flensed and the blubber rendered to make a particularly fine kind of waxy oil, for heating or lighting, for lubrication or the manufacture of margarine;* the flesh was cut up and salted for food; and the baleen—the large plates of keratin in the whale's mouth that help the animal filter food from the seawater—was processed to make corset stays or buggy whips or the rods of parasols, or any of a thousand uses that pre-Edwardian man discovered for it.

Multitudes of these magnificent, languorously moving, and tragically unsuspecting whales died each year at the hand of Europeans in hot pursuit of vast profits. Right whales and bowheads were especially numerous close to the coasts of Spitsbergen, the archipelago in the far northern reaches of the Atlantic—beyond both the remote fastnesses of Jan Mayan Island and Bjornoya, where whalers sought temporary sojourn during storms—and later in the Davis Strait between Canada and Greenland. By the eighteenth century, the Basques' technological monopoly had been broken, and French, Dutch, Danish, and Scandinavians hunters were also searching for the great mammals.

Later they were joined by Britons of the Muscovy Company, who believed (wrongly, as it happened) that they had discovered Spitsbergen and so laid claim to being the only country allowed to take whales from the coastal waters. For a while English port

---

* Whale oil is also used in the quenching of steel, the dressing of leather, and the making of both nitroglycerine and soap.

cities like Hull and Yarmouth were sending scores of vessels north—where they engaged in ugly skirmishes with their Dutch and Danish rivals, who tried to chase them away. The quarreling led the Dutch in particular to refine their hunting techniques— performing the kill from small pinnaces or lug-sailed shallops, hauling the carcasses back on board to be laid across the stern for flensing, and only then bringing the blubber back to land for rendering. Most of the whaling process was being carried out at sea—a more secure means, especially when rivals were circling, hoping to make an intercept as a whaler crept into port with its newly killed victim.

When the Americans entered the business in the early part of the eighteenth century, they were well aware of these new developments; and although the first American whaling ventures established in the late seventeenth century on Nantucket and in New Bedford and in the littler ports ranged along the southern coast of Long Island still saw much of the heavy work performed onshore, within fifty years the New England whale ships were being made large and sturdy enough and so self-sufficient that their owners could dispatch them and their crews on journeys of many thousands of miles. Rather than heading north to mix it up with the Europeans who were so bitterly and busily fighting among themselves, the Americans decided early on that their crews would head into virgin Atlantic territory—they would let the Danes, the Dutchmen, and the English have the rights and bowheads of the north, while they would concentrate on the largely untouched stocks of baleens—the fins, seis, minkes, grays, humpbacks, southern rights, and the gigantic and unforgettably grand blue whales—as well as the sperm whale, renowned for its superior oil, and which lived in what came to be called the Southern Whale Fishery.

The sperm whale, *Physeter macrocephalus*, is a creature well woven into the fabric of American literary life, in no small part due to Herman Melville and *Moby-Dick*. Melville had written in 1851 of the titanic struggle for revenge between the *Pequod*'s Captain Ahab and the ferocious great white sperm whale that in a previous encounter had so cruelly and humiliatingly savaged his leg. At the time he wrote the book, the whale fishery was at its peak, with whaling ships sailing from New Bedford, Mystic, Sag Harbor, and Nantucket bringing in as many as four hundred giant beasts each year.*

But the animal had been known in New England for at least a century and a half before this: Nantucket historians like to say that a pod of sperm whales had been encountered during a right-whale hunt as early as 1715, piquing the interest of all—for who could fail to be impressed and puzzled by so bizarre and vast a creature, with a great blunt head fully a third as long as its body, with a single blowhole through which it could throw fire-jets of water scores of feet into the air, with a pair of crescent-shaped flukes that created a devastating boom as they crashed down onto the sea, with an ability to dive two miles down into the ocean like a starfighter and stay there, without breathing, for an hour and more—this animal was bigger, heavier, noisier (emitting clicks and clacks that could be heard for miles), and more ferocious than it was possible for most seamen to imagine. Then there were later discoveries of its total utility: that its blubber could be rendered into an exceedingly fine oil for lighting and the easing of delicate metal machines, that its meat was even more nutritious than normal dark red whalesteak, that the head of this gigantic creature held a pair of pods filled with several tons—tons!—of spermaceti, a rose pink, waxy, spermlike substance that could

---

* This may sound like a lot, but in the 1960s, when Russian and Japanese factory ships were operating at full bore, as many as twenty-five thousand sperm whales were taken from the North Pacific each year.

be used to make, among other items, the purest of white candles, and that men dug holes in a whale's skull and had themselves lowered inside the huge head in barrels, the better to scoop it out; and there was the knowledge that a male sperm whale had a penis six feet long, and that as Melville recounted, a man of discernment or bravery in public or both could have an Inverness cape, with a hole for his head, made simply from the skin that covered one; and there was the discovery, wedged deep in the whale intestine, of great lumps of the famously found, floating gray greasy substance known as ambergris—whose origins had long been a mystery: it was sea bitumen, they said; it came from the roots of a marine gum tree, it was spittle exhaled by sea dragons, it was a fungus, it was man-made, the compressed livers of fishes—all of these delights gave man even more reasons to hunt down the sperm whale above all other cetacean competitors.

And so by the middle of the eighteenth century the whalers, now equipped with bigger ships, thicker sails, more capacious oil barrels, stronger harpoons, more enduring rope, and more lasting ironware, swept out from the American east into what they called *ye deep*.

Until now their voyages lasted only a matter of days, perhaps a week or two. But the more enterprising whaling men, most of them of stout Quaker stock and not given to excitement or fear, began to sail their ships as far as Brazil, or the Guinea coast, or even the Falkland Islands or South Georgia, and were away for months; there was much terror but they also spent many hours drifting idly, good for the crafting of scrimshaw. In time the more adventurous took their vessels to the south of Isla de Los Estados, and doubled Cape Horn against the vast winds and storms of those lethal latitudes known as the Roaring Forties, and with luck and good seamanship they emerged whole and passed into the whale-rich emptiness of the Pacific.

But their long sojourns in the Atlantic gave American sailors a confidence and a profound knowledge of the deep sea that was shared by few others. Whalers ranged to the farther ends of the ocean and discovered as many of its secrets as did the navigators and surveyors sent out by the maritime states: their legacy—and especially the legacy of New England whalers in the Atlantic—is profound.

4. THE PASSAGE OF GOODS

So it is scarcely surprising that when regular cargo crossings of the ocean emerged as a new means of performing transatlantic business, the Americans, long-distance specialists in this particular sea, rose to the occasion and pioneered a form of shipping that has dominated the Atlantic ever since. The development occurred in the first cold days of January 1818—and it involved the sailing, eastbound from New York, of what came to be known as a *packet ship*.

The Atlantic was thick with cargo ships already—carrying toward Europe immense tonnages of New World cargoes, most especially sugar from the various plantations of the Americas, Brazil, and the Caribbean islands, and carrying back toward the Americas the trade goods and construction supplies and items of up-to-date technology and fashion that the merchants of the colonies demanded. But these ships generally sailed only when their holds were filled up with cargoes—there was nothing regular or reliable about departures or arrivals, nor any certainty about the routes a vessel might ply: a last-minute arrival of freight bound for a hitherto unspecified port, once accepted by the supercargo, meant the ship had to divert to ensure making the delivery.

The one oceanic shipping service that did attempt to have a schedule, and even tried to keep to it, had been organized by the fledgling British Post Office almost ever since Charles II first established a mail service in 1660. It was recognized very early on that important foreign mails—official letters to embassies and colonial governors, as well as dispatches to the leading citizens of faraway places—needed to be accommodated as well as those bound for domestic destinations. Accordingly a number of postal packet ports were created in the early 1680s—at Harwich and Dover for ships carrying mails to northern Europe, at Holyhead on the Isle of Anglesey for the Irish mails, and, after its formal selection in 1688, at the remote seaside town of Falmouth in southern Cornwall.

Fast and regular sailing ships were dispatched from Falmouth to all corners of the Western world—at first with a service every two weeks to Corunna in Spain (using small vessels known as advice boats, the first of which were called the *Postboy* and the *Messenger*), and which then went through the Strait of Gibraltar for onward transmission to the rest of south and central Europe and Asia.* Then at the turn of the century, the navy's surveyor general, Edmund Dummer, proposed to the Post Office its first transatlantic service, and by 1702 he was running, as an early kind of franchise, a quartet of oceangoing sloops and brigs between Falmouth and the British-run sugar islands of Barbados, Antigua, Montserrat, Nevis, and Jamaica. From the Caribbean it was but a short step to running a service to the American mainland, and in particular New York City—this service was inaugurated in 1755, with initially two vessels, the HMP *Earl of Halifax* and the HMP *General Wall*. Once this route was in full swing, with supposedly one boat a month (though there were

---

* Wars with France frustrated attempts at a more direct route.

only four voyages in the first two years), still more vessels were commissioned into service, eventually with British routes running from Falmouth to such southern ports as Pensacola, St. Augustine, Savannah, Charleston, and, most important of all in the early years, the major northeastern American garrison city (and spermaceti-candle-making city) of Halifax.

A sporadic service, run mainly for the carriage of military mails, had already been operating between Falmouth and Halifax since 1754. Not surprisingly it had run into logistical difficulties during the War of Independence. But once that dust had settled and America had her freedom, a formal and regular service was put into operation, in 1788—with both Halifax and New York receiving mails from Falmouth—and in the case of the latter, all organized under the supervision of the presiding genius of Benjamin Franklin, colonial America's deputy postmaster general, and after independence, the new nation's postmaster general in chief.*

All of sophisticated London became swiftly familiar with the routines: on the first Wednesday of every month, mails would be made up at the General Post Office in central London for New York, Halifax, and Quebec City. A letter for Manhattan cost *four pennyweight of silver.* The leather *packets* of collected mails—from which packet boats got their name—were then put on the mail coach on the post road to Falmouth, and arrived on Saturday evening, as regular as clockwork, and were transferred to the waiting boat, which promptly made her way out of the Falmouth roads and into the swells of the Atlantic. It took on average fifty days to beat across the ocean, *uphill,*† especially if the Post Office

---

* It was his observations of the many delays to the westbound Falmouth packet ships that led Franklin to his conclusions about the nature of the Gulf Stream.

† The contrary prevailing winds of the westbound passage prompted packet boat sailors to refer to it as the *uphill* route, whereas the services leaving America for Europe were quicker, and so *downhill.*

added stops at Bermuda and Nova Scotia en route. A Londoner who posted a letter on the first of January could expect it to be read in New York City during the third week of February.

And it was not only the mails, of course: the comptroller of the Post Office, a Mr. Potts, let it be known that newspapers and magazines could be sent across the Atlantic as well. Any one of the London daily papers, such as the *General Advertiser*, the *Courant*, or the *Daily Advertiser*, set the reader back five pence a copy. The *Spectator*—still in print today—cost nine pence, and the official *London Gazette*—the most venerable of London papers, and still going strong, offering up the government's official pronouncements—was available in New York at nine pence a sheet, "to be delivered by the commanders of the several packet boats, free of all other charges."

It seems more than a little strange that it took as long as 130 years, from 1688 until 1818, for the notion of sending regularly scheduled packets of mail to be extended to the shipping of general cargoes. And in a sign, perhaps, of things to come, it was not a British institution that first came up with the notion of doing so, despite all the experience the British had accumulated. The invention of regular transatlantic cargo shipping came from a company headquartered in America.

In fact, both of the brains behind what turned out to be a truly game-changing venture were Britons, living in America. Both were Yorkshiremen, both from Leeds, and they had each come to seek their fortunes in America at the very end of the eighteenth century. By coincidence they occupied offices next door to each other on Beekman Street in lower Manhattan. By 1812, when the idea of the venture was first born, Jeremiah Thompson was a young cotton broker and owner of a small number of ships in the American coastal trade, and Benjamin Marshall—who, like Thompson, was a Quaker, as were most of the merchants in this small saga—a textile manufacturer and importer.

Both men soon found they had developed parallel interests in buying raw cotton direct from the plantation markets in the southern American states. To be sure, they had different plans for the cotton: Thompson wanted to buy his so he could market it in exchange for the fine woolen goods his father was making

A flotilla of dories, each carrying the pennant of their mother ship in the event of being lost in heavy seas, bears away to harpoon a pod of right whales basking in an unusual high-latitude calm.

back in Leeds and was trying to export to America. Marshall, on the other hand, found that he needed large quantities of cotton to send back to his family's mills in Lancashire, where they would be made into textiles that he would then ship back to New York and sell on to retailers. The two men, not entirely competitors, then decided to work in concert, and they set up offices in

Georgia, with agents in New Orleans: absent any other internal freight system,* they used their small vessels to ship cotton from the southeastern ports up to New York and then, by whatever vessels happened to be available, across the Atlantic to Liverpool.

And herein lay both a problem and, from Marshall and Thompson's point of view, a tremendous business opportunity.

The problem was one that became hugely amplified by the sudden jump in trade that followed the ending of the War of 1812 and the lifting of the Royal Navy's fitful blockade of American ports. For there were simply not enough ships leaving New York that offered cargo space across the ocean—and, moreover, no one knew when those ships would leave New York, or when they might arrive at the other side of the Atlantic.

For years it had been the custom for merchants to own their own ships: Marshall and Thompson already owned three vessels, the *Pacific*, the *Amity*, and the *Courier*, which they used for their own Atlantic cotton-shipping trade, and so they were personally well set: they did not have the problem afflicting many of their colleagues who could not find holds in which to ship goods. The particular stroke of genius of these two already highly successful businessmen, together with another Quaker shipowner, Isaac Wright—and which would ensure they were long remembered for their canny exploitation of a historically important business opportunity—was their decision to order still more ships, and to offer the space in these ships' holds to any shipper who needed it. Further, and crucially, they sent those ships out across the ocean on a fixed and regular timetable, something that had never been done before: instead of the so-called tramp ships that had operated hitherto, and which left port when the

---

* It would not be until 1829, seventeen years after Marshall and Thompson began their business, that a steam locomotive ran on the tracks of the Baltimore & Ohio, America's first freight railway.

skipper wished to, this brand-new idea of "square riggers on schedule" was offered instead.

Under this plan, a vessel of what would be called the Black Ball Line would leave New York at 10 A.M. on the fifth of every month, bound for Liverpool. A westbound ship would leave Liverpool for the uphill leg, on the first of every month. All ships carried freight of any kind, either in the hold or lashed to the decks, for any person who would pay. They also carried passengers—as many as twenty-eight in the early ships—in some fair degree of comfort.

They departed precisely on time, as advertised, whether their holds or their cabins were empty or full: they slipped their moorings in fair weather or foul, with *getting there*, and *getting there fast*, their singular priority. And in deference to the Post Office back in England, they would give to their vessels and their service the same name: they would be packet ships.

The first-ever Liverpool packet left Pier 23, in lower New York, on morning of January 5, 1818. If proof were needed that this new service was not to be dependent on the whim of weather, tide, or the master's mood, and that it would indeed function in fair winds or foul, the 424-ton, three-masted *James Monroe* cast off her moorings in a howling nor'easter of a snowstorm, and the mobs of fascinated onlookers who cheered her into the gale— church bells rang out, and cannon were fired, too—soon lost sight of her in the froth of spindrift and blowing snow.

As she rounded the seamark of Sandy Hook, with New Jersey to her starboard side, Long Island to port, and with the land dropping away fast as she was carried out to sea, she unfurled her fore topsail—and in doing so displayed to all the other ships scurrying for shelter her new signature logo, a large black circle woven prominently into the canvas. She had the same symbol on the pennant flying from her mainmast: a brilliant red streamer with a black ball, dead center.

The *James Monroe* was built for speed—this, and her fixed schedule, were the principal bait for customers—and she owed much to refinements in naval architecture that American privateers had incorporated into their cargo ships during the war, to better outrun the British blockade. She was also quite capacious: she had room for 3,500 barrels of cargo—this was the principal measure for a ship's hold in the early nineteenth century—though on this first voyage she was nowhere near full. There was, as has been noted, room for twenty-eight passengers, but only eight had signed on—at two hundred dollars for a one-way journey. Her manifest shows her holds echoing with space, since she was taking to England short commons indeed: merely a small cargo of apples from Virginia, some tubs of flour from the Midwest, fourteen bales of wool from Vermont, a smattering of cranberries from Maine, and some canisters of turpentine produced on the slave plantations in Florida. There were also ducks, hens, and a cow, allowing the obsequious stewards—mostly black men—to offer meat, fresh eggs, and milk to the understandably nervous clutch of civilian passengers. There were bales of Georgia cotton, too: the shipowners took good advantage of their own vessel to carry back to the textile mills of Yorkshire and Lancashire yet more supplies of the magical fiber from which they had first made their fortunes.

It took the *James Monroe*'s master, James Watkinson, twenty-eight days to reach the River Mersey, the ship tying up in Liverpool on February 2. Somewhere in mid-ocean she had passed (though never sighted) her sister ship, the *Courier*, battling along the uphill voyage—she took six weeks and suffered terribly. The *James Monroe*'s return was even less commercially convincing: she was damaged by a terrific storm in the Irish Sea and had to return to Liverpool for repairs. But back on New York's Beekman Street, the three Quakers, Marshall, Thompson, and Wright,

kept their north country nerve; and by 1820 all four of their
vessels were making regular scheduled runs across the ocean,
without major incident and with steadily increasing cargoes
and steadily increasing rates to be charged. Two years later they
could afford to build even bigger vessels to sail under the Black
Ball flag: The *Albion*, the *Britannia*, the *Canada*, and the *Columbia* displaced five hundred tons and sported what were said to
be the best crews on the ocean, and the slickest masters, whose
keen need to hoist all sail under even the most dire conditions
allowed them to make downhill passage as fast as possible.

Races were staged, not infrequently alarming the passengers,
who had to watch, powerless, as sails were torn to shreds in the
gales. The *Canada* once made the eastbound crossing in just fifteen days and eight hours, and even on her return made Sandy
Hook just thirty-six days out from the Mersey. "Arise and shine
for the Black Ball Line" became the watch-change cry on many
of the competing transatlantic ships of the day—the implication
being that laggardly behavior by the crew would ensure that a
Black Ball ship would beat you into port. And since crews were
paid upon arrival, this meant the Black Ballers were the first to
be handed their envelopes of cash, making them the most envied
men on the Atlantic waterfronts.

The captains, all of them hard, fierce, and wild men, driven
by personal ambition and an utter devotion to their company
timetables, soon became legends, their ships equally renowned
as the most romantic of all the ocean greyhounds ever to gather
in harbor. The crew were equally tough, known as *packet rats*,
most of them Liverpool Irishmen with prodigious appetites for
alcohol and worse, and invariably amiable scourges of the con-
stabulary of any port in which they were given leave. But they
were driven ceaselessly by their officers: one captain was known
to sleep in a special cot bolted to the quarterdeck so that he could

be sure no junior officer ever dared, while his skipper was sleeping below, to take in the sail during a storm and thus lose speed and risk delaying the schedule. It was a fierce business, entirely appropriate to the fierceness of the sea that the packets crossed and crossed again.

Before long, scores of other competitors were set up, and the shoreside streets of lower Manhattan became so crowded with waiting ships that, as Charles Dickens wrote, the packets' bowsprits "almost thrust themselves into the windows" of the office buildings across from the piers, and street traffic moved to and fro underneath them like forest dwellers beneath a canopy of branches. The Black Ballers were soon joined at the quayside by ships of the Red Star Line, the Blue Swallow Line, the London Line, the Liverpool Line, the Union Line (direct to Havre in northern France), the Fyfe Line (to Greenock in Scotland), and the Dramatic Line (with all of its vessels named after actors and dramatists). There was also, despite protests, a quite separate and competing Black Ball Line established in Britain, which caused major confusion until the courts set things right. At any one time as many as five hundred sailing vessels might be crammed side by side along the Manhattan waterfront, their prows inward, their sterns backing out into the tides of the East River like so many waiting stallions.

The British Post Office soon afterward abandoned its own packet service, as the Black Ball founders had predicted, because in London it was soon fully accepted that the Americans were dominating transatlantic trade, with fast and reliable services springing up almost daily. And a new word soon crept into the lexicon: since all the vessels belonged to a company that sent— as the stagecoach companies were already doing, in their efforts to bring scheduled order to the crossing of the country—a *line* of ships, one after another, out into the ocean, these new regular ships became known as *liners*. They were *transatlantic liners*—the first of a vast and varied company of commercial ships whose business has survived and prospered to this day.

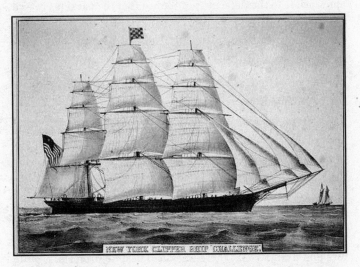

Built in 1851 at the Webb shipyard in lower Manhattan, the great clipper ship Challenge *was for a quarter of a century the apotheosis of the golden age of sailing, the exemplar of speed and grace at sea.*

Only one year and five months after the *James Monroe* first departed New York, there was another, equally significant departure from another American port. She was the *Savannah*; and though she had been built in New York City, she sailed, on May 22, 1819, out of the port for which she was named, in Georgia, bound as most of the eastbound Atlantic ships of the time were, for Liverpool. What made the *Savannah* a memorable vessel—and what prompts America still today to celebrate, albeit notionally, her springtime date of departure as National Maritime Day—is that although she had the usual three masts of a seagoing clipper, she also had a funnel—bent at the top, like a drinking straw—and below, amidships, a 72-horsepower engine. She was the first oceangoing vessel to cross the Atlantic powered by steam.

Her engine, a quarter the size of most modern vehicle engines, was designed to drive an ingenious system of paddle wheels that were supported on the two ends of an axle that ran athwartships but could also be collapsed and folded away. And though she used her engine for only eighty hours of her first journey, she managed to cross from the Savannah River to the Irish coast in a perfectly respectable twenty-three days. But she was not economical: indeed, it would be some twenty more years before the efficiency of marine engines was tweaked to the point where it made good monetary sense to abandon for good the entirely free power of the winds. Yet even in 1819 the *Savannah* was recognized as the precursor of what would eventually become a whole new way of voyaging. She might have lumbered across the Atlantic in a less-than-spectacular twenty-three days, but a century and a quarter later, steamships not entirely unlike her would make that journey in little more than three.

5. THE PASSAGE OF INTELLIGENCE

Cargoes and peoples were not the only items, however, that the fast-modernizing world required to be sent across the ocean. The yawning expanse of water was increasingly seen as a highway for information—for the exchange between the peoples on the edges of the sea and well beyond, of intelligence, news, lovers' declarations, birth announcements, reports of the passage of ships, the prices of stocks, the falls of governments, and the deaths of kings.

All of this was suddenly much needed. The world at the beginning of the nineteenth century was already changed and was well on its way to becoming the immense village it is today: the exchange of information between Philadelphia and Peterborough, or between Brazil and Belgium or Moscow and Montevideo, was every bit as important as once it had been between blacksmith and constable, or innkeeper and churchwarden. To function well, the interconnectedness of a community had always been vital: and with the world's populations mixing and melding by a slew of migrations—not least across the nineteenth-century Atlantic Ocean, between Europe and Ellis Island—a sense of global community was building, too, requiring an ever-burgeoning flow of information and intelligence.

News of King George II's death in 1760 took six wave-tossed weeks to reach his subjects in colonial America, demonstrating the desperately frustrating slowness of conveying information across the waters. Transmissions were not much faster a century later, during the Civil War, by which time the electric telegraph had been invented, easing the passage of communication across landmasses, where lines could be strung from poles. Getting the news across the sea required all manner of inventiveness: the London newspapers arranged for bulletins from North America to be sent to the outer capes of Newfoundland, for handwritten copies then to be rowed out to waiting eastbound steamboats and then hurried to the equivalent outer capes of northern Ireland, from where whaleboats would row them to the closest telegraph office for onward dispatch to London. This cumbersome process was not a huge improvement: it meant only that an item about Antietam or Gettysburg or Sherman's march through Georgia could be read in Whitehall or the clubs of Pall Mall two weeks after it occurred.

Marginal speed increases were always noted: details of the July 4, 1864, battles in Virginia were published in the London *Times* just a fortnight later, on July 16. And when President Lincoln was shot dead in April 1865, the telegraphed news, also transcribed as a handwritten letter and placed in a sealed leather packet, went by way of the steamship the *Nova Scotian*, and was carried to the post office when the ship hove to off Donegal, from where it went off to be printed and the news distributed to a shocked and dismayed London—twelve days after the shooting.

Clearly a new and faster means of communication was needed, and the newly discovered telegraph would be the key. And the key place in the formulations that followed was a barely explored, windswept island distinguished only for being the fourth-largest in the Atlantic, after Greenland, Iceland, and Ireland—the British crown colony of Newfoundland. In the middle of the century, a small group of entrepreneurs wondering how to speed up transatlantic messaging began focusing on this island—and it did so because Newfoundland offered the closest points in North America to Europe: it was a mere 1,600 miles from the harbor opening at St. John's to the cliffs of Connemara.

An undersea cable—for such had already been invented in Europe: a telegraph cable had been laid between Britain and France in August 1850, and soon afterward others were lowered into the seas between Britain and Holland, and between Scotland and Ireland—could perhaps be laid across the Cabot Strait, at the mouth of the St. Lawrence. If then an array of landlines could connect this underwater cable to the cities of St. John's and Halifax, and another line be laid to connect with Boston—why, it would need only a fleet of fast steamships to sail on a regular schedule between Newfoundland and Ireland, and messages could get from New York to London in as little as seven days.

It was at this point that the thirty-five-year-old Cyrus West Field, a wealthy scion of a paper-making family in the Berkshire Hills of western Massachusetts, stepped into this saga. The main backer of the Newfoundland plan had come to see him, in the hope of persuading him to invest. Field received the man courteously enough, said he'd think about it—and then, reading in his library that same evening, happened to turn a globe with his hands. It was moderately large globe, appropriate to a gentleman's library—and Field noticed that he was able to span his hand both between Newfoundland and Ireland and between London and New York.

And then in an instant he realized that rather than building a cable through the wildernesses of Newfoundland and Nova Scotia in the hope of saving a couple of days' transmission time for the telegraph, one could build a cable directly from Newfoundland to Ireland, under the narrowest point of the entire Atlantic Ocean. If he was able to do that, it would reduce the time for a message to pass between the two greatest cities in the nineteenth-century world from a matter of days to just a few seconds.

Field was neither technically minded nor an expert on the topography of the sea—but he immediately wrote to two men who were: Samuel Morse, who had invented the telegraphic code; and Matthew Fontaine Maury of the U.S. Navy, whose surveys of the Atlantic had established the existence of a vast mid-ocean plateau, the Mid-Atlantic Ridge. Both told Field that his idea could be made to happen: Morse had already experimented ten years previously, with cables running under New York harbor, and had written to the U.S. government to the effect that "a telegraphic communication on the electromagnetic plan may with certainty be established across the Atlantic"; Maury, though blissfully unaware that the Mid-Atlantic Ridge was a range of peaks and canyons as sharp and vertiginous as the Rockies (over which one would hardly plan blindly to drop a telegraph cable from the air, say), had written that the "plateau . . . seems to have been placed there especially for the purpose of holding the wires of submarine telegraph, and of keeping them out of harm's way."

The New York, Newfoundland, and London Telegraph Company was accordingly established in May 1854; two years later the Atlantic Telegraph Company* was similarly established in London. Both were committed to raising funds for the project. The American firm's chairman was Peter Cooper—founder of Cooper Union college in New York—a man who believed that what he was about do would "offer the possibility of a mighty power for the good of the world."

The British government was especially excited by the plan and offered to survey the route, to perhaps even provide ships to help lay the cable, and to pay a fee—providing its official messages could be guaranteed priority over all others. The Americans debated the same matter very much more intensely.

Not everyone in the New World wanted such intimate contact with the Old. Thoreau, ever the grumpy misanthrope, remarked caustically that tunneling beneath the Atlantic, as he put it, to place a communications cable would hardly be a worthy exercise if the first news to reach America was merely "that Princess Adelaide has the whooping cough." And there was still a distinct mood of postrevolutionary, post-White-House-burning Anglophobia abroad, especially in the South, where anything English was widely loathed and despised. Nonetheless, after much hard lobbying in Congress, the requisite bills passed; and on the March day in 1857 before he handed over his office to James Buchanan, President Franklin Pierce signed into law an act offering precisely the same terms as been agreed in London. Con-

---

* It counted William Makepeace Thackeray, the novelist, as an investor; Charles Dickens, on the other hand, displayed a Luddite hostility.

struction—the most ambitious construction project then ever envisaged in the world—could now go ahead.

Publicity was enormous: the papers on both sides of the ocean were filled with suggestions of how such a cable might be built—suspend it from balloons, wrote one correspondent; have it dangle just below the surface of the sea from a chain of buoys, where ships could tie up and, just like today's cars at a roadside service station, fill up with messages; Queen Victoria's husband thought it should be cased in a glass tube; still others thought—as noted in chapter 2—that there were layers of differing density in the sea that would affect how deeply various objects might sink: horses would sink lower than frogs, fat people would lie below thin, and cables would sink to only a certain depth and then hover in the ocean, like an aqueous version of today's contrails from a jet.

Scientists squabbled endlessly over how thick the cable should be—thick copper conducts the greater voltages needed for long-distance messaging, but thick copper also meant a heavy cable that might break under its own weight as it was lowered to the seabed. In the end it was decided to manufacture a cable about as thick as a man's index finger, its copper core insulated with gutta percha, then with hemp and tar, finally with steel wire armoring wrapped around it, and weighing about a ton a mile (though only around 1,340 pounds when suspended in water). In the early summer of 1857, two and a half thousand miles of this cable—340,000 miles in total length, if all its component wires were measured—were brought from the factories in London and Liverpool and coiled carefully into drums aboard two sailing vessels, the USS *Niagara* and the HMS *Agamemnon*, with half of the cable in each: about 1,500 tons weight.

In August the ships sailed in convoy to the island of Valentia, in southwestern Ireland, and a group of burly sailors dragged one end of the cable ashore through the surf into the magnificently named Foilhommerum Bay. Speeches of great portent and prayers of great sincerity were uttered. Fireworks were lit and exploded. And then, in the company of a flotilla of naval escorts, the two converted cable ships backed out into the sea, paying out cable as they went—whereupon there commenced a saga of accident, frustration, distemper, and dismay that was to last well into the following year, as it proved persistently impossible to lay the cable without it breaking repeatedly and plunging forever into the depths of the sea.

The crews tried all manner of ways to get around the problem—most notably by electing to begin not at any one side of the ocean, but in the middle, the two ships meeting at a point eight hundred miles from each coast, splicing the cables together and then sailing away from each other toward the opposing shores. But the problems they then encountered were without number—particularly a scourge of unprecedented midsummer storms, which very nearly capsized the cable-heavy British ship. And as before, the cables kept breaking and being lost. Technical failures endlessly dogged the attempts—including a celebrated moment when engineers on one ship noticed a flaw on the cable at the very instant it was being paid out over the stern wheels, and raced frantically to repair it as it slid along the deck and before it went into the water and short-circuited. They succeeded—but the cable went dead anyway.

The directors of the company back in London became increasingly exasperated as costs mounted. Some said the project was technically impossible and wanted it abandoned. The press became cynically dismissive. Poems were written poking fun at the operation. Confidence was shaken, almost to the breaking point.

But then, in the late summer of 1858, after three more failed attempts, the two ships met for one final time, spliced their cables together on July 29, sailed away from each other, and, both inexplicably and miraculously, met no problems whatsoever. The USS *Niagara* sailed into Trinity Bay in Newfoundland on August 4 and the *Agamemnon* stood off Valentia Island, sixteen hundred miles away, just a day later. The line they had sewn together in mid-ocean was still working; and even as sailors at each end hauled the cable to the already-built cable stations, where the landlines to New York and London were already waiting to be hooked up, it was still in apparently perfect order.

There was widespread rapture. On first hearing in London the news that a link had now been joined, made, and preserved in-

Cyrus Field, the transatlantic cable impresario, with Puck's famous boast from A Midsummer Night's Dream, *slightly misquoted, beside him, dominates this* Harper's Weekly *cartoon celebrating the successful laying of the first cable between Ireland and Newfoundland. Queen and president—seen in the lower part of the cartoon—were soon in busy contact with each other.*

tact, the *Times* waxed more breathless than most of its readers must have considered appropriate:

> . . . *since the discovery of Columbus, nothing has been done in any degree comparable to the vast enlargement which has thus been given to the sphere of human activity . . . the Atlantic is dried up, and we become in reality as well as in wish . . . one country . . . the Atlantic Telegraph has half undone the declaration of 1776, and has gone far to make us one again, in spite of ourselves, one people.*

The first messages were passed across it, employing Samuel Morse's now-famous code, on August 16—with Queen Victoria first offering to President Buchanan her sincere congratulations and "fervently hoping" that the new "electric cable" would cement the ties of amity and brotherhood across the seas, and with Buchanan replying soon afterward from Washington with similar verbal folderol. Soon came the first commercial message—a report from Cunard about a nonfatal collision between two ships, the *Europa* and the *Arabia*,* both of which had put into port in Canada; and then came a slew of news items. The first messages ever to be telegraphed between the two continents were either decidedly trivial, as Thoreau had fretted ("King of Prussia too ill to visit Queen Victoria"), or quite appropriately momentous ("Settlement of Chinese question: Chinese empire opens to trade; Christian religion allowed").

But it was too good to last. Very slowly, after less than a fortnight in the water, the cable started to show signs of a mysterious suffering. Its transmissions started to fade away into gibberish,

---

* The *Arabia* was the last wooden ship ever built for Cunard: she had two masts, two funnels, and two paddle wheels.

until finally they broke down altogether, the cable sending and receiving nothing. With much sadness the company directors pronounced that the cable had succumbed to some unknown submarine malady and was quite irretrievably dead.

It had lasted fifteen days. It was a failure. The one new super-continent had now fissioned, had become two again. The sea had won. Such was the public disappointment and official dismay that no further cable was to be laid for another eight years. Except that eventually, those who had kept the faith proved persuasive. In 1866, Brunel's immense new ship, the *Great Eastern*, was summoned from bankruptcy and idleness and pressed into service as a cable layer. She had her difficulties, too, despite eight years' worth of technological improvements—but eventually she sailed into the prettily named Newfoundland hamlet of Heart's Content "having trailed behind her a chain of two thousand miles, to bind the Old World to the New."

It was done. The cable worked almost perfectly, and Mr. Field from the Berkshires, though not able because of his American-ness to be awarded like everyone else an honor by Queen Victoria, was promptly nicknamed Lord Cable by the British press. His creation proved in short order so successful and then so irreplaceably vital that within the following decade the ocean's floors, north and south, became festooned with filigrees of cables. A second strand was laid four weeks after the first. By 1900 there were fifteen, including cables to Argentina and Brazil. Communication between Europe and the Americas—between every European country and every American city, north and south—became almost instantaneous and then in time a matter of routine.

Yet less than half a century after the laying of the first cable, technology offered up another advance, strengthening and quickening the electronic link between worlds. Now it became possible to make exactly the same contact across the ocean—and indeed across any points around the world, and in time well beyond the planet as well—without employing cable at all.

The first experiments with wireless telegraphy—or *radio*, as it came to be called, because this kind of electronic signal was from the Latin *radiatus*, or spread—were also conducted across the Atlantic. And by being chosen as the experimental site—quite naturally so, because the cities on its sides were the richest, most inventive, and most dynamic of modern civilization—the ocean consolidated its position yet again: it was to be a great proving ground for all the new ideas—from packet ships to supersonic aircraft—that, with accelerating rapidity, were now coming to dominate the coming technological age.

A hill on the eastern edge of Newfoundland once again, and another on the western tip of England, in Cornwall—also once again, for since the seventeenth century Falmouth had been the terminus for the Post Office packets—had been chosen to serve as the endpoints for the first test of radio, in December 1901. There had to be a waypoint in Ireland for the radio signal, too, just as there had been for the Cyrus Field's cable—but there was an added reason for this: Guglielmo Marconi, the author of all these first wireless tests, was half Irish. Though his father was Bolognese, Mrs. Marconi senior was from Ireland, of the family that made Jameson whiskey.

As many people claim to have invented radio as there are who say they created television or the incandescent lightbulb. But Marconi, who took out a crucial British patent in 1896* and who a year later began his tests with transmitters and receivers and aerials of all kinds of shapes and sizes in southern England—most

---

* U.K. Patent No. 12,039, granted to Marconi on July 2, 1896, was for "Improvements in transmitting electrical impulses and signals and in apparatus thereof."

notably between Queen Victoria's summer mansion at Osborne, on the Isle of Wight, and her son's cruising yacht of the same name sailing close offshore in the English Channel—is still most closely associated with the invention. As Thomas Edison later remarked to skeptics time and again, "Marconi is the one."

There were skeptics aplenty when Marconi announced late in 1901 that he would try to send a radio signal, already provably sent across both the English Channel and the Bay of Biscay, as well as to and from ships at sea, across the full breadth of the Atlantic Ocean. It would, they said, be as technically impossible (because of the earth's curvature) as it was morally repugnant (this from the late Cyrus Field's Atlantic Telegraph Company, which claimed a fifty-year monopoly on transoceanic telegraphy, with two years still to go).

But the twenty-seven-year-old Marconi cared little for their naysaying, even though his early attempts had been fraught with problems: an initial array of twenty aerials in Cornwall had blown down in a gale, and his experiment in using a balloon to hold up the Newfoundland aerial had failed when the balloon unexpectedly burst.

And now he was trying again. It was shortly after midnight, pitch dark, cold and windy, the start of Thursday, December 12, 1901, and Marconi was sitting before a table on top of what is now called Signal Hill, looking down on the winking lights of the port entrance to Newfoundland's capital city. The glow of a torch illuminating his notepad, he was listening intently to an earphone that was hooked to a large and curiously shaped device of valves and dials, itself connected to a wire—unseen in the dark—that ascended to a large kite, which an assistant kept flying in the strong Atlantic breeze five hundred feet above.

More than 1,800 miles away in Cornwall, on a low summit by the hamlet of Poldhu, near cliffs that fell dangerously into the raging Channel surf, another group of men, Marconi's employees, were taking turns pressing a single flat Bakelite-and-copper key on a machine similar to the one Marconi was using in St. John's. They were endlessly tapping out the three quick dots, followed by a pause, then three quick dots again. It was Samuel Morse's code for the letter S. It was early in the morning in Cornwall, still dark, with not even a glimmer of dawn in the east. It was late in the middle of the Newfoundland night. Everyone was tired.

Then, as his assistants remember it, Marconi's stern and concentrated face suddenly creased into a smile. He beckoned to his assistant, and with a broad grin handed him the earphone. "See if you can hear anything, Mr. Kemp!" And the long-forgotten Mr. Kemp pressed the receiver to his own ear and supposedly heard, above the static and the rising gale and all the other sounds electric and mechanical, the faint—and repeated—three dots, three dots, three dots. The letter S, coming through the earphone, just at the very same moment as the men in faraway Cornwall were tapping it out on that key.

It was done. It was the signal. It was the culmination. The circle had been closed: people could now finally send messages— and one day even talk—and they could do so in perfect synchronicity across thousands of miles of storm-tossed ocean, just as they might converse across an alley in a city, or a meadow on a farm.

There was some dismissive bluster. The Atlantic Telegraph Company was furious and threatened an injunction, which they hoped would scare Marconi away. Others said that he and Mr. Kemp had imagined the whole thing, and that the sounds of dots were merely the traces of stray electrons hurtling through space. But then Thomas Edison weighed in from down in New Jersey, with all his influence and authority, and declared that he believed what Marconi had said; the *New York Times* said as much a day or so later; and then the radio messages were repeated for observers, more and more accurately, and all the remaining skepticism fell away like scales dropping from the eyes, forever.

The *Times* correspondent sent a long message to his newspaper in London from Glace Bay, Nova Scotia, a year later, and had a reply from his foreign editor, in real time; and then in January 1903 a Marconi wireless station was opened on Cape Cod, near Wellfleet. Beside the stumps of one of his old aerials, there is today a bronze plaque, sheltered in a small gazebo set above the sand cliffs that overlook a typically wide Cape Cod beach, washed by the gray Atlantic rollers. The plaque says that from this place, in 1903, President Roosevelt and King Edward VII traded radio messages of congratulation; and from that moment on, wireless telegraphy, radio communication, radio telephony, and all the other present-day miracles of long-distance communication began their fantastic and improbably swift spasm of evolution.

6. THE PASSAGE OF PEOPLES

And all the while, the Atlantic's ships grew grander and bigger and sleeker and more fleet. The sturdily practical packet boats evolved first into the graceful clippers, designed for speed, and later into the iron-hulled four-masted windjammers, built for their immense cargo-carrying capacity. For no more than fifteen years from the middle of the nineteenth century, the Atlantic seemed almost to be ripped apart by the passage of scores of these clippers, hurtling back and forth at speeds unimagined just a few years before. The best of all the designers, a Canadian named Donald McKay, built some of the fastest of these ocean greyhounds: the Yankee clippers, made in Boston, were two hundred feet long, no more than thirty wide, carried twelve thousand yards of sail on three masts, had a steep stem and a graceful extended transom, and bounded through the water with unparalleled stealth and beauty. The fastest, the legendary *Sovereign of the Seas*, once made twenty-two knots; the *Lightning* covered 436 miles in a single day; the *Flying Cloud* left New York, barreled around Cape Horn in a wicked storm, and then turned up into the Pacific, making San Francisco after a total nonstop passage of eighty-nine days; and the *James Baines* took just thirteen days and six hours to reach Liverpool from Boston, and then only 133 days to speed herself right around the world. McKay's *Great Republic* was the longest clipper ship ever built, at 302 feet.

Truly, for the years that the Yankee clippers and their Baltimore cousins crossed the ocean, the vessels became objects of great awe. Parents brought their children down to the East River to gape at their stately comings and goings, and it became a contest to see who could first spot their white sails as they passed through the Verrazano Narrows. With hustlers everywhere deluging New York's public with gaily colored cards that advertised their breakneck transatlantic service, these ships became both famous and beloved—American icons, of which the citizens of a still-new country could be intensely proud. In the same way the jumbo jet became a very visible symbol of American ability, so too did the Yankee clippers.

But only for a very short while. Competition was fast coming, in the form of cargo vessels powered by steam. Even the majestic iron-hulled windjammers, which sported as many as five masts and huge yardages of sail, and could carry five thousand tons of cargo at immense speeds, made little commercial sense once the steamship had been perfected.

As soon as players like Samuel Cunard, who started a steamship service running between Liverpool and Boston in 1814, entered the market, sail was on the way to being finished. Steamship crossings could be made in less than two weeks. The new vessels were suddenly freed from the vagaries of wind and storm. Reliability of schedules—which the packets strove for, but at great risk and seldom with total success—became the accepted norm. Cargo rates started to plummet. And though some windjammers did manage to cling to their business into the new century—for some vessels, until well after the Second World War, moving bulk cargoes such as guano from remote mid-Pacific islands, where there was no chance for a steam vessel to take on coal—all of the clippers had vanished from the commercial routes by the last quarter of the nineteenth century.

The evolution from sail to steam had unintended consequences, too. The windjammers' berths in New York had long been a feature specifically of the East River, which had offered an easier location for the turning of sailing craft. The steam vessels, however—and which in time would obliterate all the competition—almost all arrived and departed from the relatively uncluttered waters of the Hudson, on the west side of Manhattan, closer to the railheads that would take their cargoes and their passengers into the American hinterland. The change led inexorably to an alteration in aspect to the fast-growing New York that has its echoes still today: the finest city views are those that now look west, to where the liners dock.

The packets and the clippers had been people movers as well as shifters of cargoes. The sailing vessels in the closing years of their careers, and steamships through all the years of theirs, and until the planes came, carried scores millions of people westward, overwhelmingly westward, and in doing so they played an essential role in the populating—indeed, in the making—of the Americas. And particularly of the United States and Canada, since both of these hitherto little-populated countries, as acts

of deliberate policy, decided they needed for a good long while a
steady flow of immigrants* from the Old World to the New.

Much of this migration, the shameful side of a vastly complex
story, and mentioned already, was of the involuntary kind, with
slaves swept up from Africa, sent under appalling conditions
across the ocean, and then padlocked into humiliating servi-
tude. Many of the others who came out at their own behest were
early colonials, from the Pilgrims of Plymouth and the settlers
of Jamestown to those who built cottages in such faraway places
as Puerto Madryn (the Welsh), Rio de Janeiro (the Portuguese),
and Halifax (many Basques, among others). Many of these were
skilled and technically able men who were invited to help build
the growing industrial revolution, to spin and weave or puddle
iron, fish or mine coal; most of those who came to America were
either from England—the young America was overwhelmingly
Anglophone, after all—or from Germany or Holland; they and
their like saw the populating of all the colonial possessions,
which ranged along the entire American coastline from Labrador to Patagonia, as part of their own personal manifest desti-
nies.

But by far the greatest proportion of newcomers were those
who came after these colonies had one by one thrown off their
foreign rulers, and who came across the sea because they saw
the newly formed nations as beacons of hope and possibility.
These were the now famously engraved *huddled masses*, people
who yearned for some respite from the grinding difficulties of
Europe—and it is the passage of these, the millions upon millions of men and women and children who came westward with

---

* *Immigrant* is a late-eighteenth-century word—which according to an obscure but prescient travel writer named Edward Augustus Kendall, who wrote in 1809 an account of his
travels in America, "is perhaps the only new word, of which the circumstances of the United
States has in any degree demanded the addition to the English language."

little but a collective and an individual sense of optimism and determination to make something of the chances that were said to be on offer in the New World, that dominate the story.

Their crossings did much to alter once again the world's perception of the Atlantic Ocean. Hitherto it had been to most an immense barrier of discouragement; now, with the payment of a modest sum for passage in bearable discomfort and the indignities of processing on the other side, the ocean was transformed into an immensely long bridge—long, to be sure, but a bridge just the same—that would take anyone bold enough to venture across into a brand-new life. In becoming the prime passageway for all these migrant journeys, the ocean became itself an integral part of a whole new world of possibility. The figures are quite staggering. While a mere one million people had arrived in America in the seventy years between independence and 1840, over the following sixty years no fewer than thirty million came flooding in—most of them northern Europeans, particularly Britons and Irish, in the years of the first great wave that lasted until 1890; and then many Italians, Germans, and Scandinavians in the half century that followed. Much the same was happening south of the equator, too: some ten million Europeans migrated to Latin America in the fifty years before the Great War, and the populations of Brazil and Argentina, which accepted particularly large numbers of migrants from Portugal, Spain, and Italy, increased massively— by tenfold in Brazil, by fifteenfold in Argentina.

And so the people came in their millions, pouring up the gangways and brought by lighters, then settling themselves uncomfortably in the ships that waited patiently at the quaysides or out in the roads. The migrants paid low "emigrant" fares for passage in the steerage—three pounds was the going rate to America for many years, though Argentina offered free passage

ATLANTIC OCEAN: COMMERCE AND COMMUNICATION

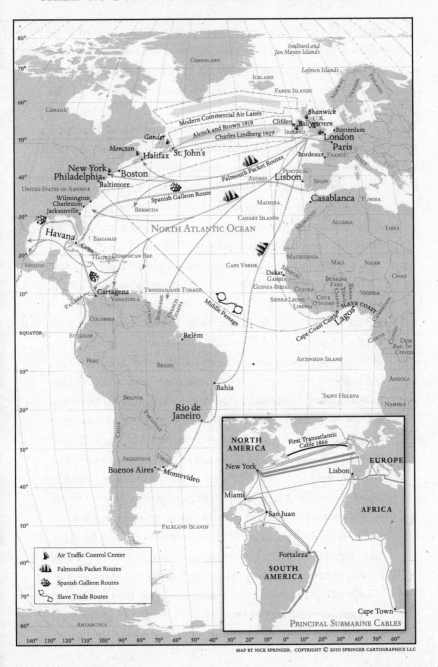

MAP BY NICK SPRINGER. COPYRIGHT © 2010 SPRINGER CARTOGRAPHICS LLC

from 1888 and handed out prepaid tickets to anyone fit and able who wished to come (a decision that Argentina was later to rue, some contend, since it brought in many less well educated migrants than the country truly needed, and proportionately fewer of the more technically able).

The migrants began their new lives at quays in Liverpool (which handled nearly five million America-bound passengers on one-way tickets between 1860 and 1914) and Glasgow, in Havre, Bordeaux and Nantes, Modano and Marseilles, Naples and Genoa, Hamburg and Bremen, and the long-forgotten port of Fiume, now Rijeka in Croatia, whence came so many of the Slavs who are inhabitants of present-day Chicago.

Conditions on the emigrant ships could be decidedly unpleasant—while the swells lived and dined pleasantly in the upper decks, those on the verge of momentous change had to put up with steerage decks that were crowded, dark, with poor sanitation and limited water, had either bunk beds with straw mattresses or hammocks, offered almost no cooking facilities, the sexes harshly segregated to lessen the temptations of turpitude, and with constant reminders from unhelpful and often hostile crew members that a ticket gave steerage passengers merely the right to passage, and perhaps a handout of bread, salted meat, and occasional chunks of pemmican or ship's biscuit, and very little else. The ship's hatches would be closed in poor weather, and so to the passengers' general misery would be added the terrible fear occasioned by being tossed around violently for days on end in often fetid near darkness—an experience quite foreign to most of the travelers, because few would ever have been near a ship, and even fewer out in the open sea. Morale invariably suffered as the passage groaned on—in bad weather, especially, good cheer could be sustained in the lower decks only by the travelers telling and retelling to themselves their imagined vision of what might lie ahead, in the promised land.

Of the numberless accountings of emigrant voyages, that of Robert Louis Stevenson, who traveled from Glasgow to New York in 1879 in a threadbare class just one level above the lowest steerage of them all, is perhaps the most famous and eye-opening. Stevenson's family was horrified and tried to delay it, but in the end *The Amateur Emigrant* was published a year after his death, in 1895, and was deemed so graphic a description of migrant misery as to be barely credible. The event that threw even more light on their situation occurred seventeen years later, with the sinking of the RMS *Titanic*, in April 1912.

For one harsh reality came to dominate the saga of that tragedy: that the lives of those in the great ship's steerage—few of whom even knew where the ship's lifeboats were—meant evidently less to the White Star Line than those of their premium passengers. It was a shocking revelation, but undeniable, for the statistics displayed a cruel truth: that while the majority of first-class passengers survived the accident, more than three-quarters of those pinned below the waterline on the steerage decks died, unrescued either because they were physically unrescuable or because few were willing to try to save them.

Maritime laws and regulations in legions—among them laws that extended well beyond those relating to ending the shabby treatment of migrants—were changed in consequence of the *Titanic*'s collision with her fateful iceberg. The irony of the coincidence of location can hardly have escaped notice: new laws regulating passage by sea were occasioned by a terrible tragedy that took place in the North Atlantic in 1912; the very system of laws and the organization of parliaments to decide and promulgate them was first created but a few hundred miles away, in Iceland, in 903—almost precisely one thousand years before.

ɪ. CASUALTIES AT SEA

Invariably it took accidents at sea to effect changes to the laws of the sea. And many of the most important recent maritime accidents took place, as with the *Titanic*, along some of the Atlantic's busiest shipping lanes. Our ability to discern this in an instant is due to a forgotten nineteenth-century polymath named William Marsden,* who while employed as secretary of the Admiralty was professionally interested in collecting and collating statistics about the world's seas. He divided a Mercator map of the world into a series of numbered ten-degree squares, known to this day as Marsden squares.

Each quarter of every year, the insurers at Lloyd's produce a casualty report—a list of vessels involved in accidents at sea and that, either through foundering, colliding, or being wrecked, were reported to have been either total losses or so seriously damaged as to require towing and rebuilding. These figures are all then plotted as black dots on a Marsden squares chart of the world, the results showing up as concentrations of accidents in all the places one might expect—in the crowded waters off Singapore, in the Black Sea, south of Sicily, in the southern Aegean.

But the Atlantic has a pall of problems on both of its coasts. Enormous numbers of accidents are reported each year along the shores of Norway and western Scotland, within the entire length of the English Channel, in south Wales, by Rotterdam, in Galicia, along the Spanish side of the Strait of Gibraltar, by

---

* Marsden was also a renowned numismatist, whose coin collection is now in the British Museum, and an authority on Eastern languages—he produced a definitive dictionary of Malay, in which he was fluent. He is remembered in naval circles as being the man who woke the First Lord of the Admiralty in 1805 to tell him of the British victory at Trafalgar, as well as of the death of Nelson.

Lagos and the approaches to Cape Town. South America, on the other hand, gets off comparatively lightly—squares 413 and 376, which include the entrances to the ports of Buenos Aires and Rio de Janeiro, show some activity—but then once the Caribbean and the North American coasts come into view, the maps swiftly turn black with pepperings of ink around the southern coast of Haiti, along the Gulf of Mexico coast from Mobile to Galveston, the length of Long Island from the Nantucket Light to New York City, and along the entirety of the St. Lawrence Seaway. Marsden square No.149, where the *Titanic* foundered, has just a scattering of dots, since accidents out in the deep sea occur only infrequently—though when they do, rescue is invariably slow to arrive and most often is too late.

Coastlines, and other ships, are what sailors fear most. Most of the infamous recent accidents have taken place within sight of land. The collision of the two passenger liners *Andrea Doria* and the *Stockholm* less than twenty miles west of Nantucket in 1956, in thick fog, became a legendary story of rescue (of the 1,706 passengers, 46 were killed) and an object lesson—resulting in yet more modified rules—in when not to rely on radar. Arguments over apportioning the blame for the costly collision between the Liberian oil tanker *Statue of Liberty* and the Portuguese cargo carrier *Andulo* off the southwestern tip of the Iberian peninsula in 1965 were so intense they had eventually to be decided by the British House of Lords, since Lloyd's insurance claims are adjudicated there (the Liberian ship lost, being found "85 percent to blame"). And the stranding and sinking of the fully laden Liberian oil tanker *Argo Merchant*, which hit a reef off Nantucket at sixteen knots while making passage from Venezuela to Boston in 1976, resulted in 28,000 tons of oil being blown out to sea, and the then American president announcing new rules regarding pollution, navigation, and preserving life on the ocean.

The crippled Italian liner Andrea Doria *lies on her starboard side after being struck by the Swedish liner* Stockholm *in the open sea approaches to New York on the foggy night of July 25, 1956. Arguments still flare over how to apportion blame for this "radar-assisted collision," in which forty-six crew members and passengers died, though nearly 1,700 were saved.*

Probably the most memorable of recent oil tanker disasters was that involving yet another Liberian vessel, the *Torrey Canyon*, which in March 1967 was steaming at full tilt toward southwest England, with 119,000 tons of Kuwaiti crude oil for the refineries at Milford Haven, in south Wales. The repercussions of her hitting, head-on, the sharp granite rocks of the Seven Stones Reef, off the Scilly Islands, were even more widespread—in terms of new laws and international agreements—than after the *Titanic*. The laconic, matter-of-fact tone of the official report, as summarized in the *Times Atlas of the Oceans*, takes nothing from the gravity of the disaster:

> At 08.40 the position was fixed by observation of the Seven Stones light vessel—it was bearing 033°T at a range 4.8nm [nautical miles]. The Torrey Canyon was now only 2.8nm from the rocks ahead.
>
> At 08.42 the master switched from automatic steering to manual, and personally altered the course to port to steer 000°T, and then switched back to automatic steering.
>
> At 08.45 the third officer, now under stress, observed a bearing, forgot it, and observed it again. The position now indicated that the Torrey Canyon was less than 1nm from the rocks ahead. The master order hard-to-port. The helmsman who had been standing by on the bridge ran to turn it. Nothing happened. He shouted to the master who quickly checked the fuse—it was all right. The master then tried to telephone the engineers to have them check the steering gear aft. A steward answered—wrong number. He tried dialing again—and then noticed that the steering selector was on automatic control instead of manual. He switched it quickly to manual, and the vessel began to turn. Moments later, at 08.50, having only

With 120,00 tons of Kuwait crude oil in her tanks, a shortcut in the sailing plan, and the ship's cook at the wheel, the California-owned, Liberian-registered supertanker Torrey Canyon was going full tilt when she struck the Seven Stones Reef off Cornwall in March 1967, causing an environmental catastrophe.

turned about 10°, and while still doing her full speed of 15.75 knots, the vessel grounded on Pollard Rock.

 A number of cargo tanks were ruptured, and crude oil began immediately to spread around the vessel. . . .

The British government eventually had to bomb the wreck with napalm—causing an additional flurry of comment, since up to this point few in the country were aware that Britain possessed the gelled-gasoline weapon then being used to such dreadful effect in Vietnam—to set fire to the spreading blaze. The court battles over the costs of the affair, and the international conferences called to consider its environmental consequences and legal and political ramifications, continued until the middle of the next decade.

Most tragedies at sea, melancholy though they may be for those involved, are events in the faraway that are invariably soon forgotten. Some—like the rescue of the men and women from the *Forfarshire* by the wonderfully named Grace Darling, in her rowboat in a storm off the Farne Islands in the North Sea in 1838—are remembered for offering up an episode of exceptional heroism. Others—like the two-masted brig *Mary Celeste*, found six hundred miles west of Portugal, sailing steadily toward Gibraltar with not a soul aboard—remain in the mind because of the mystery, in this case a puzzle amenable to so many possible causes (murder, poison, sea monster, tsunami?), very few of them good. And then there was the fate of the *Teignmouth Electron*, the tiny catamaran in which the British amateur sailor Donald Crowhurst had entered a round-the-world single-handed yacht race. He had cheated, had then found himself likely to win and thus most probably to be placed under scrutiny, and so had leapt off his craft to avoid discovery: the story lingers still, a vivid portrait of man made manically mad in the wide loneliness of a great ocean.

. . .

The vastness and imperturbable power of the sea, when ranged against the enforced solitude of a lonely sailor, can surely make for madness. It can also prompt in others a soaring of ambitions, a realization of great visions, perhaps for some the making of great fortunes. But in all these encounters with a sea so grand as the Atlantic, there has to be the presumption that the body of water itself inspires a large measure of respect and awe. If that inspiration wavers—if mankind ever begins to treat the sea with less respect than its own story deserves—then things begin to go awry. A great ocean is not a thing to regard with casual disdain: but in the manner and speed with which it is being so often traversed today, there is the rising of temptation to do just that. The consequences are myriad, and they are invariably malign.

❦ *Chapter Six* ❧

CHANGE AND DECAY ALL AROUND THE SEA

———✦———

The sixth age shifts
Into the lean and slipper'd pantaloon,
With spectacles on nose and pouch on side,
His youthful hose, well sav'd, a world too wide
For his shrunk shank; and his big manly voice,
Turning again toward childish treble, pipes
And whistles in his sound.

1. CROSSING THE POND

At the international operations center of British Airways, which takes up the entire third floor of a highly secure and discreetly marked building on an empty moor some five miles west of London's Heathrow Airport, the staff takes care to refer to every transoceanic flight as a "mission." They do so in part out of tradition; but they do so also as a reminder that, just as with today's explorations of space and with nineteenth-century ventures into godless interiors, there is never anything inherently routine or safe about their allotted task: in this case the lifting against the natural force of gravity of two-hundred-odd tons of airplane and three-hundred-odd human beings to an entirely unsustainable altitude of seven or so miles, and then propelling all without interruption for many long hours, suspended by nothing more than a lately realized principle of physics, high above a cold and highly dangerous expanse of sea.

Air travel across oceans has in recent years become to most consumers, if not necessarily to the practitioners, as tedious as it is commonplace. Its relative cheapness has rendered brief visits to the faraway perfectly imaginable to enormous swaths of society. The Atlantic, being of a width manageable for most to cross by air today without too much time or pain, is currently the most obvious pathway for millions to the most exotic of foreign fields. The Pacific is just too big; the Indian Ocean for most just too far away. So Mancunians who back in the 1970s might have considered Marbella an alluring mystery now in the first decades of the twenty-first century readily consider Miami an obvious destination for a long weekend. Parisians cross the Atlantic almost without thinking for tanning vacations in Martinique. Bored Brazilian city dwellers fly to see the giraffes and springbok near Cape Town, Belgians go in herds to sun in Cancún, Texans set off to visit the theaters in London, and Norwegians head southwest to try the slopes in Bariloche. All of this flying—together with the cargo and the courier planes trundling windowless through the nights, and with official government aircraft on routine business and military planes on secret missions—has conspired to make the Atlantic, of all the oceans, more flown across than any other.

The air-route charts present a quite alarming illustration, seeming as they do to render the ocean almost solid with passing traffic. They show in particular two great paint daubs of tracks between the American Northeast and the European northwest, so concentrated when they join just south of Iceland as to make the ocean appear almost paved, a yellow brick road in the sky. South of this thick northern superhighway are cobwebby skeins of routes linking former possessions with former masters— Mexico with Madrid, Curaçao with Amsterdam, Guadeloupe with Paris, Kingston with London, even (if to stretch a point) Havana with Moscow; while farther south still there is a thick line of the major north-south air tracks, nearly as concentrated as their east-west brethren, linking the great and growing cities of Atlantic South America with their main trading partners, old and new—Rio with Lisbon, of course, but also with Frankfurt and Moscow and Milan; and Buenos Aires with Barcelona, of course, but also with Stockholm; Birmingham, England; and Istanbul. And then even more distant, over the cold waters of the far South Atlantic, there are the lonely and half-forgotten tracks of the lesser-known city pairs: Rio de Janeiro to Lagos, Quito to Johannesburg, Santiago to Cape Town, Brasilia to Luanda.

More than thirteen hundred commercial aircraft cross into Atlantic Ocean airspace every day, and the number increases steadily, by about 5 percent each year. By far the greatest number of the planes fly across the northern part of the hourglass-shaped sea—414,000 planes checked in with its major oceanic air traffic control centers in the north during the calendar year 2006, for example. If to these are added the planes that cross from the South Atlantic up into the North and come back again, and the relatively few aircraft that cross just between the South American and African destinations, and in doing so fly over the lonely Atlantic waters to the south of the Tropic of Capricorn, then one comes up with a total figure of around 475,000 Atlantic transits every year: some 1,301 flights each day.

They do their crossing in two great waves, which in a time-lapse animation of the radar contacts look like spurts of molten gold radiating from the continents and out over the sea. First come the westbound jets trying in vain to chase the sun* by traveling generally in daylight; in contrast those going east, heading back into the Old World, fly out in the American darkness and land, by and large, early in the European morning. At any one hour of day or night there are perhaps fifty of these aircraft in flight over the sea—ten thousand human beings passing by every hour, reading, sleeping, eating, watching movies, writing, seven miles up in the sky.

· · ·

And yet from these little seven-mile-high cities in flight, only a very few of the populations will ever care to look down for more than an inquisitive instant at the wrinkled surface of the sea below, or at the thick mass of gray-white cloud that so frequently obscures it from view. These people are mostly quite careless of the ocean's very existence: it is merely an expanse to be crossed—the pond, if it is crossed quickly and casually, or something irritating and even less flatteringly named if it takes their aircraft countless hours to traverse.

The fact of inexpensive transoceanic travel has taken much the mystery of the sea away, has made us indifferent to its existence. Since the crossing of oceans has become tedious to most who do it, the oceans themselves have become the object of tedium as well. Once they were feared; they inspired awe, amazement, and mystery. Now they are to many just a barrier, an inconvenience—too large as entities properly to contemplate, too annoying as presences to warrant much care. The public attitude

---

* Not always in vain: I once traveled in the cockpit of a Concorde, leaving London just before sunset. The sun dipped into the Bristol Channel as we left; it then rose again out of the western Atlantic horizon as we reached supersonic cruising speed, hovered in front of us for all of the journey, then slipped back below the Blue Hills of Virginia as we touched down—at a local time earlier than when we had departed.

to the great seas has changed—and this change has had consequences for the great seas, few of them any good.

It has helped in particular to set the scene for what some worried few see as the endgame of humankind's Atlantic story. It is nothing very new, of course. Man has been carelessly despoiling the oceans for decades. Ever since the first factory was built beside the water, ever since the first sewer pipe was laid in an industrial port city, and ever since we started, either casually or deliberately, to spill our wastes and our chemicals into the ocean's immense and blameless sink, we have displayed a propensity to ruin it, to violate it. The land we have to live with, and so we pay it some measure of attention; the ocean, by contrast, is largely beyond our sight. It is so immense it can tolerate—or so we used to think—an immense amount of systemic misuse.

In Victorian times, though, we still thought of the ocean as vast and frightening; we still regarded it with some kind of awed respect. Not anymore. Passenger aircraft have shrunk the Atlantic's vastness to a manageable size and thereby have also shrunk our capacity to be so impressed by it. People sail across the Atlantic on their own these days, and in summertime almost as a matter of routine. The westbound sailing route from Cornwall to the Caribbean by way of the Azores is regarded as so easy of accomplishment as to be referred to disparagingly by the harder-nosed and misogynist yachtsmen as merely "the ladies' route." Some people have taken to rowing across the Atlantic, at first in pairs, then alone. One day someone with a lot of spare time and a willingness to be swaddled in many tons of grease will probably contrive to swim it. The ocean is no longer so challenging a prospect as it once was. It stands in the public imagination rather as Mount Everest once did: now that we have *conquered* it, we perceive it as somehow manageable, and on the way to being even, dare one say it, trivial.

And in lockstep with this change of perception—not necessarily caused by it, but certainly coincident with it—there has been a steady lessening, some would say an actual abandonment, of humankind's duty of care toward it. Such has already happened to Mount Everest: the base camp near Thyangboche is a slum, and the main route by way of the Western Cwm is strewn with castoffs; even the summit has as much junk on it as it does joyous flags. And now we are doing much the same to the world's seas, dealing all too thoughtlessly with them in ways that many say are threatening the seas' serenity, if not their very survival.

The oceans are under inadvertent attack, and as never before. Insofar as the Atlantic Ocean is the most used, traversed, and plundered of all oceans, so it is the body of water that is currently most threatened. Even though the central Pacific has attracted a lot of recent infamy because local gyres have swept a spectacular amount of ugly flotsam into mid-ocean patches the size of small states, it is actually the Atlantic that is in the greater trouble. It is subject to much more use—and so very much more misuse—all of it crammed into a much smaller space. It was the first great body of water to be crossed, it is now by far the busiest and is inarguably the most vital—but it has become evidently the least pristine and the most begrimed.

Yet awesome it remains, to some. In the British Airways operations center—a place of numberless computer screens and charts and weather forecasting maps and enormous panels of flashing pixels, and with dozens of serious-looking men and women* who are charged with keeping tabs on all the people, animals, and boxes of freight currently in flight around the world, and making as sure as they could that all of them were safe and on schedule

---

* Including one sitting before a very large computer screen that shows in real time images of every single aircraft in the world that happens to be airborne.

and the people as content as it is possible to be on an airplane—there is little doubt that to all of them, twenty-four hours a day, the great ocean is still held very much in awe. Great seas are not kindly entities over which to fly: if your aircraft somehow fails, where, exactly, do you put down? No pilot leaves the chocks for a transoceanic flight without remembering the first axiom in flight school: *takeoff is voluntary, but landing is compulsory.* And in the middle of an ocean it is self-evidently true not just that there is nowhere to land, but that there *just is no land. No land at all.*

Those who pioneered the practice of flying over seawater knew that all too well. Crossing a large expanse of sea perhaps didn't trouble Louis Blériot when he flew his tiny monoplane across the English Channel from Calais to Dover in 1909, just six years after the Wright Brothers' first flight at Kitty Hawk, North Carolina. For although Blériot admitted to being alone above "an immense body of water" for fully ten minutes, he also had the comfort of knowing a French destroyer was below him, monitoring his flight and ready to save him if he ditched. And for most of his thirty-seven-minute crossing, and even though he was only 250 feet up in the air, he could see the coast of France behind him and by peering ahead the white cliffs of England before him. Blériot won the thousand pounds that Lord Northcliffe had offered through his newspaper, the *Daily Mail*, and he became—not least because his boulevardier's mustache and his reputation as a barnstorming air racer—an immediate superstar, and very much the ladies' heartthrob.

But it was one thing to cross the Channel, quite another to fly across the Atlantic Ocean. Lord Northcliffe put up ten times the sum for anyone who dared try it; and though he announced it in 1913, it was not until six years later (with admittedly, a four-year hiatus for the Great War to pass) that the prize was won, and by a pair of Royal Air Force officers whose names, by a small injustice of history are still not quite as known as that of Blériot: Jack Alcock and Arthur Whitten Brown.

Jack Alcock (left) and Arthur Whitten Brown, standing beside a Vickers Vimy heavy bomber in which they and their two pet kittens crossed the Atlantic nonstop in June 1919.

The venture was Alcock's idea, conceived when he was imprisoned by the Turks after ditching his fighter plane in the sea near Gallipoli. *Why not have a bash?* he said. The pair used for the attempt a stripped-down long-range Vickers Vimy biplane, its bomb bays filled with extra fuel. In the summer of 1919, they dismantled the plane and crated it up so it could be sent by ship to Newfoundland. There they built their own runway for takeoff. They did not know where they might land—it could be a field, or a beach, or an Irish lane: it turned out to be a bog.

Plenty of others were trying for the same prize—among them

an American, Albert Cushing Read, who flew a seaplane to the
Azores, stayed there for a week, and then flew on to Portugal: it
took eleven days, and American warships were stationed under
his path every fifty miles along the proposed route. However,
Lord Northcliffe had decreed that his prize was for a nonstop
journey, achieved in less than seventy-two hours, so Read did not
win it. Nor did a Australian tearaway named Harry Hawker, who
tried it in an experimental long-range plane, a Sopwith Atlantic.
When its engine overheated, Hawker spotted an eastbound ship
five hundred miles short of Ireland and ditched; he was picked
up and went home by sea. Because the ship had not yet acquired a
radio, its crew could not tell Hawker's relatives of his rescue. Instead Mr. and Mrs. Hawker were shocked to get an official black-
bordered telegram from King George offering royal condolence
for their son's supposed loss. The better news came later.

The dashing aviators—Jack Alcock in a blue serge suit and
Brown in his Royal Flying Corps uniform, with 865 gallons of
fuel and a pair of small black cats named Twinkletoes and Lucky
Jim—set off on the morning of Saturday, June 14. They had hor-
rendous problems—up at twelve thousand feet their instruments
froze solid, their radio broke, their exhaust pipe ruptured, Brown
had to climb onto the wings to break off ice,* they became dis-
oriented trying to watch the stately heeling of the stars in order
to navigate, and went into a spin down through the clouds until
almost hitting the waves—and when they finally arrived over the
coast of Ireland, they could not find a place sufficiently free of
rocks on which to land. Finally they spotted the masts of a radio
station, circled it a number of times without at first managing
to wake anyone—it was 8 A.M. on an Irish Sunday, and the after-

---

* Or so it was said. Some have argued it would have been impossible for him, since he had
a badly damaged leg.

effects of Guinness must have trumped the callings of piety—
and settled the plane onto a field, crash-landed, and ended up
nose down in soggy black peat.

They were in County Galway, near a hamlet of Clifden. When
the radiomen awoke and realized who the two fliers were, they
telegraphed news of their achievement to London. The pair became rich and famous overnight and were knighted by the king
only weeks later. Sir John Alcock was killed in a flying accident
just a year afterward, and Sir Arthur Whitten Brown lived until 1948. They had crossed the ocean, without stopping, and they
had done it in sixteen hours and twenty-seven minutes. When
the much more showy and popular Charles Lindbergh single-
handedly flew the *Spirit of St. Louis* from Long Island to Le Bourget
in 1927, he gave due credit to the pair: Alcock and Brown, he said,
had showed him the way. Amy Johnson and Beryl Markham, who
in the 1930s separately became the first of their sex to fly the
same ocean westward, were not so generous.

The ocean is officially described by the two air traffic control
centers that have charge of North Atlantic airspace as a region
"moderately hostile to civilian air traffic"—it is vast, there are no
navigation aids and no communication relays. This means that
for a substantial portion of the journey over the ocean a civilian transport aircraft is essentially all on its own. If it gets into
trouble out in mid-sea, then it is in big trouble indeed. Such realizations have a way of inducing real awe among those whose
task it is to ferry people and goods across. What might appear to
a safely arrived passenger as no more than quotidian routine is
in fact the result of planning sessions no less intense than for
a truly white-knuckled adventure, like rounding Cape Horn or
scaling Mount Everest's South Col.

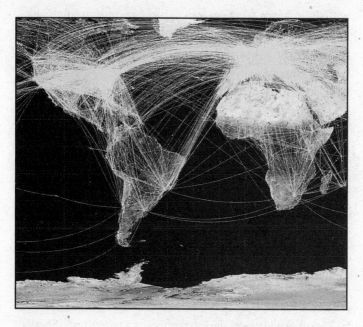

More than 400,000 commercial jets cross the Atlantic Ocean each year, as this electronic map of the air routes shows. Fair amounts of traffic pass between Europe and its former South American possessions, but between the United States' East Coast and Britain and mainland Europe, it is as though a solid bridge has been constructed, three thousand miles long.

The flight I chose to examine was one on which I was traveling home, on January 30, 2009: BA 113, an ordinary, mid-afternoon, no-excitement Boeing 777 journeyman's flight, leaving London at 3:15 P.M. and due to arrive at Kennedy Airport seven and a quarter hours later, at about 5:30 P.M. local time. The aircraft would be parked on Stand 555, it would be tail number G-YMMO, a two-year-old 777-300ER, an extended-range version of Boeing's highly regarded wide-bodied long-haul plane, equipped with Rolls-Royce Trent engines. It had just come in from Singapore and had recently performed runs to Toronto and Sydney. It was a workhorse, heavily employed on long-haul flights, and was well accustomed to flying the Atlantic.

(There were just two unusual items on this otherwise routine January day. The first was that overnight a northbound flight from Johannesburg had suffered a serious mechanical failure over Spain and had been forced to put down in Madrid. The London staff was now scrambling to send a replacement aircraft down to collect not just the stranded passengers, but also an enormous cargo of gold, something apparently quite normal on Johannesburg departures. The Madrid airport police were creating quite a fuss, however, aware that many millions of dollars' worth of bullion would be most tempting for Spanish desperadoes if word got out. And with a cell phone in every passenger's hands, it was unlikely to be a secret for long.

The other oddity was the interim report, just out, on G-YMMO's slightly older sister aircraft, G-YMMM, which had crashed on its approach to Heathrow almost exactly a year before. There was still some puzzlement over why its engines seemed suddenly starved of fuel, and the plane "just dropped," as the pilot put it, when it was coming in to land. The staff at the center were eager to assure me that even though the precise reasons for the accident hadn't been worked out—most likely ice in a fuel line, accumulated while flying over a patch of unusually cold air over the Urals—it was statistically most unlikely to happen again.)

Eighteen pages of briefing notes were handed to the captain when he and his crew checked in three hours before his aircraft's departure. The departure and arrival airports were all running normally—a scattering of lights were missing from a taxiway at Heathrow, there was construction at the end of a runway at Kennedy, nothing major. Much the same was true at the alternate arrival airports, Philadelphia, Boston, and Newark, though there were some minor navigation problems for aircraft going into Boston. As far as alternate airports en route—hooligans with laser lights were occasionally causing a nuisance by pointing them at incoming planes at Birmingham and Cardiff, there was severe wind shear and turbulence on approach to St. John's, and a strike by workers at Goose Bay, Labrador, meant that the snow had not been fully cleared from the runway, causing that particular field to be closed.

The weather during the crossing was likely to be as cooperative as expected in late January: strong southerly high-level winds at the takeoff site and until five hundred miles off the Irish coast—then clouds would set in, the winds would drop and veer to the west for most of the track, then would go back to southwesterly and freshen over Newfoundland, and then return to strong westerlies for the approaches into New York. Turbulence would be minimal; storms were unreported.

One aspect of the flight that had been already decreed by air traffic control and the planners at the airline was the transatlantic track that BA 113 should use that day. There are generally ten tracks laid out each day, five of them westbound and five eastbound—each carefully designated lanes of traffic traversing the broad width of the deep Atlantic, away from the coasts of Europe and North America, and which are shifted very slightly north and south every few hours according to the exact current position of the jet stream, and which allow the huge number of aircraft crossing the ocean to be separated safely from one another.

The westbound tracks are designated A, B, C, D, and E, and the eastbound V, W, X, Y, Z. The six hundred or so planes that head west each day—BA 113 being one of them—fly at even-numbered altitudes, separated by 2,000 feet: at 40,000 feet, 38,000 feet,

36,000, and so on. Eastbound craft operate conversely at odd-numbered levels—39,000 feet, 37,000, down to 31,000. On this day my BA 113—its call sign on the radio *Speedbird 113*—had been told to fly on Track NAT Charlie, at Flight Level 380. She would prepare to enter the critical transoceanic sector at an invisible waypoint that Atlantic aviation chartmakers had given the unlovely name of BURAK. She would make her actual entrance into the oceanic sector, sashaying elegantly into the most critical portion of the flight, at a second waypoint designated as MALOT.*

The two bodies that police the ocean at high altitude and try to maintain good order and safety for the aircraft and their thousands of daily passengers are based in Prestwick in Scotland and in Gander in Newfoundland. The first, the Shanwick Oceanic Control Centre, is an enormous complex of buildings—appropriately known as Atlantic House—situated on public housing land south of the main runways at Prestwick Airport. It has control—by way of an immensely powerful shortwave radio station sited far away in the village of Ballygirreen in southwest Ireland—over all aircraft coming to and going from the British Isles as they pass across a vast swathe of sea that extends from Icelandic waters in the north to the Bay of Biscay in the south, and onward to a line halfway across the ocean at 30 degrees west longitude.

Shanwick is usually an intense and busy place, as one might

---

* Waypoint names can sound very strange, but for that reason are five-letter nonsense words easily committed to pilots' memories. The main British Isles waypoints on the near edge of the Shanwick sector are named RATSU, SUNOT, PIKIL, RESNO, VENER, DOGAL, MALOT, LIMRI, DINIM, SOMAX, and BEDRA; those on the far side, marking the edge of Gander's responsibility—a much longer line, considering the size of territory—run from MUSVA, off the coast of Baffin Island, to VODOR, off Newfoundland. Our BA 113 entered North American airspace at HECKK, not far from the point where Leif Eriksson landed in the tenth century.

expect. But for periods in the late spring of 2010 a bizarre and eerie quiet fell on the main control room. High-altitude clouds of volcanic dust from Iceland were found to be wafting across northern Europe, and cautious bureaucrats in Brussels decided to ground most European flights and to ban nearly all air traffic across the north Atlantic. Their decision, much criticized, left millions of passengers stranded around the world, and the Shanwick controllers with precious little to do.

Shanwick's mirror opposite is across in Newfoundland: Gander Oceanic Control, by far the busiest oceanic control center in the world—its staff monitored no fewer than 414,000 crossings in 2007—handles all deep ocean traffic that passes to the west of the same thirty-degree longitude line. While Prestwick's center is located in a homely Scottish suburb, Gander occupies a series of low and unlovely structures beside a lonely former military staging post airfield among the pine trees and swamps of northeastern Newfoundland, and is remote in the extreme. Yet Gander airfield has a uniqueness beyond being far away: it is also an airport open continually, without any time or noise restriction—"A curfew, up here? You've got to be joking!"—and the airfield prides itself on being what it calls the "airlines' lifeboat," a sanctuary kept always stocked and ready on the davits for any kind of trouble that may occur in flight. "We can handle anything," the managers say. "Mechanical, navigational, unruly passengers, bomb scares, hijacking, what have you. We're trained, we're prepared. Whatever the time, whatever the need, whatever the weather, we here at Gander can take care of it."*

---

* After the terrorist attacks on September 11, 2001, nearly forty big commercial jets bound for America were diverted to Gander, and the town of 10,000 suddenly found itself playing host to an additional 6,000 bewildered newcomers. Many later spoke movingly of how hospitable the Canadians were; the Canadian prime minister later came to Newfoundland and told an audience in Gander, "You did us proud."

Speedbird 113 was due to spend about three hours of its transatlantic passage in the unreal world of the oceanic control sector,
a place commanded by the ever-fading shortwave radio signals
from Gander and Shanwick. To the passengers seated aft of the
armored cockpit doors, the ocean below is a place of utter unre-
markability—less a matter of space, more an expanse of time, a
period of necessary tedium. It is a place and a time with no markers, no fixed waypoints—other than the invisible coordinates of
latitude and longitude, which the commander up front would report by radio or by satellite data link to either Scotland or Canada
as he flew along—with no landmarks, and with no visible means
of support, other than the aerofoils and the constant low thunder
of the twin Trent engines. If things went badly wrong here—if
there was an engine fire, say, or a sudden loss of cabin pressure—
the pilot could, for most of the journey across it, either turn back
or make a turn for one of the two possible alternate airfields that
were manageably close to his chosen route—in this case either
Keflavik in Iceland or Narsarsuaq in southern Greenland.

. That would be true only for *most of his journey*, that is. There is
one relatively small sector of the transatlantic track—and which on
this particular flight was designated as a line about five hundred
miles long, an hour or so of flying between longitudes 25 degrees
west and 44 degrees west—where it would be quite impractical
to think of trying to make an alternate airport. Within this sector both Keflavik and Narsarsuaq would be more distant than the
airports either behind or ahead on the destination continent. The
only way out of a serious problem here would be to head straight
on, to keep calm, appear unruffled, pray if so inclined, and hope.
This few hundred miles is by far the most risky part of any North
Atlantic transit—and for the pilots who cross it, it is the part where
any disrespect and disdain for the ocean below fades away, where
world-weariness becomes a secondary issue, and where awe for

the vastness and unforgiving hostility of the sea beneath becomes a firm and intractable reality.

But as it happens—and mercifully it seldom happens otherwise—there was no problem whatsoever on my crossing that day. There had been little by way of turbulence or unanticipated diversion en route; the descent was as normal as the takeoff had been; the plane arrived in Kennedy Airport precisely on time; and when I mentioned to the pilot in the baggage area that I had been a little nervous crossing through the dead zone, as I called it, he laughed and said simply, it's just the place where *we have to keep on our toes.*

2. FOULING THE NEST

Yet if we return to the original point—that the casual public acceptance of transoceanic air travel has dulled us to the wonders and the beauties and the preciousness of the seas below—it is not simply the pilots in flight who need to keep on their toes. The world at large is now having to keep super-aware of the implications of flight as well, and for an entirely different reason. Aircraft in flight are dirty and fuel-hungry monsters, and because there now are so very many of them—currently some twenty thousand big commercial jets, carrying 2,200 million passengers around the world (and 100 million of them across the Atlantic Ocean) each year—the damage they appear to be doing to the fragile shroud of the earth's atmosphere, and by extension to the seas, is said by many students of the environment to be very grave indeed.

As they soar across the oceans, seven miles high and serenely beautiful to see, the planes may be out of intimate touch with land and flying all alone in the sky—but they are also leaving behind them long trails of apparently harmful gases and gray fogs of polluting particles. The Jet-A kerosene these airplanes burn emits huge quantities of the very greenhouse gases that are believed by many to contribute to the warming of the planet, mostly either carbon dioxide and oxides of nitrogen (which can increase the production of ozone in the upper atmosphere); as they whoosh steadily past, the airliners also gush out great quantities of soot and sulfates and, deceptive in their pure-white loveliness, damaging trails of condensed water vapor, too.

The amounts involved are quite remarkable. A fully laden Boeing 777 traveling from London to New York will—if burning present-day fuels—stream out fully seventy tons of carbon dioxide. A big old 747 jumbo jet, now something of an aeronautical dinosaur, will spew out 540,000 tons of carbon dioxide each year if it is exclusively employed ferrying tourists between London and Miami. Multiplying the average exhaust tonnages by the 475,000 Atlantic crossings of varying distances—a journey from Rio to Frankfurt is clearly a lot longer and more polluting than a hop between Shannon and Halifax—and given the variety of different aircraft that are recorded every year, the ocean sees more than thirty-three million tons of plane-made carbon created in its skies every year. Every one of the three hundred passengers on my flight that January afternoon bore the burden of having poured two hundred pounds of carbon emissions into the upper atmosphere. I might as well have driven across the sea, on my own, in a car made for four.

However, there are efforts ongoing to make such travel both more efficient and more carbon neutral. Engines are being more cleverly designed and planes are becoming lighter (the new and much-delayed Boeing 787 Dreamliner is half made of carbon fiber, for instance, has super-efficient engines, and is said to be able to fly long routes on a fifth less fuel than current commercial jets). There is also a great deal of research into biologically based fuels made from plants and living creatures that themselves use up in their growing the very same carbon dioxide the jets spew out when flying. If an aircraft can establish a balance between the two, between its own CO_2 output and CO_2 absorption in the fuel-growing meadows, then carbon neutrality is achieved and the owner of the aircraft—the airline, in most cases—can claim to be green, or environmentally responsible.

As a result of the new interest in preventing or severely limiting anthropogenic climate change—if indeed such a phenomenon exists, which a small number of entirely sensible scientists are still unable to accept—a lexicon of strange and exotic new words are being uttered abroad: *jatropha, camelina, babassu,* and *halophytes*—all of them plants that currently are of little use to man or animals (jatropha is poisonous to both), grow happily in marginally useful areas like near-deserts and salt marshes, and hungrily absorb carbon dioxide by the ton and produce, when pressed hard in special machines, large quantities of flammable oils.

Airlines—Japan Airlines and Virgin Atlantic being the pioneers, with the latter testing on routes over the eastern Atlantic and the North Sea—have adapted some of their aircraft engines to use the new experimental biofuels, though usually flying with only one adapted engine on a four-engine plane, for safety's sake. The initial reports said that the engines did indeed work, that they would restart if shut down (one early fear was that they might not), and that the fuels did not freeze at high altitude (which was another). Some airlines say that so-called green fuels could be employed in passenger flights by 2015; Friends of the Earth and Greenpeace promptly said they were skeptical, asserting that the only way to cut climate-threatening carbon emissions was by slowing the ever-growing phenomenon of mass flying, and to do so by at the very least imposing a massive tax on thus-far-untaxed aviation fuel.

But precious few current forms of mass human transport, or the mass carrying of cargo, can be entirely blameless when it comes to the current sin of producing carbon emissions—and not least among these, now that the world has almost entirely abandoned sail for motive power, is the shipping industry. Ships are every bit as dirty and fuel-hungry as airplanes—the surface of the Atlantic Ocean, which is even more crowded than the airways above, contributes in no small measure to the problem, too. A figure released in 2007 both by the oil company BP and by a German physics institute* suggested that the funnels of the world's entire fleet of some seventy thousand fuel-burning cargo and passenger ships pour more carbon dioxide into the atmosphere than is currently produced by every single country in Africa combined.

The head of a research group studying the effects of shipping on the environment, Dr. Veronika Eyring, has used sensors aboard the European satellite Envisat, which was launched in 2002, to plot the visible lines of clouds that mark the passage of long-haul cargo ships. The high-speed winds in the upper atmosphere make sure that condensation trails laid by aircraft are dissipated within moments of their creation. The same cannot be said of ships, however: the enormous quantities of sulfur-laden soot and other particles found in the exhaust that pours upward from ships' smokestacks—and which continues to rise, since it is so much warmer than the ambient air—has been shown in recent years to create lines of low-level clouds that can linger in the atmosphere for weeks and months.

---

* The Institute for Atmospheric Physics near Munich, in a project called SeaKLIM, carried out in conjuction with the University of Bremen.

Seen from space they are known, somewhat unimaginatively, as ship tracks, easily spotted by satellites—major east-west lines of persistent and seemingly nonweather-related clouds that stand visible in the North Atlantic, easily matched to the actual paths of shipping below. There are other tracks visible in the eastern Atlantic, running from the bulge of West Africa down to the Cape. There is an especially prominent line running between Sri Lanka and the Strait of Malacca. There is another that snakes sinuously between the great port cities of Singapore and Hong Kong.

The uncanny permanence to these clouds is due to the constant replenishment by more ships steaming beneath them—most cargo ships, even those out in mid-ocean, generally stick to accepted shipping lanes, the better to take advantage of winds and currents and in acceptance of the mathematical realities of great-circle navigation. Two of the ten main sensors aboard Envisat have proved exceptionally useful. One, known as the Advanced Along-Track Scanning Radiometer, has produced maps of the oceans showing the curious matches between low-level cloud patterns and the known habitual pathways used by cargo ships; an immensely sophisticated spectrometer, the SCIAMACHY* device, has managed to break down the emissions patterns, visible and unseen, both in scale and chemistry. The results are impressive: according to SCIAMACHY, by way of Dr. Eyring's work, the thousands of ships' engines that turn the screws of great cargo vessels around the world produce eight hundred million tons of carbon emissions annually—approaching 3 percent of all carbon emissions produced by humankind. The figure for ships' emissions happens to be almost exactly the

---

* It stands for SCanning Imaging Absorpion spectroMeter for Atmospheric CHartographY.

same as the amount of carbon laid down by aircraft—the two thus adding up to almost 6 percent of total anthropogenic carbon.

So, in addition to dealing with polluting aircraft, there are plans afoot today to make ships a great deal more efficient and environmentally acceptable, too—by all means available, except of course by cutting their numbers, something restless modern man seems incapable of doing.

One of the most effective early ways of bringing order to what was a historically ramshackle industry—an industry unchanged in its operating principles since the Phoenicians loaded murex shells in Mogador three thousand years before and shipped them back to Tyre—was that taken in the mid-1950s, when an American trucking executive named Malcom McLean hit on the idea of packing cargoes into enormous steel boxes—shipping containers. Up until then, cargoes—whether they were bags of potatoes, bales of cotton, bottles of whisky, motorcars, or machine guns—had all been loaded deep into a ship's hold by cranes, then stacked as best as their shapes and sizes would permit, by gangs of expensive, often corrupt, and rigorously unionized stevedores—in the kinds of scenes so memorably recorded by Elia Kazan in *On the Waterfront*.

The advantage of using standard-sized containers, twenty or forty feet long, and into which makers and merchants packed their own goods at the factory or the farm, was that the boxes could be put onto trucks or railroad flatcars, taken to the dockside, and loaded swiftly by specially made cranes onto the upper parts of a waiting ship as well as down in the holds. They could then be shipped to a faraway port and, without once having been opened or tinkered with or touched by interfering human hand, could be unloaded and placed on another set of trucks or railroad flatcars and taken off to the distant destination. This was the birth of what was to be called intermodal shipping, whereby a floating vessel—a ship—became just one part of a long chain of types of transport that with brutal efficiency and economy would henceforward move products from all points in the world to all others.

It was a development that may have reduced costs and enhanced efficiency—but at a stroke it also stripped ocean trading of all its remaining romance and allure. Container ships—and they are now by far the biggest vessels in the world; the biggest of all at the time of writing, the Danish MV *Emma Maersk*, weighs in at 170,000 tons and can carry fifteen thousand containers at speeds of up to thirty-one knots—must be among the ugliest of man-made creations since Le Corbusier's public housing projects. Those who retain a fondness for clipper ships, for quinquiremes, or even for dirty British coasters long rue the day that these boxy monstrosities, which must be among the most familiar emblems of today's globalized world, were ever invented. But Malcom McLean—who tried his first ship out in the Atlantic Ocean, in April 1956, running a converted U.S. Navy tanker, the *Ideal-X*, from Newark to Houston, with fifty-eight containers—knew that in the shipping industry time was everything and money was everything, and that to load a ton of cargo by hand cost nearly six dollars, while to do so on a containerized ship—indeed, on the *Ideal-X* itself, that late spring day—cost only sixteen cents. Romance may have gone out the window in a single instant, but so too did the stevedore go the same way as the supercargo, the hold vanished with the fo'c'sle, the shipping business transformed overnight from a business that involved tides and winds and gulls and sextants and signal flags and the smells of tar and sea-wet rope, into a universe of slickly oiled machines, of GPS-made, computer-calculated navigation courses, and loading cranes programmed by machine and timed to the millisecond.

McLean, whose first company was called Pan-Atlantic Steamships—and who later sold it to a tobacco company and then to
a railroad firm and in the end to Maersk, who now has a fleet
of seven of the biggest ships ever built—and who died in 2001
having accumulated unimaginable wealth—created with his
containers something that changed the world's view of the sea
forever. The containerization of the shipping industry simply
grew and grew, without particular concern for the pollution its
ever-enlarging global fleet of vessels was causing. Nowadays,
with the data from German researchers and others beginning
to impinge on the consciences of shipping company executives,
in much the same way as airline companies and plane makers
are also realizing the consequences of what they do, research is

*The world-changing idea of placing marine cargoes into same-size steel
boxes, and creating the so-called container ship, belongs unequivocally to
a former truck driver from North Carolina, Malcom McLean.*

under way to find better and cleaner fuels; and other means of hauling ships around the world's bodies of water. New rules have recently been put in place, in both the Baltic and the North Sea, limiting the amount of sulfur in marine diesel fuel, in the hopes of cutting pollution and lessening the possibility that the satellites will be able to spot the ship tracks from the clouds the vessels leave above them.

There are also some brand-new ideas. One that has gained some traction, literally, is to have an immense sail, or a spinnaker-like kite, that can be played out ahead of a great cargo vessel when the wind is right, to help tow her along even though her engines might be cut off. A German firm equipped a bulk carrier, the *Beluga SkySails,* with a computer-controlled kite-sail and in January 2008 conducted the first test sail from Bremerhaven to the coal port of Guanta in Venezuela—sticking to the tradition that almost all test runs of every kind of new maritime technology are made in the ocean where they will likely be most used: the Atlantic.

. . .

But it will be some long while before sails take over from bunker oil, and before babassu-fueled aircraft soar between the transatlantic cities. The degradation of the air above our ocean, because of our perceived need to fly and steam across it, will continue, will be remarked on as just one of the more egregious examples of modern man's weary disregard for a sea he once revered. Yet the ruin goes very much deeper than this, and literally. The visible surface of the sea, its waters shallow and deep, the creatures that live within it, and all of the seabeds coastal and mid-oceanic, have also suffered poisoning, not so much from aircraft and steamships but from the vast wafts of polluting residues that are being endlessly created in millions of factories on land.

Rachel Carson first fretted about an impending maritime catastrophe in 1960, when she wrote the preface to a new edition of her first classic work, *The Sea Around Us*, which was initially published in 1951. This may not have been the book that established her saintly reputation—*Silent Spring* accomplished that in 1962, and made her the birth mother to today's environmental movement—but it did offer the world good reason for displaying a reverence and respect toward our oceans.

If any one person can rightly be given credit for beginning the current environmental movement, it is probably the American civil servant and marine biologist Rachel Carson, whose two most famous books were The Sea Around Us *and* Silent Spring.

The first edition was a lyrical work, poignant in its innocence, adoring in its tone, never once supposing that mankind had any kind of malign intentions toward the seas, and indeed arguing powerfully for a sedulous exploitation of the mineral wealth beneath them. There is great charm, particularly, in her explanations of the world's steadily rising temperature—this was fully evident in the 1950s with much the same phenomena as today: shrinking ice caps, retreating glaciers, violent and unpredictable storms. Rachel Carson was most impressed with the theories of a little-remembered Swedish oceanographer named Otto Pettersson, who declared that all cycles of global warming have been accompanied by anecdotal evidence of great swells of deep ocean tides: he believed that "moving mountains of unseen water" beneath the sea caused "startling and unusual occurrences" in the climate experienced on earth. There was not even a hint, either by Pettersson or by Rachel Carson, that mankind had anything to do with the alteration to the climate; it was either the tides or the untoward effect of rashes of sunspots.

But that was in 1950; a decade later, although she offered no new theories for the continuing rise in global temperature, Carson did start to worry out loud about marine pollution—and in particular, this being the early morning of the Atomic Age, the pollution of the seas with radioactive materials.

The power of her prose remains undimmed. For as she wrote in her deservedly famous Preface:

> *Although man's record as the steward of the natural resources of the earth has been a discouraging one, there has long been a certain comfort in the belief that the sea, at least, was inviolate, beyond man's ability to change and to despoil. But this belief, unfortunately, had proved to be naïve. In unlocking the secrets of the atom, modern man has found himself confronted with a frightening problem—what to do with the most dangerous materials that have ever existed in all the earth's history, the by-products of atomic fission. . . .*
>
> *. . . by its very vastness and its seeming remoteness, the sea has invited the attention of those who have the problem of disposal, and with very little discussion and almost no public*

notice, the sea has been selected as a "natural" burying place
for the contaminated rubbish. . . .

To dispose first and investigate later is an invitation to
disaster, for once radioactive elements have been deposited at
sea they are irretrievable. The mistakes that are made now are
made for all time.

It is a curious situation that the sea, from which life first
arose, should now be threatened by the activities of one form
of that life. But the sea, though changed in a sinister way, will
continue to exist: the threat is rather to life itself.

She was so very prescient. The British government turned
out to be just one of the sea-besmirching villains among many—
twelve nuclear countries, according to the International Atomic
Energy Agency, did much the same thing.* Until the late 1970s
ships chartered by the British government casually tossed radioactive waste—from atomic weapons programs, from power
stations, from research projects, into a sea that, as Rachel Carson
noted, was widely regarded as "beyond man's ability to change
and to despoil." More than twenty-nine thousand tons of "highly
active radioactive waste," created mainly by the U.K. Ministry of
Defence, was dropped into the Atlantic at a specially selected site
four hundred miles west of Land's End, into waters that reached
the supposedly safe depth of nine thousand feet.

The amount of radiation emitted by the material left in what
was called Atlantic Deep, estimated at 4,000 curies of alpha activity and 117,000 of beta-gamma activity, was enormous. Lon-

---

* Most notable among the nuclear sea dumpers were the Russians, who hefted entire re-
actors into the sea, scrapped and sank atomic submarines, wrecked in remote bays legions
of barges with nuclear munitions aboard, and placed thousands of tons of power-station
waste into the ocean—mostly in Arctic waters near Novaya Zemlya. The Japanese government was also disquieted by reports saying that nuclear material was dumped at sea in the
North Pacific, near Sakhalin Island.

don did its best to soothe its alarmed citizens—especially those in Cornwall, Devon, and South Wales, close to where Atlantic tides might wash any errant material ashore—by saying that "dispersion and dilution" should ensure there was no danger, and that anyway, everything tossed overboard had been securely encased in steel barrels lined with cement. But the official balms soothed few, and it hardly helped when the government admitted soon afterward that it had actually dumped another sixteen thousand tons of only slightly less dangerous material in another zone, the Hurd Deep, not too far away in the English Channel, and that some had also been dropped into the Irish Sea and into waters off Scotland, ensuring that this isotopic gift will keep on giving for many hundreds of thousands of years.

Rachel Carson had ample reason to fear radioactive pollution; but she was blissfully unaware back then of the other substances that would come to infest the seas—not even, in those innocent times, of the weed killers that *Silent Spring* so comprehensively managed to ban from the land.

It was all so much simpler then. No doubt, like many who visited the seashore in the 1950s and '60s, she would have cursed the gobbets of tar from ships that washed their tanks off shore, and she would have been vexed at the broken floats and rotten netting that washed up among the piles of Atlantic kelp. She knew her beloved ocean was far from being perfectly clean, but its contamination had a sort of understandable ordinariness about it, tainted by a forgivable kind of pollution, of the kind you might come across in a farmyard, a wine cellar, or an auto mechanic's garage.

She had little inkling of the sinister periodic table of foul chemistry that was then to come—of the mercury that would soon be found in the flesh of almost every tuna, shark, and swordfish; of the hundreds of thousands of tons of highly toxic, highly carcinogenic polychlorinated biphenyls—PCBs—that soon found their way into the sea, killing seabirds hundreds of miles out in the ocean, contaminating seashores, shellfish, and finned fish; of the plastics that would foul beaches and entrap fish and fill the stomachs of seabirds; of the cyanide washes from gold-processing plants; of the oil pollutants from tanker accidents, shipwrecks, or drilling mishaps; or the enormous pharmaco-poeia—hormones and psychotropic drugs, antidepressants and sleep-inducing and sleep-retarding cocktails—that would slowly and steadily defy the long-standing belief in the ocean's near-infinite capacity to dilute and dissipate. This, as Rachel Carson so wisely saw, was naïveté in the extreme: the ocean was soon determined to be not so much a machine for diluting chemicals as rather a vehicle for transporting them around the planet, either by way of its waters, or by way of the fish and other creatures that live in them.

Pollution of the once-pristine and now more dirty and all-too-finite ocean is universally agreed to be a terrible thing, and of late a slew of international laws—most notably the so-called London Convention of 1972—have been set in place to direct that those who use the seas, and those whose countries border them, respect both their sanctity and their common value to the planet.

3. THE CONSEQUENCES OF GREED

Yet marine pollution by itself is not the greatest and most lasting problem that faces an ocean like the Atlantic. The sea does have a limited capacity to clean and reform itself. The creatures that live in it, on the other hand, do not. And mankind's ever-growing need for fish and other living marine animals is currently pushing one of the most fragile resources of the sea close to the breaking point.

To accommodate an almost insatiable human appetite for seafood, we are these days wantonly overfishing our seas; as a result, and astonishingly to most, we are now fast running out of fish.

A small example of just how sensitive this matter has lately become occurred for me early in the autumn of 2009, when I encountered quite by happenstance a trivial and avoidable, but, as it happens, rather interesting and very public controversy.

I had flown into London from New York on a daytime flight and arrived late in the evening. I checked my bags with the porter where I was staying, on Pall Mall, at around 10 P.M. It was a Saturday night. I was hungry and assumed I would have quite a hard time at that hour finding someplace halfway decent to eat. I strolled up through Leicester Square and into Covent Garden, walking past countless cafés and bistros, most with waiting clients spilling out into the street. And then, halfway along an alleyway, I came to J. Sheekey, a newly spruced-up edition of the seafood restaurant that I remember my parents taking me to when I was about ten, back in the 1950s. So fashionable is Sheekey's these days that I imagined it would be quite impossible to get a table, at least not without a long wait, and so I started to walk on by. Except that on a whim I turned back and ventured in, fully expecting disappointment.

Quite the contrary. The staff, looking surprised, caught unawares when the street door opened, seemed strangely relieved to see me. Their restaurant, it turned out, had tables still available. And so, unexpectedly, I was quickly seated, my glass was filled, my order taken, plates and dishes fetched and brought and cleared—and so it was that at about midnight, replete with a dozen oysters and a plate of whitebait, a fair-sized piece of sea bass, with a small dish of fennel and some new potatoes, a half bottle of Pouilly-Fumé and a cup of coffee, I strolled back down to the club. I felt good, pleasantly surprised that London, for so long a city of laughable gastronomic impoverishment, was now managing to look after its visitors so very well.

Except that when I read the papers a few days later, it turned out that there was an explanation. A few days before my visit, J. Sheekey had been publicly flayed in the newspapers for allegedly serving to its customers fish that were on a generally agreed list of overfished and consequently endangered species.

There had been a sudden flurry of interest in troubled fish and ocean fishing among concerned and sophisticated Londoners just a few days before. A documentary film shown on television had just exposed techniques that were said to be cruel, fishing that was said to be illegal, fish that were as a consequence heading for extinction, and the large population of shops, supermarkets, restaurateurs, cooks, customers, and diners who either did not know or did not care that by buying and eating such fish, they were contributing to the decline. A website had also appeared, publishing online the list of troubled fish, and the shops and restaurants that sold them—and among them was J. Sheekey, venerable maybe, but now quite publicly revealed and shamed. It is a moderately costly restaurant, the clientele by and large people who would be wanting to be seen to do the right thing— and so in droves, having seen, read, and clicked on the alarming reports, the customers stayed away. By doing so they left empty plenty of tables, and as it happened, they did so on the very night of this unanticipated visit from an innocent abroad.

But by chance, this was not to be the end of the story. The owners of Sheekey, a powerful omnium-gatherum of other fashionable London restaurants, formally complained, saying they were in fact most scrupulous in the kind of fish they sold, served seafood only from sustainable stocks, and that the website had its facts wrong. There was an unusual pause, a drawing in of breath. Environmental groups tend to have a somewhat

saintly air—verging on the sanctimonious, in a few cases—and
most are aware that they have to be extremely careful in their
various assertions of blame. An alarmed fish protection lobby
promptly huddled, and after some hesitation conceded that they
had in fact been somewhat hasty, had indeed got some their facts
wrong. They appeared chastened. They apologized—if a little
reluctantly—and they promptly restored Sheekey's to the pantheon of the blessed. Crowds of relieved ichthyophiles happily
returned, with the result that it is now all but impossible to get a
table again, especially late on a Saturday night.

It was a sorry little row, but it was one that served to point
up a reality that had until then been generally overlooked:
that many kinds of fish around the world are indeed in serious
trouble, and that it is all the fault of the endless desire for culi-
nary pleasure that currently afflicts Western mankind. We buy
or we order our food—especially our seafood, which because it
is seldom seen *in situ* and is much more of a mystery than our
meat, which tends to graze and gambol before our eyes—without paying too much attention to its origins, to the manner in
which it was caught, or to how long the populations of the fish
we favor can each be sustained. Until lately, many restaurants
were reluctant to offer much by way of information to those few
who were concerned.

Not that there is universal agreement on how good that information is. There are a large number of bodies that seek to protect
and preserve oceans and oceanic life: the Blue Ocean Institute,
the World Wildlife Fund, Sea Shepherd, the National Audubon
Society, the Monterey Bay Aquarium, the Alaska Oceans Foundation, SeaWeb, the Natural Resources Defense Council, and the
National Environmental Trust among them, all with their own
agendas and working methods, sometimes working in concert,
more often not. One can now acquire (from the Monterey Bay

Aquarium, among others) wallet-sized cards that tell you which fish it is currently prudent to eat; some of the better restaurants will identify the fisheries from which their offerings are hauled.

Different approaches are a commonplace within the environmental establishment. The Marine Stewardship Council (MSC), which was established in Britain in 1999, was an early science-based champion of sustainable fishing. It established a set of principles under which it could certify fisheries as being responsible and sustainable and thus recommended to customers: a blue and white oval logo is nowadays fixed (for a fee) to packages of fish that come from these fisheries—which currently make up about 7 percent of the world's fisheries, including, in the Atlantic, those for South African hake, Thames herring, and (as we shall see later) the inelegantly named South Georgia version of the Patagonian toothfish.

The principle underlying MSC's approach is based on promoting what it regards as "good" fish. Many American organizations, on the other hand, do their best to organize boycotts of what they consider "bad" fish (as in the campaign by the National Environmental Trust* to "take a pass, on Chilean sea bass"). Hence the "red list," which Greenpeace unveiled in 2009. This is a compendium of what it considers the most endangered fish, crustaceans, and shellfish: it contains, at the time of writing, twenty-two species, or groups of species. Eighteen of these are to be found in the Atlantic Ocean, and their endangerment stems almost entirely either from relentless overfishing or from cruelly thoughtless kinds of fishing undertaken within the boundaries of the Atlantic.

Chilean sea bass—the marketers' adroitly chosen name for the less comely-sounding Patagonian toothfish—is on the list,

---

* In 2008 the NET, based in Philadelphia, was absorbed into the Pew Environmental Group, one of many charitable bodies founded and funded by the children of the founder of the Sun Oil Company.

but is generally found off Chile's coast, in the Pacific, or else in Antarctic waters. Hoki, which without much public awareness constitutes the great proportion of fish sold by McDonald's restaurants worldwide, is also regarded as endangered, and is a small, pale-colored creature generally found off New Zealand. Pollock is usually found and fished in Alaska (the MSC regards the Alaskan pollock fishery as worthy of its seal of approval, yet it is on the Greenpeace red list, an indication of the differences to be found in this complex and controversy-ridden marine universe). And swordfish, generally caught by the much-criticized method of *long-lining*, are mainly denizens of the Pacific.

The rest of the overfished majority are found foursquare in the Atlantic Ocean: most of the fisheries for Atlantic cod, Atlantic halibut, Atlantic salmon, and Atlantic sea scallops; the albacore tuna from the South Atlantic; the bigeye, the yellowfin, and especially the magnificent, superfast, and much-valued bluefin tuna (which can command thirty thousand dollars apiece in the famous Tsukiji market in Tokyo, and is partly because of Japanese demand the most threatened grand fish of the entire Atlantic Ocean); the Greenland halibut, the North Atlantic monkfish, the bivalve known as the ocean quahog, the redfish, the tropical red snapper, most skates, most tropical shrimp found off the west coast of Africa, and the fish now delightfully known as the orange roughy, but which, before the marketers got hold of it, was known to fishermen and biologists simply as the slimehead—all these are found between Greenland and Tierra del Fuego, between Cape Town and North Cape, in the depth and shallows, warm and cold, somewhere in the hundreds of thousands of cubic miles of Atlantic waters.

Twice have I encountered the practical realities of the Atlantic's fishing crisis, once in the northwest Atlantic, then more recently in the deep sub-Antarctic South.

4. NORTH

My first encounter was well up north, off Newfoundland, where there was no specific villain other than the ineptitude of mankind in general, which in the early 1990s all but destroyed one of the great fisheries of the planet. The story of the collapse of the Grand Banks cod fishery, which I came across in the late 1990s, in a heartbreakingly beautiful but sad clutch of little communities gathered along the shores of Bonavista Bay, is a sorry tale indeed.

In the abstract, the expanse of shallow seas off Newfoundland—seas that were always portrayed, and rightly so, as rough, cold, swathed in fog, invaded by stray chunks of jagged ice and with storms so terrific and the seafloor rocks so close to the surface that the place was often lethally dangerous—long had a legendary magnificence about them. History books told us of John Cabot, who found the great silvery codfish in such abundance in these waters that he wrote that to catch them one could forget the net or the hook: a simple basket tossed from the gunwales would be filled with fish in a minute, and a mighty cod, knocked quickly insensible with a marlinspike, would be grilling on deck a minute after that. Never before had any seas anywhere in the world been so richly endowed with fish; it seemed entirely credible that oarsmen would complain that Newfoundland sea were difficult to row through, so heaving were the waters with fish; and it truly did seem possible, as others imagined, that you could probably walk from London to St. John's on the shining muscular backs of millions of cod.

The reality was not much less inspiring. I saw the Grand Banks in 1963, when making my first voyage across the Atlantic. When our *Empress of Britain* stopped briefly there, to rendezvous with an aircraft on a shallow eastern outlier of the Banks known as the Flemish Cap, the sea was at first disappointingly calm and the weather uncharacteristically clear. All changed once we got under way and a few hours later slid west onto the Grand Banks proper. We had only to cross the Banks' most easterly point, the fish-rich grounds known as *the Nose*, and the fog closed in on cue, the water became unpleasantly lumpy, and we had to ease down to a crawl for fear of colliding with any of a thicket of fishing vessels, or cutting across their nets.

The fog in these parts renders the sea curiously quiet, and I remember standing up on deck, and later out on the bridgewings, matted with moisture and shivering with cold, watching—and listening. There was the slap of the swell against our hull, the soft hiss of the bow slicing through the waves. But most noticeable were the cries and yelps of a score of foghorns, a fishermen's chorus that swelled loudly, I assumed, in those places where the cod were being found that day, and then fading, and swelling again, until finally a diminuendo, as they ebbed slowly away to nothing, and we eventually steamed off the shoals and to the south of Newfoundland, and then into the deep and relatively codless waters of the Gulf of St. Lawrence.

The blanket of mist that day was such that I never saw a fishing vessel—and such images I had of the life of a Grand Banks fisherman came most probably from reading Kipling, and *Captains Courageous*, and later, most memorably, in the 1937 film of the book, which the BBC showed on winter afternoons, and which, in a scene I seem to remember well, had both Spencer Tracy and Freddie Bartholomew fighting to stay afloat in one of the alarmingly unstable little dories that cod fishermen used to go after their prize.

That film helped it all come together for me. First there were the sleekly graceful schooners, racing up from the Massachusetts ports—back then the Americans were just as able to fish on the Grand Banks as were the Canadians, whose waters these seemed to be; the Treaty of Paris had long permitted it. Then there were the encounters with the fogs, the storms, followed by the first sightings of the shoals of tiny capelin and herring, then with the ponderously moving whales, and finally the fleet's eventual arrival at the cod grounds themselves—where they had also met up with the rougher and tougher French Canadian dorymen who had come out from St. John's and outports like Trinity and Petty Harbor and Bonavista. Then there was the dropping of the dories, no matter the weather or the height of the waves, followed by the long, wet, desperate and tiring hunt from them for the cod that lurked close to the bottom, just a few shallow feet below.

From the smoky comfort of a London cinema, it all seemed unimaginably tough and difficult. The dories were only twenty feet long, and though their prows and sterns rode high above the waves, they were designed with almost no freeboard amidships, so that their owners would find it a little easier to haul in such cod as they hooked on their lines; but then water kept crashing over the side, and whichever man was not straining at the oars always seemed to be bailing, or trying to pour cold water from his sea boots, or shuddering as yet another wave crashed down his neck and the gales blew off his sou'wester. Or else, of course, he was fishing: either hand-lining himself, or helping to pull up a longline that, between the barrels they had used as floats, might stretch five miles long across the sea and might hold a thousand hooks—each of which in the old, rich days of the fishery, might have a massive cod attached, and which needed to be brought up, freed from the barb, and slapped down into the bilge of the boat, where it joined its companions wriggling and writhing between your feet en masse.

You might eventually try to return to your schooner with a full ton of these codfish, a hundred fish, each one maybe twenty pounds, each with a huge gaping mouth, a small goatee dangling from its lower lip, an olive green back, a pale belly, and a long go-fast stripe of white along its side. Newfoundland cod, fat and heavy with a kitchen-ready amassment of succulent innards, are said by fishermen to be the prettiest of all the family Gadidae. The sight of a returning dory filled to the gunwales with them seemed for decades a most potent symbol of the enormous riches of the North Atlantic, a very visible reason for the prosperity of those who lived beside it and were fed by it.

But the mechanics of returning to your schooner in a tiny low-slung boat filled with these fish turned out to be a spectacularly difficult exercise. Even finding your ship was a challenge, especially if you had been away for hours, or maybe longer, and if during your absence the weather had closed in. Even the most powerful lantern hung from the schooner's fo'c'sle—as it was in the Kipling film, where Lionel Barrymore had hung as powerful a lamp as he could to help his men find the schooner, tantalizingly named *We're Here*—could be glimpsed from no more than a hundred feet in a thin fog, as little as five feet in a pea-souper. Then only the back-and-forth piping of the foghorns, yours and your skipper's, stood a chance of bringing you to home.

Moreover, a dory filled with fish lies even deeper in the water than usual, and the seas slopping over sides that were now almost underwater would make the craft ever less stable. Small wonder so many seamen died—in the last seventy years of the nineteenth century, 3,800 Gloucester fishermen were killed, and that from a town of only fifteen thousand—such dorymen as lived to tell went on to enjoy a camaraderie and a sense of shared pride like few other workingmen anywhere. To be a Grand Banks cod fisherman was a noble art, and only the bravest could do it.

And when they came home to port, all the bars in all the seashore towns came to know this all too well.

But then in the 1950s came the factory ships, and in an instant the picture changed.

Already the technology of fishing had been improving mightily. Hand-lining was a technique employed by only a minority of fishermen: more controversial methods like long-lining, or setting the near-invisible gossamers of floating gill nets, or even trawling along the ocean floor where the cod lived, had all hugely increased the catch. Everyone had long been happy with the Grand Banks. As more and more fishermen arrived, all was just as it had been when John Cabot came by in the *Matthew*; the world soon came to believe what he had said, and what the Basques had found in the decades following—that there was abundant fish for everyone, that for every fish caught two more seemed to spawn; prosperity for the fishermen and freedom from want for those millions who dined on fish was likely to remain an eternal reality. There were a few—which included many of the older fishermen in the Newfoundland outports, who said they *knew* their fish and their habits, and knew what was fair to take from them—who fretted that it might one day be possible to fish the stocks completely into oblivion, that disaster lurked. They were smiled at indulgently and told not to worry: the Grand Banks were a source of goodness and delight for all, and for all time.

But then had come the shipborne steam engine and Mr. Birdseye's techniques of freezing fish, then came the fish stick or what in Europe is called the fish finger and the convenience food market, and then was born the idea that fish need not to be brought to land to be processed and filleted and frozen and boxed and labeled, and that all of this could be done afloat, by a big ship that was not truly a fishing boat at all but a floating steam-engine-powered production line for the twenty-four-hour-a-day disassembly of fish and their twenty-four-hour reassembly into convenience food—and all of a sudden long-lining and gill-netting and trawling seemed the least of the challenges that an ocean fishery might face. Now it became a question of simple arithmetic: with the arrival of the factory ships, the amount of fish being removed from the Grand Banks in the 1960s became suddenly astronomical and was becoming plainly—and to use a word that in the 1960s started to ease into the lexicon, and then into the vernacular if not quite yet in vogue—*unsustainable.*

The abundance of codfish in the Atlantic is very much a thing of the past. This cheerful trawlerman was pictured in 1949 off the Lofoten Islands in northern Norway. Today's catches of Gadus morhua *are seldom so rich, nor are the fish themselves often so large.*

A ship called the *Fairtry*, launched in Scotland in 1954, was the first to start what some would call the mechanized strip-mining of the Grand Banks. Compared to the schooners and inshore fishing craft that had come before, she was enormous: 2,600 tons, and looking like a converted passenger ferry. She was also terrifyingly effective at what she was designed to do—the huge trawl net she dropped from a ramp in her stern had a mouth hundreds of feet around, and when it was pulled along the seafloor its weighted lower jaw scooped up every imaginable living thing in its path—hundreds, thousands of cod of varying ages, sexes, weights, and health, but also every other kind of bottom-feeding, bottom-living fish and crustacean, needed or not. All was sped into the bowels of this enormous ship; what was unwanted was dumped over the side; the rest was machined—filleted, salted, frozen-packed away—even as the trawl was down on the sea bottom again, hauling yet more hundreds of tons to the surface to be dealt with in the same brutally crisp manner.

The catches from this ship alone would have been astonishing. But then the Soviet fishing authorities heard of the revolution beginning, and being in the vanguard of a new Kremlin policy to distribute protein to the masses, built a fleet of similar ships, only larger still, and sent them to the Banks in the *Fairtry*'s wake. A vessel called the *Professor Baranov* was more than 450 feet long and could process two hundred tons of fish a day, all the while making frozen fish and fish meal, oil, ice, and water from its own distillation plant, and servicing up to twenty other Soviet trawlers that were lumbering across the Grand Banks like oxcarts pulling plows, scooping up even more fish than it was possible for John Cabot and all of the Basques who followed him ever to imagine.

The temptation for more proved irresistible. Within a season or two, just about everyone with a big enough net came to join the party. From the fish docks in East Germany and Korea, Cuba and Japan, dozens of lumbering and rusting strip-mine ships found their way across to the Nose, the Tail, to the Flemish Cap and to the Banks proper, and fished until they ran out of fuel and went off to bunker and carouse in St. John's. Those who lived in the fishing town of Bonavista said they could walk up to the statue of John Cabot, high on a nearby cape, and look east into the ocean and see what looked like a vast village—lights by the thousand—as the draggers, the fish factories and their trawlers, scoured the sea without surcease through every night and every day.

Fish factories that flew the flags of a dozen new countries elbowed out those who had been traditionally working the grounds for decades, and hidden in the fogs and the ferocious shallow-water storms, they settled down to work with ever more sophisticated technologies and the deployment of ever-larger trawls. The catch levels went up and up and up—until an eye-watering total of 810,000 tons of cod alone was hauled from the sandy sea-floor in 1968, the year when it all began to go badly wrong on the Newfoundland Grand Banks.

At this point the Canadian government decided something ought to be done. Too much was being taken from the fishing grounds, and for too long—a situation had been allowed to develop that simply could not go on. Government mathematicians determined that sometime between the mid-seventeenth and mid-eighteenth centuries—long enough for thirty generations of cod to have come and gone—some eight million tons of their kin had been taken each year, mainly by British, Spanish, and Portuguese hand-lining boats. But, said the same mathematicians, almost precisely that tonnage of fish had been taken during the first *fifteen* years of the factory-fishing bonanza—and, put plainly, taking eight million tons in fifteen years was the kind of figure that no fishery anywhere on the planet could possibly sustain.

A plan had to be put into operation—and with what in government terms is reasonable speed, it duly was. But though the intentions of the bureaucrats and politicians in far-off Ottawa might have been the very best, the manner in which the Canadian fishing policies of the subsequent twenty years were executed helped create a far greater disaster, and one from which few—fish, fishermen, or fishing communities alike—have ever fully recovered.

First of all, the Canadian government did what appeared eminently sensible: in 1977 it declared (in common with most of the rest of the world's coastal nations) that henceforward a two-hundred-mile-wide maritime belt off all of its coastlines would be regarded as its own Exclusive Economic Zone,* and that foreign fishing vessels would be excluded from working there. Canada's claim of jurisdiction meant that the awe-inspiring illuminated village of draggers that had been visible from Bonavista Cape—factory ships and trawlers from Murmansk, Fleetwood, Vigo, Lisbon, Pusan, and a score of other foreign ports, and which were operating as close as three miles from shore—had to leave. They could still fish beyond the new limit—which allowed them still to operate on the Nose and the Tail and on Flemish Cap—but not on the Banks.

And most of them sailed off into the sunset. The Spanish trawler fleets, squeezed by European quota rules, thought that the French territory of St. Pierre and Miquelon—the tiny euro-

---

* The modern concept of Exclusive Economic Zones (EEZs), when finally codified by the United Nations' 1982 Law of the Sea, has impressively shrunk the reality of the "high seas" on the Atlantic Ocean. For while the narrowest cinch of the ocean is that of about 1,700 miles between Ponte de Calcanhar, near Natal in Brazil, and Sherbro Island, off Sierra Leone, the two closest EEZs—between the Cape Verde Islands and Brazil's St. Peter and Paul Rocks, are separated by only seven hundred miles of truly unclaimed ocean. At its widest—EEZs included—the Atlantic high seas between Cape Town and Tierra del Fuego extend some 4,200 miles.

using, Gitanes-smoking, Calvados-drinking twin-island relic of one-time French colonial ambitions ten miles off the southern Newfoundland coast—might offer them sanctuary, and so continued to fish in the non-Canadian high-seas outer parts of the Banks. The Portuguese White Fleet—their fishing vessels still painted white, as they had to be during the Second World War to remind the German U-boats of their neutrality—did the same. Otherwise the seas emptied, and sea-bottom cod-dragging stopped.

The sudden silence that followed should by rights have given the populations of Grand Banks cod the time and opportunity to recover. For suddenly no one was doing any large-scale fishing there: no one because the Canadians who now had the sole rights to do so did not at the time have the wherewithal to fish, or certainly not as the Russians and the Koreans had done. They had neither the ships nor the will to strip-mine and vacuum their own seas as the foreigners had been doing.

However, governments had other plans. Both the federal government in Ottawa and the provincial government in St. John's decided they wanted to pump some life into the ever-sputtering economy of the country's poorest and newest province (Newfoundland had been an impoverished British possession until 1949, and since confederation with Canada had an economy that relied on little more than fish and wood pulp). In line with this vote-winning policy, they decided to start what politicians hoped would be a truly enormous Canadian-run, Canadian-owned, and Canadian-organized Atlantic fishing industry.

But then the government—and specifically a now-much-derided federal regulatory body, the Canadian Department of Fisheries and Oceans—came up with estimates of how much cod any new Canadian fleet might legitimately catch, and then managed to get these estimates wildly, almost incredibly, wrong.

They were much too high. Four hundred thousand tons of cod could be taken from the Grand Banks each year, the government said with great glee, and the newborn Canadian fishing industry, which was additionally tempted by generous government aid, not unsurprisingly took the bait. Canada's underused eastern shipyards promptly began a-welding and a-riveting, and launches down slipways became a sudden commonplace, and within a short while the departing Soviet draggers on the Banks were all replaced—by similar-sized, similarly equipped, and similarly aggressive fishing vessels that differed only in that they all flew from their jackstaffs the red maple leaf of Canada. And these boats were set to fishing offshore with a zeal and ambition encouraged by endlessly optimistic remarks from government, to the effect that plenty of fish were out there, and that the Canadian ships could carry away more or less as much of whatever species as they wished.

But it became clear soon afterward that these glowing estimates of fish stocks just *had* to have been inflated—whether through ineptitude or for corrupt or short-term political advantage no one has yet fully worked out. Some marine biologists at the time, and not a few local inshore fishermen, too, were quite certain of this and complained that calamity lay ahead—they even tried at one point to go to court and argue before the majesty of the law a case for exercising caution. But no one else was listening, and during the late 1970s and through most of the 1980s, a massive all-Canadian fishing jamboree broke out like never before.

Newfoundland became in comparative terms rich, prosperous, and now content to have been fully confederated with wise and prescient Canada. Its population was now universally happy, as the silver cod leaped in torrents out of the nearby Canadian ocean. Suddenly the old and much-maligned *Newfie* was being seen as an altogether different creature, now an admirable fellow with a fine work ethic and a newborn entrepreneurial spirit; and instead of endless miles of pine trees and sorry little centers of backwardness, Newfoundland found itself transformed, with brand-new seafood processing plants and enormous trucking operations and the opening of legions of marketing companies. This was now the face of modern, rich, better-late-than-never Newfoundland, the inhabitants suddenly being seen as a people blessed by good fortune. Someone joked that the new provincial motto should be "In Cod We Trust." An unstoppable juggernaut had been set in motion, and it seemed for a few heady years as if nothing could halt it.

But then the numbers started to decline. Early in the 1990s scientists began to publicize new figures showing that the number of cod being caught on the Banks was decreasing savagely, and that the number of cod that were spawning—a critical figure for the future—was going down as swiftly as a punctured balloon. The government, aware of the economic boom it had helped create for Newfoundland, tried to keep on smiling, telling all who would listen that all was well. In 1992 its own marine scientists, those who had gotten the numbers so badly wrong a decade before, suddenly sensed the consequences of their errors and suggested limiting the annual catch to no more than 125,000 tons. Politics then got in the way: ministers tried to placate the juggernaut by ignoring the figures and setting their own target at almost twice the level: 235,000 tons. Even this they saw as politically risky; officials had to explain that though the new suggested level might be a long way down from the wonderful 810,000 tons that had been caught back in 1968, it was actually no more than a measured reduction, a figure both sensible and prudent.

But far from being sensible and prudent, it was actually quite irrelevant—for during that early fishing season there came grim and unanticipated news from the sea: that try as they might, Newfoundland fishermen all of a sudden couldn't catch anywhere close to even a tenth of that tonnage of fish. And then it dawned: something terrible and unimaginable had happened. The cod, quite simply, had run out.

The trawlers went out, dropped their nets and cranked open the mouths, dragged for their allotted hours through the fishing ground, and pulled everything back up—and discovered that the trawls were coming up empty. The inshore fishermen sailed their little boats around inside the twelve-mile limit, baited and dropped their lines over the well-known fishing holes—and watched in dismay as their hooks came back up, clean and shiny and quite lacking in cod.

All of a sudden the truth hit everyone square on in the face. Everything that had taken place since that two-hundred-mile limit had been put in place and the foreigners had been thrown out was shown to have been no more than a wild party, with shots and snorts of unyieldingly bad numbers leading inevitably to the partygoers suffering all the symptoms of a really bad trip. It was a party that came to a crashing end much too soon, and the hangover started the moment the shutters came down.

And so the government, entirely stunned, had no option. It closed down the fishery. In June 1992, almost five centuries after John Cabot had told of a corner of the sea brimming with the most beautiful and edible of ocean fish, all of them had been caught by man, and the sea had been rendered quite barren. It was said Newfoundland's waters had maybe 1.5 million tons of spawning cod; now those remaining in the bays amounted to perhaps sixty thousand tons—essentially nothing. The seas are now just empty. The Grand Banks is now an ex-cod fishery.

And thus it has remained ever since. There have been experiments to restart the fishery, but they all eventually sputtered out. And as I soon found when I drove up along the Bonavista peninsula, stopping at outports such as Catalina, Port Rexton, Newman's Cove, Trinity, and the northerly town of Bonavista itself, where John Cabot's statue stands gazing out to sea—small amounts of cod can still be found in the bays and inlets all around. But fishermen are absolutely forbidden to take them—anyone caught with a codfish will be slapped with a heavy government fine. Some argue that allowing a catch of one ton per fisherman each year might make sense—but the government, perhaps in embarrassed recompense for having made all too many mistakes in the past, refuses.

Some of the processing plants have closed, or now work short shifts with such other fish as can be found and legally taken; some thirty thousand Newfoundlanders have been put out of work. There was a lassitude, a terrible sadness to the place—shuttered shops, boarded-up factories, padlocks on chain-link fences around plants that used to bustle with workers.

Blame for the collapse of the cod fishery is spread around liberally. Some in government blame the warming weather, which they admit no one can do anything about; others assert that the ever-hungry harp seal eats spawning cod, and since politicians can do something about that, many urge the eradication or culling of the harp seal colonies. Inshore fishermen blame the trawlers and the statisticians. Offshore fishermen are angry with the government for snuffing out their livelihood and offering them little in return—even though unemployment insurance payments in Newfoundland are generous, and critics suggest that the fishing industry in this corner of the Atlantic is all too generously subsidized and instead should be allowed to stand or fail alone.

But these arguments are all trivial compared with the one reality: that not so very long ago the northwest Atlantic Ocean yielded up a marvelous bounty—and mankind's greed and a fatal propensity for short-term thinking caused that bounty to disappear, most likely forever. An entire shoreside community has fallen victim, too. Whether that is the truly great tragedy, or whether the decimation of the population of cod on the Grand Banks is the matter for greater grief, is a question that goes to the heart of our relationship with the seas around us.

John Culliney, a marine biologist working in Hawaii, once remarked that the oceans, "the planet's last great living wilderness," perhaps present a frontier where man had "his last chance to prove himself a rational species." Here off Newfoundland, mankind's apparently utter dereliction of duty toward stewardship of the Atlantic suggests little reason for optimism.

5. SOUTH

And yet in the far south Atlantic, matters appear to be in rather better shape. The fishery created in 1993 in a huge area of British-administered sea around the island groups of South Georgia and South Sandwich—its 850,000 square miles making it the largest remaining part of what was once the formidably grand British Empire—is currently one of the most policed and efficient in the world. Most of the Chilean sea bass to be found on northern restaurant menus comes from there, most of it certified approvingly by the world's fish protection organizations.

In truth, like most people I had long been unaware of the simple existence of this body of British-run sea. At least that was the case until one February day in the early 1990s, when I had a most unexpected encounter, and heard tell a revelation of a most curious string of circumstances. But a little background is necessary.

In my university days in the 1960s, I briefly shared rooms with an exceptionally bright young man named Craig, who took a degree in classical Persian, achieving what by all accounts was a stellar first-class mark. He was promptly recruited by the British Foreign Office as a diplomat—and maybe given other duties by the clandestine services, too. He was sent, not unsurprisingly considering his linguistic skills, to various of Her Majesty's legations dotted around Southwest Asia. We remained friends, and from time to time I received letters and cards telling of his postings in places like Amman, Jeddah, Jerusalem, and Tehran. He told me once that he reasonably expected that, providing he did not blot his copybook, he could end up as British ambassador to Iran and cap his career with all the honors and decorations that so senior a British diplomat would likely accrue. I think I last heard from him in the mid-1980s, when he was in a branch of the Foreign Office that dealt with Palestine: his career seemed then to be entirely on track.

But then came one cloudless summer morning in February twenty years later, when, on a Russian cargo ship, I was steaming into the approaches to Port Stanley, the capital and main harbor for the Falkland Islands. I was on the bridge and by curious chance heard a call on the VHF radio's Channel 16, asking if I happened to be available.

The deputy governor of the Falkland Islands presents his compliments, squawked the caller—and would I care to join him for luncheon? Naturally I said yes, though I had not the faintest inkling of who either the governor or his deputy might be. Moments later a small launch appeared, manned by a pair of soldiers; they tied up by the ship's gangway, saluted me aboard, and we sped back to land, the colonial flag streaming behind us in the breeze.

It was Craig who was waiting at the quayside. He was heavily bearded now, looking a little older than I remembered, but he was as warm and welcoming as always. We strolled off to his government-reserved table at the Upland Goose, the small hotel made briefly infamous during the Falklands war ten years before, and we ate a lunch that, inevitably, considering this was the sheep-rich Falkland Islands, had been composed around one of the hundred known variations on the theme of roast mutton. We then ordered coffee and brandies, went out into the garden, and sat in the watery South Atlantic sunshine to reminisce. It was then that I asked the obvious question—except that Craig stopped me, holding up his hand.

He knew it was bound to come up, he said. He knew what I would want to ask. *What on earth is this man, a Persian scholar and Farsi speaker, someone with a glittering diplomatic career so firmly in his sights, now doing in a place like this?* We both were more than a little embarrassed. But Craig said he had prepared himself for one day meeting his old friends, and had decided that if he ever did, it would be best to tell the truth.

It turned out that some years before the office had posted him to the embassy in Rangoon. He was to be Head of Chancery, a senior position from which he would be groomed—even in a mission so far away from his normal patch—for what was looking now as his all-but-inevitable ascent to diplomatic stardom. Everything was entirely on course—except that Craig, who by then was in his late forties and still unmarried, proceeded to become involved with a similarly middle-aged and single Burmese woman. In normal circumstances this would scarcely have mattered—but it so happened that Britain's then ambassador to Burma turned out to have a deeply felt opposition to any of his staff entering into relationships with what he called "the natives."

A formal letter was then written to London requesting Craig's removal—and since it was a complaint from a head of mission, London was duty bound to take notice. So the poor man returned

home, there to commence a spiral of absolute career destruction. There were postings now to such overlooked and in those
days nonrelevant offices as Luanda, Mogadishu, and Ascension
Island. "And now this," he said, and rather sheepishly handed
over a card from his wallet. It was covered with tightly packed
script.

There was the familiar lordly emblem of British government
service, then his name, and then his title—a classic illustration
of the axiom that the longer the title, the less enviable the job.
He was indeed HM Deputy Governor of the Falkland Islands and
Dependencies, as the radio message had said, but in addition he
was *Assistant Commissioner and Director of the Fisheries of South Georgia and the South Sandwich Islands.*

But it was his reaction that I then found most surprising. "I'm
the first director, the first one ever," he declared, rather proudly.
"And you know what? You'd think I'd be bitter and angry at what
happened—but it is actually quite the reverse. I am completely
loving it. This, down here, is just the purest paradise." And he
then proceeded to tell, breathlessly, how the diplomatic service these days was primarily devoted to such dull matters as
trade—but down here, he was living entirely out in the clear,
cold, endlessly fresh air; he had access to an official boat; he was
able to travel to some of the most spectacular island scenery in
the world; he had come to know the location of the best breeding grounds for such exotic creatures as wandering albatrosses,
right whales, and countless kinds of penguins. He was never
compelled to wear a suit; he seemed to encounter only people
who were fascinating, obsessed, passionate, and adventurous;
and was currently able to help create one of the best-run fisheries on the planet.

"Five years ago I wouldn't have known one end of a fish from
another. I couldn't tell a krill from a kangaroo. I lived in offices.

I went to endless policy meetings. I fretted endlessly about what the London office might think. But now every single aspect of my job has changed. I'm still paid reasonably well. I am still a British diplomat. Burma, for all the short-term misery it caused, ended up doing me a big favor. It got me sent here. And these past two years in the South Atlantic have turned me into one happy, happy man."

I could surely see it. He was quite radiant with pleasure, brimming with delight. He still wrote in Farsi and had a collection of Persian classics in his study. He would always love that part of the world. But now he had found something very different, and found it just as entrancing. Had he stayed on in the South Atlantic, he would no doubt have remained a happy and fulfilled man. But the truth is more somber: he fell ill only a few weeks after our unexpected Falkland Islands encounter, was brought back to England by air, and died not long afterward. We buried him in a village in Rutland, on a blustery March day.

His Burmese girlfriend, who had moved to London and with whom I was in touch for many subsequent years, wrote in the late 1990s to say Craig would have been gratified to know what he and his successors had achieved in the South Atlantic. When she lived in Rangoon she had, as she cheerfully put it, little enough interest in fish—the Burmese had other matters of deeper concern. But in due course, and through her brief life with my old friend, she had also become entirely fascinated with the doings of the sea, and now a total convert, she championed the sanctity of the oceans with a huge enthusiasm.

* * *

The South Georgia and South Sandwich fishery has a reputation very different from that of the fishery eight thousand miles away to the north in the selfsame ocean, off Newfoundland. The Grand Banks may be ill-famed as a world-class fishing catastrophe, a marine monument to greed and carelessness, but the managed waters midway between Cape Horn and the Cape of Good Hope have in recent years become one of the great environmental success stories of the world, a story of caution, restraint, responsible solicitude—and of relentless patrolling by big ships, with guns.

Yet only in recent years. *Caution, restraint,* and *solicitude* were words seldom traditionally applied to the creatures that live in the waters off South Georgia. Right up until the 1980s the harvesting of fur seals, elephant seals, penguins, sperm whales, and right whales was an immense and highly profitable industry, and had been almost from the very moment Captain Cook found the "wretched, horrid and savage" island of South Georgia in the late eighteenth century. Horrid or not, by 1912 the inhospitable, entirely glaciated main island sported no fewer than six enormous whaling factories,* and the decimation of whale populations—the humpback, most notably—became an almost unstoppable phenomenon. British and Norwegian whalers processed more than thirty thousand blue whales in one year, 1929. Now these majestic and gently amiable creatures, the largest animals on earth, are reduced to a population of less than two thousand.

The British jurisdiction over the islands might eventually have helped curb some of the greater excesses of destruction—except that by 1925 factory ships had been invented, and for the next sixty years there developed a pelagic industry over which no one, however well intentioned, had any jurisdiction at all—and so Russian, East German, Korean, and Japanese ships began a free-for-all in the South Atlantic that resulted in the near-total

---

* It was an attempt in 1982 by an Argentine scrap-metal dealer to dismantle one of these disused stations—though refusing to have his teams' passports stamped by the resident British magistrate, on the grounds that Argentina did not recognize London's sovereignty over the islands—that led to the invasion of the Falklands and to the subsequent brief and bloody war to restore British rule.

demolition of many of the region's more fragile species, fish and cetaceans alike.

Yet this mayhem—publicized by groups of ever-more-vocal marine environmental bodies, who capitalized on the widespread public sympathy for the fate of the southern ocean whales—did in time prompt the British government, particularly in the aftermath of the Falklands War of 1982, to set about changing the rules. By the late 1980s London had decided to provide the organization and manpower to try to make sure that whatever fish were to be taken in future, at least from waters over which it had authority, would be taken sensibly and with great caution. Whaling, sealing, and penguin hunting were banned; it fell then to London to make sure that the fish with which the southern seas teemed—some of the local species existing in almost as great an abundance as the cod once had in the seas off Newfoundland—were never to be put similarly at risk.

The array of creatures in the southern Atlantic is rather different from the populations in the north. There is, for example, an abundance of krill, the tiny shrimplike creature eaten by baleen whales and still much favored by Russian, Ukrainian, and Japanese fishing fleets. It is either canned, made into paste, or sold as frozen blocks for the feeding of livestock, or else it is disguised and sold to humans, mostly unawares. There is the icefish and the rock cod, both of which flourished off South Georgia but were fished to near extinction by Eastern Bloc trawlers in the early 1980s. And there is the Patagonian toothfish, which for some reason escaped the notice of the Russians and the East Germans—until, that is, about 1988. That was shortly after this rather large (up to seven feet long), long-living (a toothfish can live to fifty), exceptionally ugly, and exceptionally tasty fish was *rebranded*, given the newly invented name of *Chilean sea bass*, and began to find itself appearing on the menus of white-linen fish restaurants in North America and Europe.

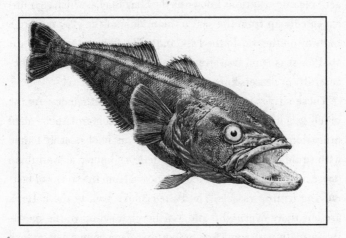

The alarming appearance and the inelegant name of this enormous crea-
ture—the Patagonian toothfish, Dissostichus eleginoides—*required*
some adroit massaging by the seafood industry to make it palatable. It
now appears on menus as Chilean sea bass, a name invented for it in
1984.

The words *Chilean sea bass* emerged in the lexicon in 1984; the first appearance in South Atlantic waters of fleets of Russian fishing vessels specially equipped to take away large quantities of them—their Latin name is *Dissostichus eleginoides*—was four years later, during the austral summer of 1988. Since then this fish has become so popular, so desperately wanted by restaurateurs around the world, that it has come to be referred to by journalists as the *white gold of the southern oceans*. To those custodians of the southern oceans who remember what happened on the Grand Banks, this development led to some dismay and no little concern.

Toothfish are generally hunted in the waters around South Georgia and the shallow banks around the extraordinary jagged volcanic eruptions known as the Shag Rocks, which soar unexpectedly up from the sea in the middle of the gray nowhere, halfway on the way to the Falklands. The fish are caught in the shallow seas from trawlers, or in deep waters rather more successfully by vessels known as long-liners.

These ships are possessed of a bloodless efficiency. As the name *long-liner* implies, very long wires—some of them eight miles long—equipped with thousands of hooks, each baited with squid or sardines or a cheap delicacy known as Namibian horse mackerel, are streamed overboard from the sterns of fast-moving fishing vessels. The baited hooks sink to the bottom, are left there overnight, and are then hauled up in the morning—usually with some four or five tons of enormous fish on each line, which are passed through rollers that detach the hooks; the fish have their much-prized cheeks automatically removed and are blast-frozen and sent into the refrigerated holds.

There are two problems with this kind of fishing. The first is technical and particularly tragic: before the baited hooks sink to the seafloor they tend to attract the attention of seabirds, and biologists have long noted with alarm that tens of thousands of birds—many of them petrels or great albatrosses—manage to get themselves caught, whereupon they are dragged down into the sea by the weighted lines and their sinking hooks, and so drown. Fishermen are now being asked to attach colored streamers and bird scarers to their lines to prevent damage to the rare and wonderful albatrosses in particular; encouragingly it is a solution that is said to be working—at least when fishermen do as they are asked.

But this illustrates the second and more serious problem: many fishermen pay no heed to such requests, because much of the toothfishing that goes on near South Georgia is, or until very recently has been, downright illegal. The seabirds continue to die at the hands of the fishing pirates; and the fishing grounds more generally are being put at risk of being depleted, just as they were thirty years before in Newfoundland.

This is why powerful ships armed with big guns have in recent years been sent to these waters, to scare away and deter illegal fishing—something never done in Canadian waters. And the reason the South Georgia toothfish is now fast recovering, and why the fishery is being seen as an exemplar of what fish management could be, is that this hard-hearted policy appears, at least in its early years, to be working.

I came across the sharp end of the policy on a recent visit to South Georgia. I was on a former Russian research vessel, which on this occasion was taking a small number of bird-watchers to see a breeding colony of wandering albatrosses on a rocky outcrop to the southwest of the main island. We were sailing along in the open sea, beyond sight of land, cruising at maybe ten knots, when suddenly a vessel appeared at great speed astern of us on our starboard side and our bridge radio started chattering urgently.

"Unidentified Russian vessel two miles ahead of me, report your name and business," said a crisp, no-nonsense, English-accented voice. "This is the British warship HMS *Northumberland*. State your name and business, and your reason for being in these waters, please, immediately. Slow down for possible boarding and inspection."

And so we had to heave-to, were compelled to identify ourselves, state for the record that we had neither the will nor the ability to fish, and outline our purpose for being in what were not, as we supposed, international high seas but rather British territorial waters. Had we been deemed suspicious, marine boarding parties were ready in their Zodiac speedboats to come and grapnel their way up our sides; had we decided to run, we might soon have been discouraged by a shot across our bows.

But all turned out quite otherwise. By chance the captain of the navy ship was an old acquaintance of mine, and once he had established our credentials he asked if he might "put on a bit of a show" for our passengers as recompense for asking their ship to stop. All promptly delighted in a fifteen-minute display of marine gymnastics, the brand-new navy vessel wheeling this way and that at high speed through the swells, throwing up Niagaras of spray and leaving a wake a mile long. Finally, fun over, she sounded three blasts on her whistle and took off into the sunset, vanishing over the horizon in minutes. Another sector of sea had been declared at least briefly poacher-free; yet more shoals of toothfish in the depths below were left safe for another night.

That was in the early 1990s, when everyone was still jittery in the immediate aftermath of the 1982 war. Matters are rather more stable these days, although the navy is there just in case. All of the fishing vessels licensed to operate in South Georgian waters are obliged to keep a transponder constantly reporting their position to the government; and as additional precaution there are even more potent combinations of protective enforcement: Hercules aircraft fly out from the Royal Air Force base on the Falkland Islands, eight hundred miles away, and spy satellites are programmed to report any vessels below that appear to be fishing illegally (and this includes nighttime operations of the vessels known as squid-jiggers, which use thousands of lures to attract squid by the million, and which can be easily seen from space because they use batteries of brilliant white lights beamed down into the water to attract them).

There is also a high-speed and very long-range fishery patrol vessel, orange-painted to stand out against the ice and equipped with an Oerlikon heavy machine gun, whose crew will chase, board, and arrest any miscreants and will readily confiscate or sink any errant vessels. The MV *Dorada* did just that in 2003: it helped chase the *Viarsa*, a Uruguayan-flagged toothfish pirate with nearly $4 million worth of illegally acquired fish in her freezers, halfway around the southern world, finally arresting her and her crew off the coast of South Africa.

There is something of a historical symmetry to this one story. The *Viarsa* turned out to be owned by a syndicate of Galicians, from northern Spain—the very people who pioneered fishing in the deep Atlantic five hundred years before. There were reported to be some twenty other such ships, all owned by the same people, but registered in places like Belize and Ghana, Argentina and Panama, and all involved in the same mission: that of plundering the far oceans for fish, whatever the risks and all but certain of great rewards. The irony is that the same imperative that first brought northern Iberians to the unregulated foggy desolation of the Grand Banks in the sixteenth century now eventually also brought them—though this time illegally—to the ice-cold but now highly regulated waters of the sub-Antarctic in the twenty-first. On both occasions it was an imperative prompted by a seemingly insatiable desire to take an infinite amount of fish from seas that the Galicians—today the second greatest fish eaters in the world, after the Japanese—believe contain a nearly inexhaustible supply.

6. A DISRESPECTED SEA

That inexhaustibility might have been true in the sixteenth century. It is most certainly not true today. The fish that jostled for space in the days of Columbus and Cabot, Vespucci and Francis Drake are all savagely diminished today by a sorry conspiracy of deliberate depletion. Small wonder, with delusions of perpetual abundance still so popular—despite the evidence—and with such a ceaselessly vast worldwide appetite for fish, that alarms are now and at last being sounded.

Some say that all fish are in worldwide danger. Many who decry the eating of meat for environmental reasons argue that fish should be shunned with equal vigor too, because the fish in the seas are in every bit as much danger as the buffalo once were on the high American plains. Not a few predict that all commercial fishing, worldwide, will have been essentially extinguished before the current century is halfway done.

Certainly the oceans are changing under the malign influence of landsmen so utterly careless of the sea. We must all know anecdotal evidence of this—some new, some not so new. In the 1960s, for example, I used to visit a remote sea loch on the northwest coast of Scotland, and would occasionally take a boat out as far as my courage would allow, sometimes sheltering in squalls in the lee of a squat green island called Gruinard. Locals told us not to go too close. Once I did by accident and saw notices on the shore warning that it was unwise to land since half a century before the island and its surrounding waters had been deliberately infected with anthrax, a wartime experiment whose effects had lingered longer than anyone expected. They thought that the sea was big enough to wash the toxins all away: no one imagined the reverse would happen, and that the seas themselves would become poisonously disarranged.

Also back then, but along the arm of another quite distant sea loch, we would spend happy hours walking the shores, stopping every so often to gaze down through the pellucid waters of the rock pools at the brilliantly colored undersea gardens, the

waving fronds of vivid purple anemones offering brief protection from the sun to nervous scatterings of scarlet crabs and
tiny stranded fish. But all has changed. Careless visitors have
come in legions since, and the isolation of these shores has in
recent years been quite destroyed. The clear pool waters that I
have seen on more recent visits are now frothed with blowing
foam, and I believe it is not my imagination that suggests that
there are fewer creatures to be glimpsed these days, and none
of them so brilliantly colored as in my childhood memory.

And yet again, farther to the south, but still in a part of western Scotland buffeted by gales and rinsed by the rushing sea,
there is yet a further sign of careless ruin. Where we once used
to lie in solitude on the machair and watch the sea otters and
the basking sharks offshore, and marvel at the gray emptiness
of huge seas all around, there is now a long series of floating
platforms, made of wood and buoyed up by floating blue barrels: fish farms, with pumps endlessly whirring and lights
endlessly flashing and oil-dripping speedboats scurrying to
and fro carrying barrels of feed to the thousands of trapped
animals within. The waters within the cages are a constant
fury of movement and bustle: the fish below are frantically
jostling for space—not jostling as they were reported to be in
the North Atlantic five centuries ago, a joyful consequence of
their freedom and fecundity, but because they were now being
pinioned in huge numbers behind submarine wire fences, and
being kept there, fins worn, muscles weak, infections spreading, until big enough to be netted out and sent off by truck to
the markets in the big cities of Europe.

There is change and decay all around the sea. On the very
winter day I write this, yet another melancholy report of ocean-
depleting news comes in: this time it seems that the rising
acidity of some tropical ocean waters, supposedly caused by the

solution in seawater of an excess of man-made carbon dioxide, is shown to be robbing certain fish of their sense of smell, rendering them unable to detect nearby predators. It is not just that we appear to want to eat every fish in the ocean that we find appetizing; we also seem to want to give a helping hand to other fish-hungry monsters and thereby thin the population still further.

One cannot but hang one's head in shame and abject frustration: we pollute the sea, we plunder the sea, we disdain the sea, we dishonor the sea that appears like a mere expanse of hammered pewter as we fly over it in our air-polluting planes—forgetting or ignoring all the while that the sea is the source of all the life on earth, the wellspring of us all. The Atlantic, first to be found, to be crossed, and to be known, is by far the most polluted, the most plundered, the most disdained, the most dishonored of the world's oceans.

To compare the man-made collapse of cod in the North Atlantic with the seemingly sensible current management of sea bass in the South hints at a way in which man may at last be changing his ways. But it is by no means a perfect comparison. The lamentable decisions that were made by the Canadian government in the 1980s were made in a democracy, and by a government that felt understandably obliged to meet the short-term requirements of the Newfoundland fishermen, who also happened to be voters. There are absolutely no voters in the South Georgia waters. There are no permanent human inhabitants. The colonial government can manage the fishery there with impunity, doing as it thinks prudent, never having to tip its hat to any interested party—other than the fish themselves, one might say.

But nonetheless, a growing human resolve to change our ways is becoming slowly apparent—and it appears likely that it will be the Atlantic Ocean that will provide the test bed once again for the way that new resolve is calibrated. It is the ocean that will also be bound to demonstrate the consequences if we fail.

And what, one is bound to wonder, might those consequences be? Could the sea somehow contrive, in some unimagined way, to resist our unending misuse of her, and in some fashion or other start to strike back? What price might mankind have to pay, if after decades of his misuse and carelessness, the Atlantic determines to do just that?

Chapter Seven

THE STORM SURGE
CARRIES ALL BEFORE

———◆———

Last scene of all,
That ends this strange eventful history,
Is second childishness and mere oblivion;
Sans teeth, sans eyes, sans taste, sans everything.

---

1. THE ICE DEPARTS THE SCENE

Some peculiar things are happening in the North Atlantic Ocean, and no one is quite sure why. The changes present themselves in many guises: here is an illustration of just one of them.

Early in September 1965, I was in East Greenland, waiting for a group of Inuit fishermen to pick up our university expedition party from the shores of a wide fjord called Scoresbysund. We had been working up on the ice cap for some months and had now come down to the seashore to wait, as previously arranged, for the first stage of our long journey home. We waited, and waited. Three days passed. But the boat never came. Eventually, over our shortwave radio, we heard the explanation: two weeks of relentless easterly gales had unexpectedly driven billions of tons of Atlantic pack ice in from the Denmark Strait, blocking passage along the two-hundred-mile-long fjord to all navigation except big icebreakers. It was something that happened once in a while back then; no one had told us. And the boat for which we were waiting was certainly no icebreaker, being no more than twenty feet long and made of wood. She had a crew of three and was named the *Entalik*.

So there we were—stranded, trapped, and potentially in quite a bad way. Each of the six of us was in good shape, but we were almost entirely out of food. We had several hundred pounds of a specially formulated low-temperature-spreadable van den Berg's margarine, ten boxes of Weetabix, and, bizarrely, a single carton of bay leaves—all left over in a cache from when our expedition had first landed, three months before. Yet we were far from helpless: we had a radio and a gun; and more by luck than any particular proficiency at marksmanship, someone managed to bring down a barnacle goose and, though I hesitate to admit it today, a ragged and yellow old polar bear. We ate both; the goose was as tasty as geese are known to be; the bear was as stringy on the inside as he was mangy on the outside, and his thighs were infested with scores of a disagreeable inchlong kind of flatworm known as *Planaria*, which we had to pry out with our Swiss army knives before stewing the meat over the Primus stove. We took care not to try the bear's liver, which Greenlanders had long warned us would be highly toxic, being crammed with too much vitamin A.

Eventually we received a radio message from the crew on *Entalik* saying they had managed to struggle through the pack ice for three days and nights and were now as close to us—about a mile—as they were ever likely to get. As long as the ice did not move too much, they could stay where they were for perhaps another day, but then they should head back home. The winter was beginning to close in: at those latitudes, a little above 70 degrees north, the vanishing of the autumn sun came earlier and earlier by several minutes each day, and the nighttime temperatures had started to plummet, and even in the daytime it was chilly enough for small flurrying snowstorms.

Our choices were limited. We had already missed the last ice-breaker of the season, which would have taken us back to Denmark, and which sails twice annually from a tiny settlement thirty miles along the north side of the fjord. But if we now failed to reach the security of the settlement, and were trapped here on the south side all winter, we would clearly be in a very serious situation. The cold would begin to be intense, and it would soon be totally dark. We would probably starve.

If we were going to have any chance of reaching home, we had to walk out to the boat, trek out across the sea, over the ever-shifting ice floes, and hope to find *Entalik* at the ice edge. We would have to leave immediately, before the boat, faced with the risk of being stuck fast for the winter, turned about and headed for port.

So we struck our tents, gathered our essential belongings, and tied them to our backs; we roped ourselves together for safety, and wearing crampons and carrying ice axes at the ready, we clambered over the pressure ridges that had formed where the floes met the beach, and then started to walk out over the ever-shifting, ever-tilting mass of ice floes, jumping from one to another over leads of black and ice-cold water that we knew to be a thousand feet deep, at least. Not that its depth made any difference: a minute in the freezing water and a man would be dead, long before he was halfway to the bottom.

It took many hours, and there were many close calls, but we eventually made it to the boat; and though she was to be stuck in the ice for a while herself, as her crew had feared, and though we

had to shoot yet another animal—this time a musk ox—in order
to eat, and despite more rafts of difficulties that are not wholly
relevant to the story,* we did make it out to the Scoresbysund,
and within a week we did get back to England. Moreover, the four
of the six of us who were students made it in time for the start of
classes that term, though just barely.

.   .   .

Of the six, five went on to become professional geologists, and
two used to go back to that same part of Greenland almost every
year. They have come to know the ice, rocks, and animals—and
the weather—of this utterly remote corner of the world almost
as well as do the Greenlanders who still live there. And over the
years they have noticed that some things have changed, rather
strangely.

Outwardly, some are of a kind that would be wholly expected.
The settlement is now a little larger—it was peopled by four hundred then; it has five hundred now. It has changed its name, no
longer memorializing a British whaling captain named William Scoresby: it is now Ittoqqortoormiit, the Greenlandic word
meaning *the big house*. The icebreaker from Copenhagen still
calls twice a year; but now there is a twice-weekly air service in
the summertime, too. *Entalik* is long gone, but in its place there
is a bigger and much sturdier boat. The young men who crewed
back in 1965 are old-timers now, and there is a party of strong
young hunters and sailors who take the new craft into the far-
thest recesses of the fjord system—at 220 miles from the Atlantic
to most distant feeder glacier, the longest in the world—taking

---

* We had to charter a small plane to ferry us in groups back from the settlement to an airfield in western Iceland. The pilot, a heroic figure named Bjorn Palsson, was killed shortly
after so bravely helping us—and when I visited Iceland to research this book and mentioned
his name, all remembered him, even though forty years had passed. At the time none of us
knew he had long been famous in Iceland as a "rescue pilot," the kindliest of daredevils.

supplies to those who have chosen to live even more remotely, far from the open sea.

These young men who work the boat were the ones who noticed the most significant difference in their surroundings. When first they started their job, they could assume that if they took out their boat late in the season, there would inevitably be days when they ran into storm-blown pack ice, just as we did in 1965. The fjord would close up on them with the same billions of tons of floes, and navigation would become difficult if not impossible. The first blow-in usually occurred in late September—a signal, regular as clockwork, that the winter was about to begin, the water to freeze and Scoresbysund to become fully welded into the Denmark Strait ice pack, as usual. Very occasionally the ice would blow in some weeks before; in our case the ice that impeded our passage home had come in the first week of September. Very, very occasionally ice was seen in the fjord at the end of August. But this was rare, and when it did occur, as to the men on the old *Entalik*, it was very unwelcome indeed.

But what then started to happen—and so far as the Greenlanders remember, it started to happen in the middle of the 1990s—was that the first appearance of blown-in floe ice in the fjord began to be delayed, pushed back until later and later in the season. There was never any ice in August, which was a relief. But the blow-in seldom occurred in early September, either—and that, even if reason for good cheer among the boatmen, was rather odd. For season after season at the end of the twentieth century and then on into the twenty-first, the fjord remained ice-free until very late in the month. Indeed, there might be no ice at all until the very end of September, which was hitherto unheard-of.

And then the yearly close-up, the moment when the ice in the fjord becomes so thick and permanent that all the local boats are hauled up onto shore and turned upside down for the remainder of the dark winter. By then the sun was well down in the sky, and it was half dark for most of the day. Such blackness normally coincided with the onset of thick ice, but now there was open sea outside, and waves, and the sounds of splashing and trickling water that were usually associated with all-day sunshine. The Arctic winter might have set in, but the Arctic waters were not as cold as they used to be; and the ice was not there in the amounts that used to be seen. By the end of the new century's first decade, everyone began to realize that what at first seemed just peculiar was now a trend. The Arctic, or at least this small part of it, was really getting warmer.

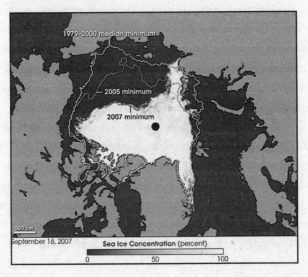

The steady melting of the Arctic ice sheet since 1979 is returning much of the north polar region to blue water during the northern summer. Scoresbysund in East Greenland—the largest fjord system on earth—is the two-pronged inlet seen on the lower east side of the island. Despite being solid with floe ice in the 1960s, its fjords are now entirely ice-free until late winter.

Thus has climate change arrived at a small Atlantic settlement in East Greenland. The rise in temperature in this corner of the ocean is quite gentle, perhaps—not as impressive as the breakup of country-sized ice sheets that lie eight thousand miles to the south, off the coast of Antarctica. The economic consequences for East Greenland will not be as profound as that of the melting of the polar ice pack north of Russia, which in 2009 allowed cargo ships to traverse the so-called Northeast Passage between Archangel and Vladivostok. But it will alter the voyaging habits of the Greenlanders—the late onset of winter will lengthen the time they hunt for narwhal, walrus, or seal; it will extend the time they can devote to fishing; and it will affect the dates when they can expect to see the icebreaker coming up from Copenhagen at the beginning of the season—for, yes, the late freeze-up in the autumn is also matched by an ever-earlier breakup of the ice in spring, as one might expect—and the date on which the final one will go home to Europe. Consequences both profound and dangerous are said to be likely from the coming warming of the world. In remote places like Ittoqqortoormiit lesser changes will be coming, too, and not all of them in this corner of the world are thought to be entirely bad.

A whole series of hitherto unanticipated events like this—some of them small and mattering only locally, others very large and of worldwide concern—are currently being observed all around the world's oceans but are most keenly observed in and around the Atlantic. We know all too well what many of these changes are. Some of the creatures that inhabit the seas and the coasts—from whales to codfish to polar bears—are now and in disturbing numbers dying, or being involuntarily culled, or are having their livelihoods threatened and their habitats altered. Some ocean currents are changing their routes, their size, and their strengths. The ambient temperatures of the world's seas and the air above them are both increasing, and, according to some, are increasing much more rapidly than at first supposed, which is said to be especially troubling.

The patterns of weather in places are being disrupted, and dramatic storms are becoming more numerous, their intensities steadily and ominously worsening. Ice caps, glaciers, and hitherto persistent snowfields are melting, and they are doing so swiftly, and this immense translation of solid into trillions of tons of liquid water is compelling the oceans to rise to levels that may well threaten all of our coasts and many of our cities.

There is much alarm about all this. Not a few see widespread doom in the oceans' ills, and there has been a good deal of apocalyptic fretting. It is widely held—but far from universally—that mankind's industrial excesses are greatly to blame for this, and that unless humans change their ways, the world, and the seas that give it life, will shortly be in the direst trouble.

2. THE APPORTIONMENT OF DOOM

One truth is inescapable: the Atlantic Ocean as an entity will one day disappear. The continents that surround it and give it its current distinctive shape will themselves change their appearance; under the influence of great forces, they will move around the surface of the globe, and the Atlantic's waters will be forced to go elsewhere. Whatever well-publicized other troubles may be afflicting the ocean today, they will have no bearing on this. The length of the Atlantic's existence as an ocean is entirely independent of whether its temperature rises or falls, or in which direction its currents go or whether its toothfish or its polar bears live or die. The length of the Atlantic's existence has nothing whatsoever to do with man, who has no influence on how long it survives, and man will have long ceased to exist by the time the ocean is ready to vanish.

But the *relationship* between humankind and the ocean is a different matter entirely. That is something that surely depends to a very significant degree on just how solicitously humankind treats the oceans under its care. A noisy argument is in full spate right now, just as this chapter is being written (and will probably be continuing as it is read, and well beyond) over the degree to which the bad habits of man play a significant role in the circumstances of the sea.

We know the sea is in trouble. We know that man plays at least some part in the wreaking of that trouble. Examples abound, most recently in the Gulf of Mexico—officially defined by the seers of Monaco as part of the Atlantic, even if it doesn't look as though it should be—which in the early summer of 2010 was being catastrophically polluted by a drilling rig that exploded and sank in waters a few miles from New Orleans, a city that was itself still recovering from the trials of Hurricane Katrina five years earlier. A volcanic torrent of oil from a ruptured seabed pipe a mile below the Deepwater Horizon platform spread to coastlines from Texas to Florida, despoiling and polluting and killing. Eleven men died in the explosion itself.

The event—which was both predictable and preventable—promptly confounded the growing lobby of those one time skeptics who were starting to think that undersea oil exploration was safe, or at least safe enough. But to those others who recalled other great tragedies in the Atlantic—the 1988 loss of the Piper Alpha North Sea platform, with its appalling casualties, being the most egregious—it was a catastrophe that simply served to confirm another belief, that oil drilling at sea was a business inevitably and ultimately harmful to sea and man alike.

However, there was a third group—and a very large one. These were people who were quite beyond persuasion, and who believed that the world's modern industrial needs must in any case trump such petty concerns. To this group the loss of the rig and the pollution it caused, though tragic, were of rather less consequence. They were something that, as environmentalists shudder to hear, just *went with the territory*.

The melancholy events in the Gulf serve to pose the question one further time: what is the truth? Are the legions of troubles of the sea truly the result of the mischief of man? And further: could it be that when giant hurricanes like Katrina are spawned, or when the polders of Holland are indundated or the beaches of Africa are clawed away and villages founder beneath the waves—could it be that the sea is showing signs (at least to those who like to anthropomorphize it) of somehow *striking back*? Or are all the ocean's troubles entirely cyclical and natural-born, are the storms and sea-level rises parts of cycles, too, and is the ocean more than likely just to remain ponderously aloof to man's fleabites of bad behavior?

This is where complication and controversy begin to appear. I recognize all too well that it would nicely serve this book's purpose to be able to show that man is fully or even wholly responsible for the ocean's ills, and I would clearly like to be able to do so. But I also know that there is a vast body of competing claims on the topic—with people of great distinction and good faith arguing that of course man is responsible, while others of equally stellar reputation and good faith claim that to suppose such a thing is the height of arrogance, and that man is far too puny and crabbed to be of any importance at all to an entity as vast as the Atlantic Ocean. Ever since 1995, when the Intergovernmental Panel on Climate Change uttered its historic declaration that it saw "a discernible human influence on the global climate," the debate has assumed totemic importance, its adherents and opponents battling for hearts and minds as though hierarchs of some newfangled religion. Politics, somewhat unhelpfully, is now a party to the argument, too, muddying the issues still further, adding new and louder voices to what is already a cacophony.

That all being said, there are now a handful of proven verities, certain hard truths about the present situation of the seas that even the most ardent deniers of change have generally come to accept.

The first is at once the most simple and the most profound: the world is getting warmer, and the temperature of the oceans, and especially the temperature of the Atlantic Ocean, is currently rising at a sudden and an alarming rate. And this is going to have consequences for many of those who live on or near the sea and derive their sustenance from it. Whether these consequences are temporary or permanent matters little: the importance is that they are going to affect everyone, and not just the narwhal hunters of Ittoqqortoormiit.

The specific central facts of the argument appear to be these, with three sets of observed data seeming to be unassailable (which does not mean they pass unassailed, as we shall see). First, it is measurably clear that during the last quarter century, the average temperature of the atmosphere at the world's surface has been increasing, and has risen by an average of 0.19 degrees centigrade during each decade. Second, observations from ships, aircraft, satellites, and scientists on the ground have led to the conclusion that the ice sheets and ice caps in the Arctic Ocean, on Greenland, and on the continent of Antarctica are all losing mass; and that since 1990 the glaciers and ice caps elsewhere that have been slowly shrinking for half a century have suddenly started melting very rapidly. And third, according to satellite observations, the world's sea level has been rising at the rate of 3.4 millimeters every year for the last fifteen years, and the rate of rise is increasing.

Beyond these three facts, a number of other less certain—or more contentious—assertions and predictions are made, and by an overwhelming number of climate scientists. First, the global sea level is estimated to keep on rising, and by 2100 it will have risen by more than one meter, perhaps by as much as two. Second, this rise in global sea level is linked to the melting of the ice caps. Third, a series of so-called tipping points are now fast being approached, and if the observed warming trend continues (which is itself by no means a certainty), then changes to all manner of the world's features and phenomena—rain forests, monsoons, hurricane frequency, desertification—will occur and may become irreversible.

The fourth point, made by many and now believed by most, is that all of these developments are occurring at the same time as a dramatic rise in the tonnage of carbon dioxide and other so-called greenhouse gases—which in essence sit in the upper atmosphere and prevent the escape of heat from the planet—that are being spewed out by the chimneys and exhaust pipes of industrial mankind. These emissions, which have all stemmed, one way or another, from the burning of fossil fuels, have increased by no less than 40 percent since 1990.

The fifth point, made by most but disbelieved by many, links all of the unassailable facts about warming, melting, and rising to this last well-known fact: that the increase in mankind's carbon emissions is not merely coincident with the increase in global temperature, but is the ultimate cause of it. And at this point the two lobbies part ways, decisively, noisily, and with often passionate discourtesy. The one lobby insists that this is so; the other casts all manner of doubt and claims that the world's money—vast quantities of which are being earmarked to pay for a lowering of mankind's carbon emissions to decelerate the rate of warming—could and should be far better spent elsewhere. Population, most of these climate skeptics say, is the major problem (although recent data have shown that population may be beginning to peak, and maybe shrink back) and other vast troubles—diseases, lack of water, poverty—need to be addressed first, long before attention is given to what they say is the utterly unprovable linkage between carbon emissions and global warming.

3. UP SHE RISES

There are many predicted consequences of the warming of the world. Some of them are restricted to the land, as with the increase in droughts and the expansion of deserts. Most, however, are bound up with the future of the seas, and of them two are becoming paramount: the rise in the level of the sea and a slew of possible changes in the world's weather.

The rise in sea level is perhaps of the greatest immediate interest, not least because the millions who live beside the sea are often vividly aware of when and if it is happening. There are two causes for the phenomenon, which is a very real and (at least in human terms) rather long-term trend: since 1870, when data used to be collected from mechanical tide gauges, rather than by today's satellites, the world's seas have risen by some eight inches.

The first cause of this rise stems from a simple law of physics: that as the ambient temperature increases, water expands. The warming sea, in other words, is becoming not so much taller as more *bloated*.

Though this thermal expansion is expected to contribute about 40 percent of the global sea-level rise—perhaps more than half of it, say some—it is a rather difficult concept to conjure with. Some argue that the basins that hold the seas will grow bigger in hotter weather and thus keep the level the same. Physicists who support the bloating idea counter by explaining that water expands more than rock, and so their assertion is correct. One has to take the word of science in matters such as these.*

It is much easier to grasp the other reason for sea-level change, one calculated to be responsible for the other 60 percent of the rise, and which concerns the physical form of water that has much of the high-latitude and high-altitude world in its grip: ice. As long as the world's land-based ice—its glaciers, ice caps, permanent snowfields—remains frozen, then all will be well, or at least all will be stable. But if a large proportion of this ice melts, as it has been doing for at least the past twenty years, and if the melting continues to accelerate, as it has started to do in recent years, and if all the locked-in water becomes unlocked and flows down into the seas, then there will be trouble—or at least there will be instability. That's because the world's seas will become fuller, and their levels will go up and up, and they will do so for perhaps a very long while, and possibly unstoppably.

Since ice is the key to the stability of the seas, the Atlantic Ocean is the key ocean to watch. Of the world's three great oceans, the Atlantic sports by far the largest amount of ice. A glance at a map shows why the Atlantic is such a catchment area for polar cast-off ice and why the other two oceans are less so.

The Pacific's connection with the Arctic, for instance, is pinched off by the sixty-mile-wide Bering Strait; and though the

---

* Global-warming skeptics, who often claim that scientists are not necessarily paragons of rectitude, received something of a boost in November 2009 with the leak of thousands of emails from a noted climate research center in England that alleged—though never proved—that researchers had been performing "tricks" with their statistics, and at the same time grumbling about the freedom-of-information laws that opened their work to public scrutiny.

Bering Sea has plenty of winter floe ice, the glaciers of Alaska and Kamchatka and far northern Russia produce relatively little new ice as their gift to their local ocean. The South American Andes spill some icebergs into the South Pacific by way of Chile—though most of them melt into high-level lakes, from which rivers course principally through Argentina and into the Atlantic. The South Pacific then (technically, at least, though an atlas might suggest otherwise) barely connects to the Antarctic icefields at all—the frozen continent lies hundreds of miles to the south of where the delineators at the International Hydrographic Organization in Monaco declare the South Pacific to end. And the Indian Ocean, which is mostly a Southern Hemisphere ocean, has no physical connection to the Arctic either. It has a southern boundary that, just like the Pacific's, ends many hundreds of miles shy of the Antarctic coastline.* The South Pacific and the Indian Ocean seldom if ever see icebergs and seldom if ever sport ice floes, though of course the presence or absence of ice floes makes no difference, other than seasonally, to the level of the seas.

The Atlantic Ocean, by contrast, is powerfully and intimately linked to bitterly cold polar waters, as well as to lands that breed icebergs in huge numbers. Both in its far north and its far south, the Atlantic gets more than its fair share of solid and spectacular ice.

In winter the open waters of the North Atlantic are littered

---

* The truth of the ice-free Indian Ocean assertion rests largely on the agreed location of the northern limit of the Southern Ocean (the line where the southern Indian, Pacific, and Atlantic oceans end). In the official ocean-defining document SP23, referred to in chapter 2, the International Hydrographic Organization places this limit at 60 degrees south. The government of Australia objects, however, demanding that the Southern Ocean be seen to extend all the way up to its own southern coastline. Icebergs seen close to Australia's Heard and McDonald islands are thus, by the strict application of the IHO's definition, actually in the southern Indian Ocean.

with icebergs, borne on the currents to the waters well to the south of Greenland—as the *Titanic* so disastrously came to know. The North Atlantic also becomes choked with ice around Iceland—as any Fleetwood fishing trawler that is sent out for cod in wintertime well knows, too. And there is an unbroken passage of wide sea north of Iceland that drives directly to the North Pole, unchecked by any land at all, allowing pack ice—with occasional trapped bergs from Arctic-bound glaciers—to drift into the ocean proper, there to be joined by thousands of Greenland icebergs.

But Greenland—the biggest noncontinental island in the world, currently home to fifty-seven thousand people and almost three million cubic kilometers (700,000 cubic miles) of ice—is the real key. All of its ice is currently melting or ablating, at varying rates, and hundreds of well-lubricated glaciers slither off the ice cap either directly down into the Atlantic from the island's east coast or else indirectly, by way of the Davis Strait and the Labrador Sea, from the huge glaciers on its west.* Greenland feeds new meltwater into the ocean without cease: it has become like a gigantic faucet turned on full blast, with the bath filling fast and no one on hand to turn the water off.

The Atlantic's attraction for new ice is not a northern phenomenon alone: its southern waters are similarly specked with icebergs—thanks mainly to a peculiar accident of tectonics. A sunken mountain range known as the Antarctic Peninsula juts northward from that continent directly into the heart of the southern Atlantic—it almost reaches up the southern tip

---

* There has been much hysteria over the melting of the Greenland ice cap and the supposed speed-up of its glaciers—a widespread belief once championed by former U.S. vice president Al Gore. But lately the glaciers have slowed again, and to levels last seen in the twentieth century, which has removed some of the politically convenient drama from the situation. Most climate scientists still believe, however, that slow and steady melting will continue.

of South America. Then by a quirk of its geology, it swivels, as South America does also, until both headlands end up facing the east. Here the two sets of cliffs, with Cape Horn in Chile to the north, and the British possessions around Elephant Island to the south, help create the infamously lethal body of water known as the Drake Passage. On maps it resembles an iron-plate exit wound from an eastbound bullet: on the Pacific side the entry is smooth, but in the Atlantic the walls are bent upward untidily, looking like an immense and ragged-ended funnel in the sea apparently purpose-built for hosing materials into the deep ocean.

And hose them it surely does. Through the passage storms a cocktail of fierce westerly winds, immense currents of ice-cold water, and melting icebergs in prodigious quantities. These great citadels of ice are swept at impressive speeds directly into the southern Atlantic, making their way south of the Falkland Islands and close to the island groups of South Georgia and South Sandwich. Icebergs in the waters of the southern Atlantic are a danger that the few vessels steaming here regard with endless wariness. But more than that: from the moment they enter the water, they increase the level of the ocean. And if thousands, millions of them are poured into the sea from the land, the level of the sea would begin to notch up to accommodate them, millimeter by dangerous millimeter.

The presence of alien ice is thus very much an overwhelmingly Atlantic phenomenon—and with increasing air temperatures, the melting of the ice, and the rising of the sea, the sea-level problem is first and foremost an Atlantic problem too. The oceans are connected, of course—many oceanographers refer to *the world ocean* and regard the named seas more as inventions of mankind—and the Atlantic's problem will become the world's problem in very short order. But the symptoms of the change will be noticed—and are being noticed—in the Atlantic Ocean first.

One very large rise in the Atlantic's level, a long while ago, led to social changes on the order of those which some fear today. About 8,700 years ago, during one of the many previous global warming periods, an ice barrier penning in the waters of Lake Agassiz, a huge glacial lake in what is now central Canada, broke apart. An immense quantity of freshwater—equivalent, some say, to fifteen Lake Superiors—coursed wildly into Hudson Bay and out into the open Atlantic. The ocean's level rose more than one full meter in a matter of weeks. Archaeological evidence suggests, if somewhat tenuously, that the rise in sea level had effects all around the Northern Hemisphere, with farmers as far away as the Black Sea promptly leaving the coastline and moving to the much safer hillsides to begin tilling the soil, where, as it happened, hunter-gatherer cultures already existed. Tension, one imagines, swiftly resulted.

Elsewhere today there are similar signs of popular concern over the consequences of sea-level change and climate alteration. The government of the Maldive Islands received a great deal of press attention in 2009 for holding a cabinet meeting underwater, with all its ministers wearing frogmen's suits, to demonstrate their vulnerability to a rise in water levels, even though they will be far smaller than the changes that followed the Agassiz flood;* the government of Nepal held a similarly well-advertised meeting at Everest Base Camp to illustrate how the melting snows and icefields of the Himalayas were ruining the country's crops and flooding its villages. Generally, though,

---

* Much is made of the plight of those impoverished low-lying countries that are likely to be submerged. The Maldives in the Indian Ocean and Tuvalu in the Pacific are most frequently mentioned, and some small islands off the coast of Bangladesh have begun to disappear already.

the first to feel the effects of the ice melting will be those living in and around the Atlantic. There may be no modern-day equivalent of an ice-dam collapse imminently expected, but other changes—the weakening of the Gulf Stream and the gravitational effects of being close to the North Pole among them—are conspiring to make the sea-level rises in the North Atlantic even more impressive, and the response of those who live beside the sea commensurately so as well.

Take the port of Rotterdam. Take Holland, indeed. Take, in fact, all of the Netherlands. Perhaps no country anywhere around the Atlantic is so bound up with the sea—a quarter of it is below sea level (the Dutch word *Nederland* meaning *low-lying land*), and ever since the 1920s reclamation and the construction of dykes and dams and flood control measures, and the polders—the reclaimed land—they protect have been central to the making of the country, literally so. So crucial to the nation's existence is the protection of the polders that a model of politics has been created to recognize it: the Polder Model is shorthand for *consensus politics*—that no matter how profoundly you might differ on other issues, if anything threatens the polders, all Dutchmen know that arguments cease, as the integrity of the polders comes first.

Catastrophic storms mark the remembered dates of Netherlands history much as wars of liberation and the reigns of king do so in other, drier lands. So while America has her 1776 and 1865 and 1941, and while Britain has 1066 and 1688 and 1914, the Netherlands has her 1170 (the All-Saints Flood, when the Zuider Zee first became salty), 1362 (when the Great Drowning of Men occurred, with twenty-five thousand people swept away by an immense storm surge), 1703 (the fantastically lethal Great Storm, which also affected England, and stirred Daniel Defoe to write a book about it), and 1916 (when the concerted nationwide attempt to stanch that winter's inflow of North Sea waters—an attempt that continues to this day—was first launched). There were appalling storms in January and February 1953, too—a spring tide and a northwesterly wind combined to whip up the seas and breach the dykes and seawalls and drown nearly two thousand people. The reconstruction in the years that have followed have made the Dutch population determined that no such thing ever happen again.

Which is why the Netherlands, above all others, is currently moving fast to ensure that the rising ocean beyond its massive seawalls does not wash the nation out of existence. Maps are now published that show how deeply the country would be inundated—if no precautions were taken—by only a modest rise in sea levels. If the ocean were to rise by a single meter, almost the entire Dutch coast, from the German port of Bremerhaven in the north to the French channel port of Calais in the south, would be at risk of flooding. Floodwaters would extend miles inland, as far as the cities Breda, Utrecht, and Bremen. Half of the country's fields would be salt-water-logged and unusable. Tides would wash in and out of great Dutch cities like Amsterdam, The Hague, and Rotterdam.

But the Dutch are not going to allow any such thing to happen. Sea defenses—enormous movable lock gates and barrages—that protect the polders and prevent storm surges from moving up rivers are of course being strengthened and their levels raised. But the major cities are doing rather more—with Rotterdam, Europe's busiest port, and a city of seven million, much of which lies below sea level even now, in the forefront. The city fathers have decided that rather than fight the incoming waters, they would be better advised to create a long-term ability to live with them—accommodating the waters to create a kind of new northern Venice, and making sure that with clever engineering, they keep this one from sinking.

And so they are encouraging the deepening of existing canals and the widening of existing rivers; they are building enormous water storage tanks beneath all new office blocks and parking lots; they are encouraging the making of grass roofs and enormous and endlessly thirsty public gardens; they are creating children's parks that are designed to be usable in dry weather, but which can be instantly transformed into shallow lakes for water sports when it rains or when the tides turn nasty; they are extending the great shipping docks and container terminals farther downstream along the Rhine and the Meuse, and in perhaps the most obviously sensible accommodation to the predicted new water levels, they are building in the old docks large numbers of structures that float. Just for now there are experimental pavilions being built on top of pontoons; before long, say the burghers, there will be housing estates and shopping malls, all floating gaily on the waters, no matter how high they rise.

Most other great cities are more conservative in outlook and more strapped for cash, and are simply throwing up modern versions of defensive earthworks. London, sitting uncomfortably in its basin of clay, will be affected mightily by the rising seas, but has not yet ventured into the kind of experimental water world that Rotterdam is planning. It expects its estuary towns all to be fully awash, it worries about its nearby atomic power stations—almost all of which are built beside the sea because of their need for cooling water—and it frets about water flooding into the underground railway system. But it is doing little—it can afford to do little, and has in any case lost the bold infrastructural vision of a century before—and may suffer more than most. The only defense London has now is the Thames Barrier, still a wildly futuristic-looking array of movable submerged gates and plates placed across the river at Greenwich. Designed in the 1970s to hold back storm surges, it has been raised more than a hundred times since it was built, and the rising of sea levels means it is sure to be raised with much greater frequency in the near future. But then what? When the barrier was constructed, the rate of sea-level rise was both constant and predictable; now it is accelerating, and there is less and less predictability about how the water at the river's mouth will behave. There is talk about a new barrier and what happens if none is in fact built. Lurid images are being published showing the Houses of Parliament awash, substations at Canary Wharf in a riot of short-circuiting sparks, the dean of St. Paul's splodging down the nave in his rubber boots, and the wheel of the London Eye reflected in its own lagoon. The city, already a byword for urban damp, is suddenly alarmed at the prospect of evolving into London-on-Sea, with all that implies.

New York, too, is thinking on similar defensive lines. Unlike London, it sits on stable geological features that rise well above sea level—but it was tunneled into and bored through until it resembled an ants' nest, and all the tunnels lie well below sea level. A storm surge coming into New York harbor could flood the subway lines without difficulty—even now huge pumps remove fourteen million gallons of seepage from the tracks and tunnels every day. But far more goes on underground than subways: the telecommunications cables and fiber-optic lines alone are vital for the running of the world's financial industries: soak them in the water and the world starts to fall apart. Small wonder the authorities have begun buying new pumps and creating new and hidden drainage schemes to keep the water away from all the high-technology equipment belowground; and new committees of specialists are sprouting like mushrooms, all bent on keeping New York from drowning on the day when the big water comes.

The city has nearly six hundred miles of shoreline, and since climate modelers believe that for technical reasons the Northeast of the United States will suffer rather greater sea-level rises than elsewhere if the Greenland glaciers melt quickly, these miles are suddenly being regarded as very vulnerable indeed. So from Paramus to Elizabeth, from Raritan Bay to Throgs Neck, plans for strengthening the docks and anchorages are being laid, emergency evacuation plans are being dusted off—and plans to build two huge flood barriers are being openly discussed. One would be sited a few hundred yards on the seaward side of the Verrazano Bridge, while the other would stand across the entrance to the Arthur Kill, between Staten Island and New Jersey. Engineers have already worked out their costs and the benefits: the politicians, though, are still waiting to be convinced.

There are currently some forty climate-change-related construction schemes under way in oceanside cities around the world, most of them in the Atlantic. Central to all of these preparations—whether they are advanced or not, whether they incorporate revolutionary designs or not, whether they are likely to work or not—is the assumption that the worst will happen to each city *when the weather gets bad.* The climate experts in all of these cities loudly assert that the weather is already getting bad—that as the world warms, as the ice melts and as the sea levels rise, so, and by way of a series of complicated physical changes that are still not wholly understood, is the weather becoming dramatically worse as well.

So the vulnerable cities are not merely going to slide slowly and elegantly under the sea, millimeter by millimeter. They are going to perch on the edge of inundation until a winter's night some years or decades hence, when the a storm rages itself into an uncontrollable maelstrom of fury, and a battering of huge waves breaches the dykes and the levees, and water courses into the city center in torrents, destroying all before it. Violent weather, added to higher water, turns an alarming development into a serially lethal one. And violent weather, it is said, is becoming much more common.

4. HERE IS THE WEATHER FORECAST

But is it? Is the Atlantic weather changing? Are there now climatic reasons for supposing that the ocean, much abused, is about to wreak its sweet revenge? We may pride ourselves in our modern sophistication—but the very fact that we ask such a question, that we are now so anxious, so mired in self-recrimination, transports us back to much the same level as the Mayans and the Caribs, who asked exactly the same questions centuries ago. Are we making the gods angry? was the question they asked. Is the ocean striking back at us? we inquire nervously today.

Anecdotal evidence suggests that something meteorologically untoward may be happening. In 2009, in the beach surf off Rio de Janeiro, for instance, there were sightings of Magellanic penguins, borne two thousand miles north from their customary homes in Patagonia. That caused much alarm and puzzlement. Biologists summoned to their aid thought they must have been following shoals of anchovies that had been swept north by changing currents and winds: Brazilian newspapers reported with incredulity the sight of bikini-clad young sunbathers carrying the creatures home to their refrigerators, with both women and birds frightened and helpless (and the penguins dying).

On the other side of the sea, and in another hemisphere, the impoverished republic of Liberia has recently suffered wave after wave of storms that have begun gnawing away the country's coast, causing hundreds of houses in some small communities to tumble into the sea and vanish. One larger town named Buchanan* recently appealed for funds to start building seawalls to halt the encroachment. The Liberian government has warned that unless the international community helps, the town and many like it will have to watch helplessly as streets are inundated and its people are forced to move inland, and that the country will be compelled to alter its very shape in order to accommodate the changing sea.

Then again, in Denmark there are other strange symptoms: the average national wind speeds appear to be gathering up, and in the Danish countryside sales of wind turbines are reported to be climbing because the more frequent gales seem suddenly so commercially alluring. In Cape Town huge forest fires have raged close to the city center, and the country's national flower, the King Protea, has been almost wiped out locally, as has its chief pollinator, the sugarbird. The violent rainstorms that doused such blazes a decade ago now no longer happen: the weather in the Eastern Cape has changed, say the locals: *the seasons*, one was quoted by the BBC as saying, have *run amok*.

And then, of course, there was Hurricane Katrina. The devastation caused by this Category Five storm, which was born in the Atlantic Ocean off the Bahamas on August 23, 2005, and struck the south coast of Louisiana and Mississippi six days later, was appalling. Though it did not hit New Orleans and was only a Category Three storm when it made landfall, nearly two thousand people died in and around the devastated river city, and damage to property was in the scores of billions of dollars, making it the costliest natural disaster in American history.

The government was severely criticized for its inept handling

---

* Many Liberian towns were named to honor an American president—Monrovia, the capital, was named for President James Monroe. Buchanan might be supposed to honor President James Buchanan, but it was in fact named for his cousin Thomas, the country's first governor when it was still an American colony.

of the storm's aftermath—but that overshadowed the stellar achievement of the one government agency charged with predicting the storm in the first place. Katrina was so classic in its development and construction that the National Weather Service, with uncanny precision, predicted just about everything about it. The NWS bulletin issued by the Baton Rouge, Louisiana, office just a few hours before Katrina made landfall remains a textbook instance of when official language can be far more chilling than the purplest of literary prose:

URGENT—WEATHER MESSAGE

NATIONAL WEATHER SERVICE NEW ORLEANS LA
1011 AM CDT SUN AUG 28, 2005

DEVASTATING DAMAGE EXPECTED

HURRICANE KATRINA: A MOST POWERFUL HURRICANE WITH UNPRECEDENTED STRENGTH, RIVALING THE INTENSITY OF HURRICANE CAMILLE OF 1969.

MOST OF THE AREA WILL BE UNINHABITABLE FOR WEEKS, PERHAPS LONGER. AT LEAST ONE HALF OF WELL CONSTRUCTED HOMES WILL HAVE ROOF AND WALL FAILURE. ALL GABLED ROOFS WILL FAIL, LEAVING THOSE HOMES SEVERELY DAMAGED OR DESTROYED.

THE MAJORITY OF INDUSTRIAL BUILDINGS WILL BECOME NON FUNCTIONAL. PARTIAL TO COMPLETE WALL AND ROOF FAILURE IS EXPECTED. ALL WOOD FRAMED LOW RISING APARTMENT BUILDINGS WILL BE DESTROYED. CONCRETE BLOCK LOW RISE APARTMENTS WILL SUSTAIN MAJOR DAMAGE, INCLUDING SOME WALL AND ROOF FAILURE.

HIGH RISE OFFICE AND APARTMENT BUILDINGS WILL SWAY DANGEROUSLY. A FEW TO THE POINT OF TOTAL COLLAPSE. ALL WINDOWS WILL BLOW OUT.

AIRBORNE DEBRIS WILL BE WIDESPREAD . . . AND MAY INCLUDE

HEAVY ITEMS SUCH AS HOUSEHOLD APPLIANCES AND EVEN LIGHT
VEHICLES. SPORT UTILITY VEHICLES AND LIGHT TRUCKS WILL BE
MOVED. THE BLOWN DEBRIS WILL CREATE ADDITIONAL DESTRUCTION.
PERSONS, PETS AND LIVESTOCK EXPOSED TO THE WINDS WILL FACE
CERTAIN DEATH IF STRUCK.

POWER OUTAGES WILL LAST FOR WEEKS . . . AS MOST POWER POLES
WILL BE DOWN AND TRANSFORMERS DESTROYED. WATER SHORTAGES
WILL MAKE HUMAN SUFFERING INCREDIBLE BY MODERN STANDARDS.

THE VAST MAJORITY OF NATIVE TREES WILL BE SNAPPED OR
UPROOTED. ONLY THE HEARTIEST WILL REMAIN STANDING . . . BUT BE
TOTALLY DEFOLIATED. FEW CROPS WILL REMAIN. LIVESTOCK LEFT
EXPOSED TO THE WINDS WILL BE KILLED.

AN INLAND HURRICANE WIND WARNING IS ISSUED WHEN SUSTAINED
WINDS NEAR HURRICANE FORCE OR FREQUENT GUSTS AT OR ABOVE
HURRICANE FORCE ARE CERTAIN WITHIN THE NEXT 12 TO 24 HOURS.

ONCE TROPICAL STORM AND HURRICANE FORCE WINDS ONSET, DO
NOT VENTURE OUTSIDE!

So what exactly was Katrina? Was it merely the name given
to the storm by the Weather Service—as names have been given
since 1953—*while in fact its real name should have been "global warming"*? Or is such a claim, which was first made by a well-known
columnist of the *Boston Globe* as the storm first hit, simply another example of the "unadulterated garbage," as an Australian
climatologist put it, that afflicts a debate that has now become
both very public and very political?

The questions that have loomed large ever since the end of the
2005 Atlantic hurricane season—which was especially ferocious,
with two more storms after Katrina that were indeed much
stronger, and record-breakingly so—is whether ocean warming is making hurricanes more numerous, whether it is making
them individually stronger and more lethal, and whether per-

haps it is making them both. And if ocean warming is the fault of humankind—then are we making hurricanes more deadly and more commonplace? Is it all, in other words, our fault?

In 2005 sharp battle lines were drawn over this issue—in a battle that happened to coincide with the extraordinary ferocity of both Katrina itself and that year's hurricane season generally. The 2004 season had been formidable, too: four titanic storms had hit Florida that summer, causing some $45 billion in damage. Now another season had brought death and destruction to an even greater degree. Something seemed to be going awry—to some a trend seemed to be developing.

Starting with a ripple of wind in the African savannah, monstrous and enduring hurricanes occasionally form over the Cape Verde Islands in the eastern Atlantic. Those few that make landfall in the Caribbean or the Americas—such as Andrew in 1992, or Bonnie, illustrated here, in 1998— can be dramatic and lethal.

Not surprisingly, press interest in the possible link between storms and man-made global warming grew wildly as the extent of the mayhem was fully realized: indeed, the climatologists' famous "hockey stick" graph, which had long been touted as showing a recent near-exponential rise in atmospheric warming, was itself almost duplicated in appearance by a graph drawn by the author Chris Mooney in 2007 that showed the number of articles in the reputable American press discussing the possibility of the link: it appeared to rise exponentially, too.

An Atlantic hurricane—a counterclockwise rotating windstorm, more accurately called an Atlantic tropical cyclone—is a surprisingly fragile creature. The place and manner in which it is conceived and born, the uncertain progress it takes toward its maturity, the direction and speed at which it then moves across the ocean, the ways in which it grows and then achieves its greatest strength, the mechanics of its decline and its subsequent staggering progress to extinction, are all the result of the tiniest and most subtle fluctuations in the condition of the ocean and the winds that feed, direct, and sustain it.

Very basically, hurricanes—the word is originally Carib; a hurricane is specifically an Atlantic phenomenon only*—are created in the northern summertime, usually between June and November. For them to form, very warm subtropical seas must be overlain by relatively cold air, such that any moist air that rises from the sea is cooled rather quickly. Many hurricanes are first spawned in the shallow waters off the eastern Caribbean; a number of often very serious hurricanes are born much farther

---

* In the North Pacific such storms are named for the Cantonese combination for great wind, *da-feng*, hence *typhoon*; while in the Indian and South Pacific oceans *cyclone*, the generic scientific name for a circular storm, remains good enough. With hurricane science so currently fashionable, new words are being cheerfully invented: a student of historic storms is now known as a *paleotempestologist*.

away, in the shallow waters of the eastern Atlantic, around the Cape Verde Islands. The conditions in what are known as these cyclogenetic regions are essentially the same: plenty of warm water below, nicely cool air above, and the rising up of water vapor and its subsequent unusually fast refrigeration.

This rapid cooling—which causes clouds, and rainfall, and the release of latent heat by the mass of air—can under certain (and still not fully understood) circumstances result in the creation of great and powerful disturbances in the vertical columns of air—the making of invisible phenomena that a glider pilot or a weather balloon would readily recognize as very strong eddies and thermals.

Beyond this column of air, the pressure gradients in the cyclogenetic latitudes, where these columns of fretful air are born, happen to generate winds, most usually trade winds, that blow from the northeast. These winds contrive to nudge or tickle any unstable column of air into movement—and this movement, under the effect of the Coriolis force, on some infrequent occasions causes the column to start rotating, turning always counterclockwise in the Northern Hemisphere. The prevailing trade winds then steer this fragile and vaguely rotating column westward across the sea—and providing that the water below is warm enough, so that the air ascending into the column is adequately moist, and providing that the upper atmosphere is sufficiently cool to condense it into cloud and rain, so more disturbance will be funneled into the rotating column and it will become supercharged with thermal energy that, when translated into kinetic energy, will cause the winds caused by its rotation to go around faster and faster. Once in a while, fifteen times or so each season, this whirling mass of air and cloud will develop into a proper storm. Depending on the speed of its maximum sustained winds, the storm can become classified as a hurricane, and if there is enough warm water to fuel it, it will spiral up through the five official categories of vigor and potential danger until it becomes a thing of awesome proportion and power.

The ultimate key to the growth of a hurricane is the warmth of the water over which it passes. One of the reasons Katrina became so vicious was that when it passed westward from its birthplace above the Bahamas, it slid right over one of the feeder currents of the Gulf Stream, a narrow underwater river in the Gulf of Mexico known as the Loop Current—and in August 2005 the Loop Current was unusually warm. Its difference from normal may have been minimal—measured only in fraction of degrees—but for something as sensitive as a developing hurricane, it was enough to make an enormous difference. The additional fuel that this slightly warmer water provided powered the relatively modest Katrina storm right up to Category Five strength. This was the development that prompted the National Weather Service to issue its famously dramatic Sunday message, and the storm surge it created, and its eventual landing—though by then rather weaker—led to the dreadful catastrophe that followed the next day.

If warm water is the key, and rising sea temperatures result in more warm water, then the correlation would seem obvious: warmer waters mean fiercer hurricanes, and possibly more of them. But the science is not that simple, and the correlation—at least in history—seems unconvincing. There is no certainty, for example, that any kind of trend has become truly apparent. In the short term the highly active hurricane seasons of 2004 and 2005 were followed by two years of below-average activity, then by a year, 2008, that with sixteen named hurricanes was only moderately severe, and then by 2009, which was about as violent as a vicarage tea party. In the medium term—in the years since 1995—there have been rather more hurricanes, and a very large number of stronger ones. But the very long-term statistics—and there is a project named HURDAT that seeks to find all available data on all Atlantic hurricanes since 1851—is tending to show less of a trend and instead a number of cyclical patterns.

Many climatologists argue for the importance, in any discussion of sea temperature change, of what is known as thermohaline circulation: this involves the sinking into the cold depths of the saltier water that is formed by the evaporation of the sea at its warm surface, and the drawing into the ocean of warmer water to replace that which has been sunk. There appears to be a cycle—known as the Atlantic Multidecadal Oscillation—that is linked in some way to changes in thermohaline circulation. The years since 1995 have seen rather more intense thermohaline circulation than usual—though it is within the boundaries of its oscillations measured in the past—and so some believe that instead of witnessing a trend, we might be witnessing a normal cycle, with the oscillation today in one of its routine warm phases. This does not mean that warming is not happening: but the fact that it may be being superimposed on a cyclical phenomenon makes for more complication than is comfortable. (And of course the warming is happening, and it could be affecting the thermohaline circulation, rather than the other way around.)

Moreover, even the most ardent believers in anthropogenic climate change acknowledge that superviolent storms like Katrina cannot in themselves be ascribable to global warming—only if there are a very large number of such catastrophes could such a correlation be certain, and there is still precious little data to support that. All that can be stated with certainty is a very obvious reality: that recent Atlantic storms have been lethal and costly not because there are more of them, *but because more people have settled and more expensive buildings have been built in the places where the storms have happened to strike.*

So the best short-term solution to the regular destruction of so many Gulf and Atlantic coastal communities perhaps needs to be stated again: it requires not so much any need to cool the world, but to persuade people not to come to live in those places where, habitually, the world goes mad. There are many excellent reasons for wanting to limit carbon emissions, but preventing storm damage to American coastal communities is not one of them. The communities should never have been built. Strip the vulnerable coastlines of Florida, Louisiana, Alabama, Mississippi, and Texas of great mansions and sprawling oil refineries and strip malls and country clubs and casinos, and suggest to the inhabitants that they move inland and away from the hurricane corridors—then to a degree the human problem solves itself. The tropical Atlantic Ocean and its neighbor seas are capable of very great violence—perhaps greater today than ever before. Until they can be permitted or persuaded to calm down, the best immediate solution is simply to keep their waters and their winds at arm's length. So long as the ocean is still *going on*, then down in hurricane land perhaps mankind should be thinking about *going away*.

5. THE LITTLE-KNOWN SEA

The warming of our oceans has its most visible effects on matters great and familiar—on Rotterdam, on hurricanes, on penguins or anchovies. But the rise in temperature, however caused, also seems to work its way into more unfamiliar worlds—and one of them illustrates the notion that it is probably best if we leave the seas alone, because we know much less about them than we think. For there is much currently expressed concern over whether global warming will have a particular effect—whether for good or ill we do not yet know—on a creature that turns out to be probably perhaps the most plentiful species on our planet, and yet one of whose very existence we were quite ignorant until 1986. That was the year when this creature was first discovered, and it was found in the Atlantic Ocean.

The sea teems with tiny drifting beings, plankton, which are suspended, wafting, moving aimlessly among the placid similitude of the world beneath the surface. Where they are and what they do there depends much on the nature of the water on which they waft: on whether it is warm or cold—an attribute that depends on the one hand on latitude, and on the other hand on depth, for they drift inside a three-dimensional universe, whether it is very salty or less so, whether the pressure is high or not so high, whether the sea chemistry is benign or strange, whether it is light or dark—for no light at all finds its way below a thousand meters, and it is perpetually pitch-dark except for vague glimmers from the blooms of bioluminescent creatures and the tiny orange firefly glows from brave beasts that flourish beside the scalding thermal vents. Yet in every zone, from the oxygen-rich splashiness of the coastal waters to the near-freezing blackness and iron-crushing pressures of the deep abyssal trenches, there is almost invariably life, and most of it is microscopic, and most of it is still unknown.

Many of the tiny creatures that inhabit the oceans' well-lit upper waters emit gas or gaseous compounds. One hard-shelled algal beast, *Emiliana huxleyi*, emits dimethyl sulfide, which some believe contributes to the unique aroma we call the smell of the sea.* But most, being photosynthesizing animals, suck in carbon dioxide, make carbohydrates, and in immense quantities turn out oxygen. Maybe 70 percent of the planet's total oxygen comes from such seaborne organisms: one of these was discovered in 1986—a blue-green algae hitherto not known to exist, and which was given the name *Prochlorococcus*.

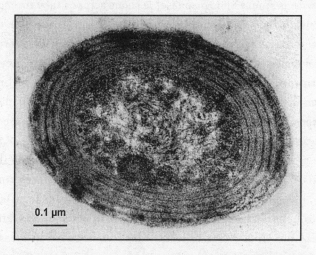

0.1 µm

Quite probably the most abundant living creature on the planet, the cyanobacterium Prochlorococcus *was first discovered in the Atlantic's Sargasso Sea in 1986. These minute creatures employ their chlorophyll-b to produce as much as one-fifth of the world's atmospheric oxygen.*

---

* The sea does not smell out at sea: a sailor notices the aroma only on approach to land, where it is caused by the reaction of dimethyl sulfide and seaweed, and should more properly be called *the smell of the shore*. Where the gas is emitted in mid-ocean it rises into marine clouds and is distributed around the planet, where it joins the complex matrix of elements that go to make up life. To add further to the heroic reputation of this tiny creature, the trillions of *E. huxleyi*'s cast-off shells that fall to the ocean bottom end up forming calcareous deposits, like chalk.

A young researcher at the Massachusetts Institute of Technology, Penny Chisholm, first found the creature in the Sargasso Sea. She and Rob Olsen, her colleague from Woods Hole, were on a research vessel sailing from Cape Cod to Bermuda with, as an onboard trial, a machine normally used in hospitals to assay blood and known as a flow cytometer. The principle of this device is simple enough: a laser is shone across a tube through which a fluid flows at speed—blood in hospitals, seawater on Penny Chisholm's boat—and detectors pick up the light scattered and deflected by any tiny particles, invisible to the naked eye, suspended in the flowing liquid. The two researchers had no idea the machine would even work on the boat; and if it did, they expected to find numerous examples of a particular blue-green algae they already knew existed.

What they did not imagine was that the device would show the existence of millions and millions of even tinier creatures, tiny, oval-shaped living entities, around six microns in diameter, one two-hundredth of the width of a human hair. But these creatures were not simply tiny; once examined under electron microscopes, they were found to have incorporated into their minute workings a type of chlorophyll that permitted them to absorb carbon dioxide and to extract from the seawater a tiny amount of oxygen, which then escaped into the atmosphere.

Taken individually the amount of free oxygen that any one of these algae might produce is microscopically insignificant; but Penny Chisholm calculated that *Prochlorococcus* existed in such unimaginably large numbers—one hundred thousand of them in a single cubic centimeter of water, perhaps a trillion trillion of them in total—that they were quite probably the most common creature in all the world, and would in total produce immense quantities of oxygen.

They prefer to live in the warmer seas, essentially wafting

about in the oceans between 40 degrees north and 40 degrees south, or south of a line connecting New York and Lisbon in the north, north of another drawn between Buenos Aires and Cape Town. There they lie, contentedly at the bottom of the food chain, waiting to be eaten by tiny shrimp that would then be consumed by small fish, and on and on, up until the hungriest predators of all, mankind. Or one perhaps should say they probably lie at the base of the food chain, for while it is difficult to imagine anything smaller existing in the sea, Dr. Chisholm felt that *Prochlorococcus* was an example of how nature had once again displayed its infinite capacity, as she put it, to *humble* the world of science, and could readily do so again. Before 1986 we did not know that such a creature existed; now it is recognized as perhaps the most common being on earth—or rather in the ocean—and it plays a central role in keeping land-based creatures alive.

To dramatize this creature's importance, it can reasonably be claimed that one in every five breaths any human being takes contains oxygen created out at sea, and quite specifically by *Prochlorococcus*. We now know it exists, and it goes without saying that if anything disastrous were ever to befall it, the survival of all beings that require oxygen would be placed at risk. In the two decades since *Prochlorococcus* was found, a great deal of research has been done to see what might harm it, and how. Specifically, researchers have been trying to determine whether the warming of the seas might limit its ability to absorb carbon dioxide and frustrate its propensity for creating oxygen.

It turns out that *Prochlorococcus* seems thus far happily resilient to the warming of the planet. It likes warm seas and flourishes in them. Any increase of the sea's temperature might well cause the range of *Prochlorococcus* to expand into the newly warmed waters, to push beyond the present day's 40-degree-latitude lines—and that might have its own effect on not just the outward flow of oxygen into the atmosphere, *but on the absorption of carbon dioxide already in it.*

It is tempting—but entirely fanciful—to imagine that such a development might balance some of the expanded emissions of greenhouse gases that are so troubling humankind today. An expansion of the range and population of *Prochlorococcus* might well turn out to be a component of the earth's self-regulating mechanism, so crucial to James Lovelock's famous Gaia theory—which holds that the world is to be viewed as a self-contained living being, able to change its own ways and to deal with its changing circumstances. This curious animalcule might be even more precious than at first supposed: not merely supplying the air that we breathe but somehow dealing with our most dangerous pollutant. But this is an idle thought: there is no evidence; a lot of research still needs to be done.

And yet all this concerns a being we were entirely unaware of twenty years after man first went to the moon. Those who have long claimed the sea to be far less known than outer space seem suddenly to have a special brand of wisdom.

· · ·

The great forces that created the Atlantic in the first place will in time—in a very long time, in human terms—also destroy it. The forces, part of the tectonic mechanisms of the planet, are better understood today than when they were first revealed in the 1960s, but they still present something of a mystery. They are difficult to appreciate in part because they are so complex, but also because of the time scale involved: we are around to witness only the tiny incremental movements and shifts by which the world changes its topography, even though those tiny shifts can often be, for mankind, catastrophically lethal and terrifying.

The earthquakes, eruptions, and tsunamis that have shaken the world during the two thousand years mankind has been

able to chronicle them have seemed like gargantuan affairs—
death and destruction on what for humans is a titanic scale have
been rained down by events that are now a familiar part of history: Lisbon 1755, Krakatoa 1883, San Francisco 1906, Tang-
shan 1976, Sumatra 2004. Seen in a planetary context, events
like these have barely any significance. They are tiny shape-
shifts that come to assume real importance only when millions
of them, over millions of years, have taken place. The Sumatran
tsunami of December 26, 2004, may have killed a quarter of a
million people, and may have passed into human history as one
of the greatest natural disasters of all time—but it moved the
sea floor south of Sumatra no more a few meters northward, and
the sea south of Sumatra is many thousands of miles wide. It
would take a million years' worth of submarine Indian Ocean
earthquakes before this corner of the world would appear to
have changed its appearance even minimally.

It is a fortuitous accident of tectonics that the Atlantic is seismically the least vulnerable of the oceans. The Indian Ocean is
scarred by subduction zones and faults, and it was no surprise
to the geological community that the 2004 tsunami originated
there. The Pacific is almost entirely surrounded by volcanoes
and is rocked by ceaseless earthquaking from Japan to Alaska,
from California to Chile, from Kamchatka to New Zealand. But
the Atlantic, by contrast, has as its geological centerpiece only
the Mid-Atlantic Ridge, which is certainly opening up and dis-
gorging lava all the while—but does so in a somewhat lethargic,
somnolent manner, and by the standards of the neighbor ocean
can hardly be called seismically violent. When Anak Krakatoa
was born off the coast of Java in 1930 it appeared with terrible
violence and drama; when Surtsey was born off the coast of Iceland thirty-three years afterward it proved spectacular to see
but was more a boisterous oozing than a calamitous detonation.

That is not to say there has been an absence of memorable activity in the Atlantic. Much has happened, and the recent occurrences faithfully and fully recorded, more so than elsewhere because sophisticated, organized, scientifically curious, and technologically able man has been living on the shores of the Atlantic for very much longer than around the other oceans.* There are many early records of violent seismic activity in the eastern Atlantic between Portugal and the Azores, for instance, beginning with a record of flooding in the Tagus in the winter of 1531, and huge waves in the sea nearby that wrecked scores of ill-prepared fishing boats and sailing vessels. Then there was the enormous earthquake that all but destroyed Lisbon on November 1, 1755: it is said to have sent massive sea waves to Madeira and Agadir, as one might expect, but also to have caused destruction as far away as Martinique, in the Caribbean.

The question of whether destructive tsunamis are likely to travel across the Atlantic has prompted some recent concern, ever since the Indian Ocean wave of 2004, which spread rapidly and killed many, from Bengal to Sri Lanka and beyond. The records show few credible accounts of long-distance tsunamis being generated in the Atlantic—the Lisbon event is probably the only one. The Grand Banks earthquake of November 1929, triggered by a magnitude 7.2 earthquake south of Newfoundland, has been studied in great detail—great waves of sand and water, known as turbidity currents, swept down the undersea canyons and severed many of the submarine telegraph cables, the precise thirteen-hour sequence of their failures being recorded by the sudden loss of signal—but it seems not to have created much seismic excitement beyond the St. Lawrence Estuary. Similarly,

---

* This is not to diminish the claims of both India and China to having some early interest in and aptitude for earth sciences, but until very lately modern developments in seismology and vulcanology have been nearly exclusively Western.

the immense explosion in Halifax harbor in December 1917, mentioned in chapter 4, did transmit a number of rolling tsunamis—but they lasted only a few minutes and never made it into the open ocean.

A three-hundred-mile long sand deposit running along the east coast of Scotland, between Dunbar and Inverness, is thought to have resulted from a famous submarine landslide off the Norwegian coast eight thousand years ago. And all sorts of mayhem is thought to have been caused on the far side of the ocean when Lake Agassiz collapsed, but no physical tsunami evidence for it has yet been uncovered, even though researchers are hoping to find fossil sandbars on the west coast of the Labrador Sea. Until then the rather tenuous suggestion, already noted, that Black Sea farming patterns changed as a result of the meter-high rise in sea level remains the sole suggestion of transoceanic impact of the great Agassiz flood.

The concern over the possibility that destructive superwaves might be able to cross the Atlantic has come about in part because of what happened in the Indian Ocean in 2004; but in rather larger part it has also spiked because of an item of wild speculation that appeared in the press in 2000—and which held that New York City was at risk of inundation because of an impending landslide on the island of La Palma, in the Canary Islands. News stories in some of the more excitable press—and a lengthy documentary film shown by the BBC—had it that a block of basalt the size of the Isle of Man was about to fall off the western side of the Cumbre Vieja volcano, and that the American president should take immediate note lest they be caught unprepared for the devastating effects of a wave that would race westward across the ocean at five hundred miles an hour and, when it struck, would drown major American cities beneath a wall of water scores of feet high.

It later emerged that the researchers who first informed the press and who helped with the BBC film, though based at the University of London, were funded by a large Chicago-based insurance brokerage, Aon Benfield, which would doubtless welcome a public made skittish by the publication of reports of ever more bizarre threats—MEGA-WAVE TO ENGULF MANHATTAN!— to world serenity. The seismological community generally has poured scorn on the reports, has said that the mathematical models used were outdated and wrong, that the chances of such a landslide in La Palma were vanishingly remote, and that tsunamis have little history of traveling across the Atlantic Ocean, though admittedly for reasons unfathomed. The researchers retired to lick their wounds; the BBC issued something of a retraction; and most recently the European Space Agency said it would conduct a survey of the Cumbre Vieja volcano to ascertain its stability and presumably try to reassure the world that New York is not about to be drowned, certainly not imminently, and probably never.

The volcanoes of the Atlantic are also generally more benign than those elsewhere. To be sure, there are vicious examples, most of them in the Caribbean. On Martinique, there is most notoriously Mont Pelée, which erupted on Ascension Day in 1902, and killed almost all the twenty-eight thousand inhabitants of the town below with its rolling clouds of red-hot ash and superheated air. A prisoner in an almost airless cell survived, and joined the Barnum & Bailey circus. Patrick Leigh Fermor wrote *The Violins of Saint-Jacques*, a novel imagining that the celebration ball going on when the volcano erupted was swept in its entirety into the sea, such that the orchestra, still playing gamely, can be heard to this day by fishermen sailing by overhead.

Some others are more discommoding than catastrophic—the very geologically similar complex of the Soufrière Hills on the British colonial possession of Montserrat, for example, erupted in 1995, killing many fewer people but ruining the island capital of Plymouth and forcing its abandonment. Dust from the eruption in 2010 of Eyjafjoll in southern Iceland severely disrupted air transport all across Europe. And back in 1961, the entire population of the South Atlantic island of Tristan da Cunha—another British possession, with some 250 inhabitants—had to be evacuated to England after their volcano erupted, directly threatening the little settlement of Edinburgh-of-the-Seven-Seas.

Fewer than three hundred people—seven families, all interrelated—live on the mid-oceanic volcanic island of Tristan da Cunha, 1,800 miles west of the South African coast. Generally the British possession is isolated and alone, the islanders always fretting that their volcano might erupt again, as it did in 1961.

When this eruption began all the Tristanians, elderly women and babes-in-arms among them, took off in longboats to Night-

ingale Island, twenty miles away, and sheltered off the beach
awaiting rescue—the Atlantic Ocean seemingly offering them
safer asylum than the solid land on which their ancestors had
chosen to settle. But two years later, once the mountain had settled itself down again, most of the islanders elected to return.
They live there still, proudly offering themselves to passing ships
as *the most isolated inhabited island in the world.* The volcano may
growl and steam, the sulfurous gases may produce widespread
illness, the islanders' isolation may bring all the disadvantages
of inbreeding, and the economic trials of the inhabitants may be
endless and legion, but in this otherwise unvisited nook of the
Atlantic, mankind clings on with limpet-like tenacity, as if to try
to remind the ocean just who claims mastery.

. . .

Some of the islanders on Tristan, and the technicians at the
weather station three hundred miles farther south, on yet another British colonial sibling, Gough Island, might have noticed
something else rather unusual in recent years.

The prevailing winds in both places—but most especially on
Gough Island—are from the west. In Gough they are usually very
strong: the island, which lies at just about 40 degrees, 31 minutes south, is very much in the Roaring Forties, and the westerlies here do indeed roar, without cease.

Or at least they used to. During the last thirty years or so the
climate in these latitudes has somewhat altered. The westerlies
do not blow so strongly or so often, and are now not so perpetual
as they still seem to be just a few score miles to the south. It is
as though the Southern Ocean Super-Gyre, the climatic forcing agent that is ultimately responsible for the high wind belts
around the Antarctic, and which sailors know as the Roaring
Forties, the Furious Fifties, and the Shrieking Sixties, has recently shifted southward, toward the pole. The reason for this,

climatologists insist, is the human-induced depletion of ozone in the atmosphere above the western Antarctic: it seems that the winds might have slipped down toward the ozone hole, as it were, to fill the gap the ozone has left behind, confirming the age-old principle of *nature's abhorrence of a vacuum.*

The effect of this southward shift of the Atlantic winds has been most unexpected: it has caused warm and salty water to dribble out into the Atlantic Ocean from the Indian Ocean, through some hitherto unknown deep-sea phenomenon known as the Agulhas Leakage. This warm and saline seawater seemingly enters the North Brazil Current—an exceptionally complicated northbound current that flows along the coast of Brazil toward the Caribbean. It is believed that this water could then enter the birth waters of the Gulf Stream and change its strength, temperature, salinity, and direction even more than it is being changed today.

Thus is a further complication—and one almost certainly initiated by mankind, if the filling-the-ozone-gap theory is correct—being added to the mix that is the Atlantic Ocean. The weather patterns around the sea will change still further—though for good or ill, no one has any current idea. All that is certain is this: with new and fiercer hurricanes forming off Cape Verde, with volcanoes erupting in Montserrat, with the sea level of Rotterdam rising and the ice in East Greenland melting, with black smokers and white smokers generating more heat and dull red light and so nurturing the clouds of thermophilic bacteria near the Mid-Atlantic Ridge, with Surtsey building itself up yet again, with Eyjafjoll erupting dustily, with Iceland still splitting apart and the cables running across the Grand Banks in danger of being broken once more, with *Prochlorococcus* expanding its range and burping out yet more oxygen into the air, and with, as now, Indian Ocean water leaking across into and warming and making more salty the seas near Gough Island, Brazil, and the Caribbean—with one or all of these things now happening, and with much wondering over whether humanity is able to accommodate them or whether they signal the beginning of the end of man's relationship with this most vitally important of seas—it is clear that some exceedingly strange things are happening in today's Atlantic Ocean, and no one is quite sure why.

EPILOGUE:
FALLS THE SHADOW,
FADES THE SEA

———

All the world's a stage
And all the men and women merely players:
They have their exits . . .

The tiny beacon they call the *lighthouse at the end of the world* will one day meet up with another of its kind that presently stands ten thousand miles away on the far side of the globe. And when that happens, lighthouse striking lighthouse with the slowest and gentlest of collisions, the Atlantic Ocean as we know it will cease to be.

The final moment of the Atlantic's existence will come in about 170 million years. It will be brought about by an episode of highly improbable-looking tectonic gymnastics, in which the tip of South America snakes itself down and around the entire continent of Antarctica, then heads back up northward and collides with the tip of the Malay Peninsula somewhere in the region of Singapore.

A great deal of mathematical modeling had to be done to reach this vision of the world's future appearance. Much of the calculating has been worked up by a Texas-based group that specializes in paleogeography and tectonic futurism, headed by Christopher Scotese. Another group based in England, known informally as The Future Is Wild, and with more obviously commercial ambitions, is hoping to find in Hollywood and the publishing industry a market for its carefully modeled visions of the planet's geological and biological future. Both groups have devised scenarios for the coming couple of hundred millions of years: both agree that the supercontinent whose breakup gave rise to the Atlantic—Pangaea—will one day re-create itself,* and they have agreed to name it Pangaea Ultima. Precisely how the continents that currently exist get to this point is a matter for scholarly argument, but there is agreement that in the end the world will have one continent, it will be surrounded by one sea, and all of the oceans that currently exist, the Atlantic included, will have long since been consigned to history.

However, at least at this moment the very opposite seems to be happening. The Atlantic, far from heading toward history, is getting much bigger and wider. The lines of volcanoes and rifts along the Mid-Atlantic Ridge are continuing to spew new mantle material to the surface, and the convection currents below are continuing to shift apart the seabeds on either side of the ridge, like conveyor belts moving in opposite directions: the Americas are being moved ever farther west, and Africa and Eurasia slide ponderously to the east. This is a process that all geologists be-

---

* There is a widespread belief in the geophysical community that there is only a finite amount both of land and water on the planet, and that both are constantly reordering their shapes and relations to one another, and in a cyclical manner. The regular reordering of the oceans even has been named the Wilson Cycle, after the Canadian father of the plate tectonic theory, J. Tuzo Wilson. The believers hold that continents, too, break apart and reform themselves every 400–500 million years, and that the earth as presently configured is in a state halfway between the maximum spread of its continental fragments and their next re-formation into a single massive one.

lieve will continue for maybe another five million years, perhaps much longer. It is after this that the mathematical models start to diverge.

One group predicts what it calls *extroversion*, a process whereby the continents appear to peel open like blooming flowers, only then to shift back upon themselves and eventually coalesce into one. In this scenario, the Atlantic continues to open ever wider; the Pacific is slowly squeezed closed as the two American continents pivot around Siberia toward a collision with East Asia; and Africa, India, and Antarctica move as one around and up toward South Asia's various peninsulas and islands; until Pangaea Ultima is finally made, and for the time being the world halts in its tracks, with its gigantic new landmass surrounded by an even more gigantic, newly shaped sea.

The other group of modelers are the supporters of *introversion*, and in this scheme, which is somewhat more complicated, the Atlantic, after a period of expansion, would suddenly begin to shrink because subduction zones would form along the eastern seaboards of both North and South America. Lines of volcanoes would start to erupt off New York City, Halifax, and Rio de Janeiro—though both places, and all other centers of human habitation, would by this time have long since ceased to exist—and the ocean floor would start to vanish beneath the American continent. Europe and Africa would at the same time continue to collide, eventually squeezing the Mediterranean to nothing; Baja California would slide northward; Antarctica would head north. In some two hundred million years from now North America would collide with Africa, and South America would slide itself around the tip of southern Africa, heading northeastward until the moment when it collided with Southeast Asia.

It is this scenario that involves, at least in theory, the tantalizing prospect of the collision of two lighthouses.

. . .

The tip of South America, where the Andes plunges into the two oceans that the mountains separate, is a place of vivid and lonely beauty, and of lives lived under the impress of endless strong and stirringly cold westerly gales. Cape Horn—a low brown island less distinguished in appearance than its history suggests—dominates the perception of the region, but there are also the snow-trimmed peaks of Tierra del Fuego, the dusty plains of Patagonia, the windswept estancias where sheep huddle against hedges and where bushes of *calafate* tremble in the storms; the corrugated iron-roofed *frigorificos*, where farmers and their gauchos would bring their lambs for slaughter and shipment, the bleached bones of long-dead whales lying along the shores of the Strait of Magellan, the bleached spars of long-lost clipper ships lying in the bays to which they ran, fatefully, on failing to double the Horn—these are what makes the southern extension of the Andes so frightful, and alluring.

Lying twenty miles off the mainland's eastern tip there is an island—Isla de Los Estados, the locals call it, the Spanish translation for Staten Island, since this elongate jumble of sharp peaks and deep valleys and stunted beech trees and sphagnum bogs and the ruins of old prison camps was first discovered and named by the Dutch, for the states general that directed the expedition. This is the *other* Staten Island; and while the one settled by the Dutch in New York is now a successful sprawling suburb with half a million people, Staten Island in South America has no one permanently living there at all. It is desolate, forbidding, quite inhospitable to man. A succession of lighthouses built there were abandoned because of the gales; even a sturdy military prison erected there in 1899 lasted only three years before being damaged by storms, provoking riots and escapes. Nowadays it has been declared a wildlife refuge for its colonies of Magellanic penguins, and a small detachment of Argentine sailors is sent there on forty-five-day rotations: they roundly dislike it, with its foul weather and inhospitable terrain.

Jules Verne had a peculiar lifelong fascination with Isla de los Estados, though he never visited. He wrote his final novel—*Lighthouse at the End of the World*—about a rollicking piece of gangsterish behavior on the island, and it centered on the dousing of the light and the luring onto the island's rocks of passing merchantmen. A century later a Parisian navigation enthusiast named André Bronner, in a delightful flash of Gallic madness, rebuilt the last of these lights, which had fallen down after being abandoned. He said later that he realized how important this single dim flicker of a light had been to all the great clippers and steamships making their way around the Horn in darkness; he developed a blinding obsession with the romance of this tiny gleam in the wilds of nowhere and managed to raise money from rich Parisian friends to build a replacement.

It took Bronner and seven similarly deranged colleagues two long midsummer months to build the new light. He took pâté and cognac and cases of decent burgundies to sustain the party, and employed a composer to write a *symphony at the end of the world*, which was played on the wild March day in 1998 when the light was handed over to the Argentine navy. The sailors who man the base now attend to the light, a modest affair that is powered by solar panels, requires little maintenance, and like most of the Staten Island lights that preceded it, is as a warning to Cape Horners all but useless. The predecessors were too small, and for some curious reason were all built behind obscuring mountain ranges. Bronner's new version has been rendered of limited value by the advent of GPS navigation, which keeps even the most delicate vessels on secure passage for a transit of the Cape.

So far as this story of the Atlantic's life span is concerned, the

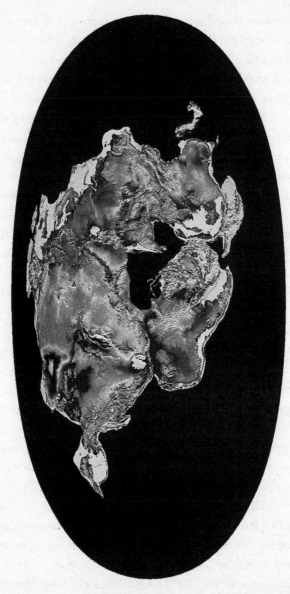

Two hundred fifty million years from now the continents will have co-alesced into another Pangaea, and the only internal body of water will be the stagnant relic of the Indian Ocean, the Capricorn Sea. The Atlantic Ocean, 440 million years after its birth, will have vanished clear away.

symbolic importance of the Staten Island light goes well beyond its usefulness. For the headland on which it is built—the island's northeasternmost cliff, just below Mount Richardson and Pickersgill Point, which stand as testament to the early British explorers of the region—is likely to be the first part of the Americas to collide with Asia, once the world is done with its predicted orgy of moving.

If the mathematical models are correct, in a little less than two hundred million years, the site of the light at Mount Richardson will be slowly edging toward the place where a light now stands at the most southerly point of the Malay Peninsula. This is Raffles Lighthouse, built in 1854, and lighting both the entrance to Singapore harbor and the Strait of Malacca. But when Raffles meets Richardson, when Singapore meets Staten Island, then, and finally, the long and slowly squeezed waters of the Atlantic Ocean will have been compelled to go elsewhere. The maps that Christopher Scotese creates show a small inland sea bordered by India, Arabia, East Africa, Argentina, and Sumatra; but this is hardly a sea, will in all likelihood not last, and has the melancholy distinction only of holding the last captive molecules of what was once the oldest and—in terms of the civilizations around it—the grandest ocean on the planet.

The Atlantic Ocean was born 190 million years ago; and given the mechanics and the timing of its likely death, it will survive as an ocean for maybe another 180 million years. Its total life span will thus be getting on for 400 million years—years almost entirely given over to gigantic geological dramas, to climatic phenomena on a scale barely possible to imagine, to the evolution and extinction of thousands of kinds of animals, birds, fish, plants, and single-celled beings, and of all stages in between.

For maybe 200,000 of those 400 million years, humankind

existed and flourished on the shores of the ocean. He and she and their kin first populated the east of the sea, then swept around and across the landmasses on the far side of the world before appearing on the west side of the same ocean some thousands of years later. Humans were powerfully afraid of the ocean for centuries, assumed it represented the edge of the known world and was populated by terrifying monsters. They ventured into it timidly and retreated from it swiftly—and then finally they crossed it, from east to west, in the eleventh century of modern recorded time, and in doing so found that far from being the edge of all the world, the Atlantic was now a bridge to an entirely new one.

It took four more centuries to find it properly. But once the existence of that new world had become a clear and undeniable certainty, once it was accepted that the water just crossed was indeed a newly known ocean, then this body of water, three thousand miles wide in the north, four thousand in the south, and rather less than two thousand at the ocean's waist between Africa and Brazil, became the central stage for all manner of humanity's most stupendous endeavors and amazements.

The ocean became, in a sense, the cradle of modern Western civilization—the inland sea of the civilized Western world, the home of a new pan-Atlantic civilization itself. All manner of discoveries, inventions, realizations, ideas, the mosaic of morsels by which humankind advanced, were made in and around or by way of some indirect connection with the sea. Parliamentary democracy. A homeland for world Jewry. Long-distance radio communication. The Vinland Map. The suppression of slavery. The realization of continental drift and plate tectonics. The Atlantic Charter. The British Empire. The knarr, the curragh, the galleon, the ironclad, and the battleship. The discovery of longitude. Codfish. Erskine Childers. Winslow Homer. The convoy system. St. Helena. Puerto Madryn. Debussy. Monet. Rachel Carson.

Eriksson, Columbus, Vespucci. The Hanseatic League. Ernest Shackleton. The Black Ball Line. The submarine telegraph cable. The Wright brothers, Alcock and Brown, Lindbergh. Beryl Markham. The submarine. Ellis Island. Hurricanes. Atlantic Creek. Icebergs. *Titanic. Lusitania. Torrey Canyon.* The Eddystone Light. *Bathybius. Prochlorococcus.* Shipping containers. NATO. The polders. The Greenland ice cap. The United Kingdom. Brazil, Argentina, Canada. The United States of America.

All these, and a thousand things and people and beasts and events and occurrences and people, go to make up today's Atlantic. They act as a reminder of the immense complexity of an ocean that has been central and pivotal to the human story. They are all now part of a new continuum of study that has come to be known in recent years as Atlantic History, a discipline now widely taught, and taken so seriously that there is now a history of itself, *a history of a history,* so critical has the idea of an Atlantic identity become to both the contemporary and the future world.

But such grand ideas, though necessary meat and drink to the academic world, can be elusive, fugitive concepts to those who merely like to stand on an Atlantic cliff top and contemplate the awful majesty of the sea rolling and unrolling away to the horizon. To them—to us, I would rather say, for this account has been a story told for those who regard the sea less as a concept than as a capricious and wondrous confection of water and waves and wind, of animals and birds, of ships and man—I would offer just a final story. It is a story of a forgotten man, and his small and lonely struggle with this sea, and a struggle in which the sea, as always, won. It involves a shipwreck, and a rescue, and a lonely death.

. . .

Perhaps we all have a secret wreck story to cherish—a saga to think about in the warmth of a much-eiderdowned bed, maybe, while a cold rain is lashing down on the windowpanes, and the trees are thrashing, and one prays silently to oneself for any sailors out on a wild night like this. My own came to me in a book I once read. It was just such a think-of-the-sailors night—cold and exceptionally wild—and I was staying on a lonely *estancia* in southern Patagonia, huddled beside a log fire of baronial dimension, a hot whisky to hand. I was reading by a dim library light the extraordinary story of wreck and ruin that had taken place half a century before on the far side of the Atlantic Ocean, five thousand miles away.

It was the story of a heroic rescue that had taken place on a distant Atlantic seashore where legend had long said the rescue of shipwrecked sailors was quite unthinkable: among the rocks, reefs, and utterly waterless sands of that corner of South-West Africa known as the Skeleton Coast.

The first of the ships that had been wrecked, and which was the ultimate cause of the drama that unfolded back in that southern summer of 1942, was the MV *Dunedin Star*—seven years old, a 13,000-ton refrigerated cargo carrier, handsome without being graceful, Liverpool built. She had a crew of sixty-four and twenty-one passengers, most of them Londoners escaping the bombs.

She was on passage southbound. It was a Sunday night, November 29, when she, imprudently hugging the coast to avoid prowling German U-boats, struck the Clan Alpine Shoal (marked on the Admiralty charts of the day with the ominous letters PD, *position doubtful*). The collision ruptured her hull below the waterline, and the captain had little choice but to beach her. He managed to get out an SOS call, and then the power failed. Before the lifeboat motor broke down some forty-two of his passengers and crew made it through the treacherous surf and onto the utterly inhospitable shore. The rest were compelled to stay aboard.

Over the next few days four ships arrived to help with the rescue. One, a Walvis Bay tug named the *Sir Charles Elliott* (named for a colonial grandee), also ran aground. Two of its crew were drowned trying to swim ashore. One was the first mate, a Scotsman named Angus Macintyre; his body was never found. The other was a Namibian, Matthias Koraseb, who is buried ashore: it is his ghost that supposedly haunts this wilderness, his cries said to be the howling winds.

The other three ships tried gamely to help those left onshore, and while the surviving men hunted for driftwood and tried vainly to fish, the women and children huddled out of the sun in a makeshift shelter. The crews offshore tried to float rafts of food and water toward the coast, but most of them were lost, swept northward by the fierce current, or else upended and lost in the raging surf. Then one by one the rescue ships, running out of food and water themselves, left, their distressed captains flashing messages of good luck by heliograph.

Air force planes next tried to help, at first dropping food and water from the air—but all the early packages promptly burst, leaving the survivors aghast as the precious potential water supplies exploded uselessly into the sand. Two of these planes, heavy Ventura bombers loaded with supplies, then landed near the party, both getting bogged down in the dunes. After four days of digging one of them managed to lumber out and get away—only to crash into the sea half an hour later. Its crew survived, managed to swim ashore, and had to be rescued themselves.

Unknown to all, another rescue party of police and soldiers was making its way painfully by land from Windhoek, five hundred miles south. The conditions were appalling: the sand and the salt pans with their fragile crusts meant that on some days the eight-vehicle convoy made only two or three miles. But slowly, carefully, the rescuers crept north—until finally, after twenty-six days of unimaginable suffering in the blistering and waterless heat, the party was reached. All of them, even an infant who had gone temporarily sand blind, were alive; and all of them reached safety, arriving at a military hospital down south, appropriately, on Christmas Eve.

The story was suppressed for the rest of the war: the colonial authorities were keen to keep the German navy ignorant of Allied military dispositions along the West African coastline. The drama was never related in full until 1958, when a South African naval historian named John Marsh found the official papers and wrote *Skeleton Coast*, the book that so captivated me in Patagonia all those years later.

So I decided there and then that one day I would travel to the Skeleton Coast—a place so named because of all the skeletons, of both men and the vessels in which they had wrecked—to see if I could find a trace of the *Dunedin Star*. Some years later I found a commercial ship to take me east from Patagonia, by way of the Falkland Islands, South Georgia, and Tristan da Cunha, and eventually to Cape Town; from there I flew to Windhoek in Namibia, and finally in a two-engined Cessna flew up to a tiny tented camp in the middle of the northern desert, close to the Angolan frontier.

The Skeleton Coast's waves roared away in the distance. The place was said to be entirely deserted: just a scattering of seal colonies, packs of predatory jackals, endless miles of dunes, morning fogs rolling in from the ocean, the eternal cold surf. Armed with maps and a GPS fix of the wreck site, I set off the next day with two local guides. We traveled in a large and very battered old Land Rover, a car that had twin gearboxes and was equipped with differential locks and the ability to inflate its own tires and all the other necessities of deep desert travel. The night we left was pitch-black, except for a brilliant carpet of stars above us. It was also cold, and until we reached the sea completely silent,

except for a faint moaning of the wind and a very distant rumble
of the ocean.

After winding and bumping through the sands for hours, hauling ourselves up and over mountain spurs and ridges, sometimes
following half-imagined tracks left by earlier wanderers, but usually driving on virgin, sea-washed beach sand or on spurs of blistering granite, we came to a place I recognized. It was a headland
named Cape Fria, with an immense colony of fur seals, stinking
and noisy, and surrounded by a cordon of yellow-eyed jackals
that were busily engaged in carrying away the weaker seal pups.
The cape had been a landmark, mentioned in the book, since the
Dunedin Star had stranded no more than fifteen miles from here.
But the men and women in their shelter, who had no radio, never
knew how close they were to it and thus to a potential source of
food, for a seal is an easy thing to hunt and very nutritious when
cooked. It was probably a blessing they did not know: in the sear-
ing heat, and without water, they never could have walked to the
cape. It would have remained tantalizingly out of their reach, doing terrible damage to their ever-flagging morale.

Heat waves were now beginning to rise from the desert. It
was over ninety degrees, the summer air harsh and dry. The fog
over the cold morning ocean had now vanished, and the fast-
climbing sun was a coppery disk against an almost white sky.
We passed over thousands of ghost crabs that scuttled in great
armadas down to the water's edge. There were flocks of seabirds,
scores of skeletons of stranded whales, occasional wooden cable
drums and buried carboys and spars of wood. And then, half an
hour north of Cape Fria, a couple of miles after we had passed
the northern tip of a long and brilliant white salt pan, I spotted
something: two objects that began to loom larger and larger in
our field of view.

One turned out to be a half-buried metal cylinder, much corroded, about forty feet long, its upper part eaten away, a metal rod spearing its way up from its midsection toward the sky. To the northeast and southwest of this enormous object were rows of much-decayed wooden boxes, a number of what appeared to be hatch covers, small caches of bayonet-mounted lightbulbs (of a design once peculiar to Britain, I knew), and a scattering of bottles. All told, the site was about three hundred feet long.

The other object, three hundred more yards away, was a small forest of about fifty wooden spars, driven deep and hard into the sand and forming what might seem like a series of crude rooms. If covered with fabric—sailcloth, say, or mariners' tarpaulins—this could easily be the rudiments of some kind of sanctuary. It was finding this that made me suddenly believe we had probably made it.

I had some map coordinates, given to me by a man in Windhoek who had a lifelong fascination with the story of the *Star*. I held his sheet of notepaper up and switched on the GPS I had brought with me. The device spent a moment or two connecting to its network of satellites above, and then its screen suddenly fixed on a number—18°28'South, 12°0' East.

It was exactly the same as the number scrawled on my piece of paper. This—the shelter, the cylinder (most probably a boiler, or a fuel tank taken as cargo), the scores of still unbroken lightbulbs—was indeed all that remained of the wreck. It was sited fully two hundred yards in from the surf but where it had originally been beached—a reminder of how the western coast of Africa is slowly moving in toward the sea, just as the sea on the ocean's other side, in the Carolinas, at Cape Cod, is with equal relentlessness encroaching on the land.

We spent a couple of hours at the site, just sitting, quite mesmerized. The wind picked up steadily during the morning, and though I spoke some thoughts that I imagined were fairly profound into a recorder, when I played them later the howling of

the gale and the hissing of the blown sand against the micro-
phone made them almost impossible to hear. But I could hear
myself saying what I feel to this day: that it was incredibly moving, being at a place where so many people, after such privation,
almost didn't make it—but did.

One is not supposed to take anything from sites like this. But
I did, and for what I thought was good reason. I found among the
jetsam a tiny glass bottle, the kind of thing that I imagine one of
the elderly lady passengers on the *Dunedin Star* once carried in
her handbag, filled with *sal volatile*, in case she ever took a turn.
It was quite empty now, of course; but it had a working screw cap
and an elegantly frosted surface; and I had a purpose for it, were
I to complete the expedition I had come here for, were I to find
just one final thing.

We nearly got into big trouble. My guide was so exultant at
having discovered the wreck that he drove at a quite madcap
speed back along the beach. There is no road; the hard-packed
sand of a carefully chosen part of the beach is an ideal substitute.
But "carefully chosen" is an important caveat. Too close to land
and the sand becomes deep and dry and the wheels will spin
uselessly, and you have dig yourself out. Too close to the sea and
the sand turns to syrup, and your wheels turn toward the ocean
and you can be stranded there, caught perhaps in the onrush of
the tide, just like a thousand boats before.

And this is just what happened. The driver was heading fast
along the beach, but the oh-so-carefully selected strip of hard-
packed roadway became unexpectedly more narrow as a low cliff
rose beside us on the left, the landward side. The tide was rushing in from the right, hungrily. We stopped just as the roadway
ran out altogether, and spray from the oncoming waves started
dashing against the seaward windows.

The driver swore, lustily. We already knew the radio didn't

work, so if we were trapped, we couldn't get help. He threw the gear into reverse, and shouted to us all to pray. A shower of gray and dirty water suddenly rose in a huge curtain in front of the car as the wheels bit uselessly into the wet porridge below—until suddenly one of the tires, possibly one of the rear tires, bit into a small patch of hard sand—and the car shot backward.

Now he was moving—but he had to keep it moving, keep the car going backward, both straight and very, very fast. The water was coming in swiftly now, covering the packed sand—but seeming for one precious moment not to have mixed in with the grains, not to have broken through the surface tension, not to have begun to change its consistency and its viscosity. And so the car shot backward across the water, as if it were walking on it, almost miraculously—and then after five minutes of steady reversing like this we bumped over the low cliff of sand that had caused the problem, flushed out a small shower of scuttling ghost crabs—and were, quite mercifully, safe.

The driver, his hand shaking, wiped the sweat from his brow. We sat in silence in the heat, the windows open to catch the on-shore breeze, gazing out to sea. Beyond the breakers the green Atlantic growled on, restless and eternal. It had a sort of calm smugness to it, someone said. A smugness as if to say that yes, we had escaped its clutches, but it knew it would claim another victim, give it time. And then another, and another. Sufficient numbers of men would always be drawn to occupying their business in its great waters, deep or shallow, for one to be entirely certain of that.

· · ·

We made it, eventually, to Rocky Point. It was here, sixty miles south of where the *Dunedin Star* had wrecked, that the *Sir Charles Elliott* had also stranded, and where two of her crew had drowned. I had been told there was a grave, little known and seldom seen.

The remains of the tugboat are still visible, just. The breakers churn the water offshore into a ceaseless maelstrom of white,
but at certain moments it is just possible to spy two slender pin-
nacles of black rising defiantly two or three feet above the sea.
That is all: two corroded stanchions, or aerials, or parts of the
superstructure, which break clear of the Atlantic for just a second or two between waves. Twenty years ago you could still see
the bridge and parts of the funnel; but that is all gone now, and
these forlorn relics will probably last for only another year or so.

Onshore nearby and in sight of the wreck, on a sandspit protected by a shallow lagoon from the full ferocity of the sea, is the
grave. It must be one of the most remote and least visited grave
sites in the world, and sad to say it is quite irredeemably ugly, a
boxlike structure of four courses of red brick, and with a large
brass plaque sloping up to the north. Its stark lack of beauty is
tempered somewhat by the mass of whalebones that cover it,
some of them washed there by the sea, most others placed by the
occasional visitors.

The tomb is first and foremost a memorial to Matthias Koraseb, who came from South-West Africa and whose body was
found and buried here, in the sand of his birth. But also mentioned on the plaque is Angus Campbell Macintyre, whose body
was never found: he had been the first mate of the *Elliott*. In the
old Scottish tradition of leaving a stone on a mountain cairn, I
had long wanted to leave something on this grave, if ever I was
able to visit it.

I am not at all ashamed to say that I was very moved, standing there in the Atlantic wind, listening to the growling of the
surf, beside this lonely little grave. And so, although I knew at
the time that this would be a sentimental gesture, I wrote a note.
It said quite simply, *Thank You for Trying. Now Rest in Peace.* I signed
it and dated it, then folded it up as small as possible and pushed it

down through the narrow neck of the little smelling-salts bottle that I had taken from the wreck. I screwed the cap tightly shut and placed the little token beneath the stones and the old whale-bones and pieces of driftwood that covered the grave, and left it there. A message in a bottle; I hope it stays in place for many years.

Here on a lonely beach on Namibia's notorious Skeleton Coast is the grave of Matthias Koraseb and the memorial to the never-found Angus Macintyre, who died in a failed attempt to rescue the survivors from the SS Dunedin Star in 1942. This book is dedicated to Angus Macintyre.

Angus Campbell Macintyre was a Scotsman. He was born beside the North Atlantic Ocean and he died, during the commission of an act of great kindness, far away from his home, in the South Atlantic Ocean. What cruel symmetry, I thought. As

I leaned on the grave, looking out to sea, I thought back to that morning, nearly half a century before, when the liner on which I was crossing the same ocean for the very first time—and which was steaming across to the New World, and from a Scottish port—stopped to play a small part in a similar mission of mercy for someone who was also in distress upon the sea. On that occasion the scales were tipped in our favor: the mid-ocean rendezvous was safely made, the crisis averted.

But on this occasion, and as is so often true, the advantage had been with the sea: the wild ocean currents had swept mankind away as if he were no more than spindrift and foam. The man has gone; his ship has almost gone; in time the grave will be submerged by the ocean's rising waters, and all solid memory of the small event will have been washed away. I hope the dedication of this book to his memory will serve as something, to someone.

But whatever the fate of Able Seaman Macintyre, the ocean in which he lies will long remain. In one form or another, in one shape or another, and perhaps known by other names than as a memorial to Atlas, its waters will always exist, so long as the planet exists. They will always be present, gray and heaving, washing and waiting, extending out to the deeps, stretching across to the far horizon and then far, far beyond. Mankind may come and go, but an Atlantic Ocean of some kind will endure, will always be at the end of the beach or down at the base of the cliffs. It will always be in motion. It will always be present. Whether seen or unseen, heard or unheard, it will be imperturbable and irresistible, and as the poet has it, it will be quite simply *be there*, always just minding its business, always just *going on*.

ACKNOWLEDGMENTS

It goes almost without saying that researching a book that took me to places as distant and varied as Morocco, Brazil, Argentina, Newfoundland, Monte Carlo, Namibia, and Norway, as well as to such islands as St. Helena, Greenland, Tristan da Cunha, Bermuda, Muckle Flugga, and a score of others, was enormously interesting and the greatest of fun. But it was all made very much more rewarding by the kindness and assistance of a great many people, both Out There and Back Home; and though I remain haunted by the fear that I might have left out someone who put me up in his seaside house for a long weekend or loaned me his entire personal library of antiquarian works on maritime history, I hope in the few lines that follow to thank at least most of those without whose help the writing of this book would have proved quite impossible.

The idea of my writing about the Atlantic was in fact born on the shores of a competing sea, the Indian Ocean. One evening, on the terrace of the Galle Face Hotel in Colombo, the British writer and diplomat Tom Owen Edmunds, the Galle Literary Festival's chief organizer Libby Southwell, my wife, Setsuko, and I were gazing into the sunset and chatting idly about the relative richness of the historical associations of the world's great seas. As the evening wore on, it became abundantly clear that, lively though the Indian and Pacific oceans might be, in terms of the making of the modern world, the Atlantic could lay claim to having played a role that was infinitely livelier. So to Tom and Libby—now married and happily established in Islamabad—my thanks for assisting in the inspiration; and my gratitude to Setsuko, too, for seeing matters through from that point to this, and for her farsightedness, kindly patience, and endless practical support.

I soon discovered I was not alone in sporting a fascination for the Atlantic. Anne-Flore Laloë, at the University of Exeter, was at the time writing a thesis on the history of European and American intellectual connections with the ocean, and she jumped at the chance to be helpful. We met first at the National Maritime Museum in Greenwich, in the company of two researcher-curators on staff there, Claire Warrior and John McAleer. These three experts then established the beginnings of a paper trail for me, suggesting directions, books, libraries, and all the other paraphernalia I would need for my subsequent journeying, and I am grateful to them beyond measure. Anne-Flore, now Dr. Laloë, remained an enthusiastic supporter throughout, sharing her thoughts and discoveries as she assembled her own monumental work.

Deborah Cramer, in Boston, and Richard Ellis, in New York, have each written a number of critically acclaimed books about this ocean specifically and about the seas more generally and the life within them. Both gave unstintingly of their time and advice, and I am delighted to acknowledge their generosity of spirit and depth and breadth of their maritime knowledge.

Ted Nield, an old friend at the Geological Society of London, has written extensively about the formation and brief existences of the supercontinents, and was able to lead me down the maze of winding passageways that runs for the millions of years between Ur and Pangaea and up to the present day. John Dewey, a former Oxford geology professor now with emeritus standing at the University of California–Davis, kindly offered his thoughts on zircons and the origins of the earth, and his advice was augmented by insights from Stephen Moorbath, still at Oxford; Bruce Watson at the Rensselaer Polytechnic Institute in Troy, New York; John Rogers at the University of North Carolina–Chapel Hill; and by robust rebuttals from the much-respected British critic of plate tectonics, Joe McCall.

Chris Scotese, known for creating legendary tectonic visions at his PALEOMAP Project at the University of Texas–Arlington, offered his valuable time and expertise, too; we happily use, on pages 41 and 446, his impeccably crafted images of the imagined surface of the world as it probably once was and in time may well be.

Among others who took the time and trouble to help, I most gratefully mention: Amir Aczel in Cambridge, Massachusetts; David Agnew and Martin Collins, who, from their offices in London and the Falkland Islands respectively, gave wise counsel about the state of South Atlantic fisheries; Lesley Bellus and the staff of Wilderness Safaris in Windhoek, Namibia, who helped me with accommodation and logistics on the Skeleton Coast; Renee Braden, archivist of the National Geographic Society, who provided me with a wealth of early cartographic information; Kent Brooks of the University of Copenhagen, who advised about the ice conditions off the East Greenland coast; Penny Chisholm of MIT, to whom Amir Aczel introduced me, and who gave me the latest information on her phytoplanktonic discoveries; Charles Clover in London, who writes about the environmental impact of ocean fishing; Simon Day and Bill McGuire in London, experts on the possible impending collapse of the great Cumbre Vieja volcano in the Canary Islands; Susan Eaton in Calgary and Gregory McHone of Grand Manan Island, New Brunswick, who together helped me to understand the notion of congruent margins—the geological "fitting together"—around the North Atlantic coast; Chris Ehret of the University of California–Los Angeles, Curtis Marean of Arizona State University and Sarah Tishkoff of the University of Pennsylvania, who research the origins of humankind in southern Africa; Paul Falkowski of Rutgers University, who made a series of very specialized undersea Atlantic maps available to me; Dennis Feltgen and Chris Landsea of the National Hurricane Center in Miami; Ed Hill, director of the National Oceanographic Centre in Southampton, England; Ian Hogg of Tsawwassen, British Columbia, whose company I enjoyed in the South Atlantic and who later kindly read chapter 4, and as a former Royal Navy officer with a keen knowledge of warfare, made critical comments; Susana Lopez Lallana of Cordoba, Argentina, who made useful contacts for me on the Isla de los Estados; Paul Marston and Richard Goodfellow of British Airways in London, who arranged a detailed preflight briefing for me on the progress of Speedbird 113 between Heathrow and Kennedy airports; Captain Christopher Melhuish, USN (Ret.), former commander of the USS *Constitution* and now a civilian policy planner with U.S. Navy Fleet Forces HQ in Norfolk, Virginia; Eyda Merediz of the University of Maryland, who has written about the Canary Islands; David Morley, colonial administrator of the British territory of Tristan da Cunha; Iain Orr, a former British diplomat and now biodiversity champion in London; Alex Roland of Duke University, a specialist in the commercial shipping history of the North Atlantic; Jenny and Murray Sayle, now of Sydney, Australia, whose marriage essentially originated when Murray singlehanded his way most of the way across the ocean in a small yacht, comforted through the storms by Jenny on the radio in Newport, Rhode Island; Patricia Seed of the University of California–Irvine, who gave great help on early Portuguese navigators in the South Atlantic, especially Gil Eannes and the attempts to double Cape Bojador; Kirsten Shepherd-Barr of Oxford University, who led me into the byways of Faroese literature; Athena Trakadas, of the National Oceanographic Centre, Southampton, who explained in absorbing detail just how purple dye was extracted from the murex of Essaouira; Captain Robert Ward, a director of the International Hydrographic Organization in Monaco; and Mary Wills of the University of Hull, whose field is the study of the suppression of the slave trade.

I must thank Sir Richard Gozney, HM Governor of Bermuda, for his kindness as well as for the hospitality generously offered both by him and Lady Gozney on my visit to the Crown Colony at the time of the 400th anniversary of the island's European settlement.

Carol Zall, of the BBC/WGBH program *The World*, was very supportive of my travels. Cybele Tom, then of Oxford University Press in New York, gave me many useful pointers when I first conceived of this book, and its structure to a considerable degree reflects the wisdom of her words. And, as always, my indefatigable oldest son Rupert Winchester, in London, was on hand to help whenever I needed something looked up, or checked, or an errand performed: my debt to him for help with all of my recent work is immense.

If this book eventually merits any favorable attention—and of course, any errors or infelicities are entirely my own—then it will be due in large part to the irreplaceable skills of my New York editor, Henry Ferris. This is the third book on which we have collaborated, and though he is a decidedly tough editor, his thoughtfulness and courtesy have conspired to make this most necessary of processes much more than merely bearable. Traditionally in the creation of a book it is the research and the writing that are the fun parts, with the editing the time for payback. Not so with Henry: I find these days that I look forward to receiving his notes, however vividly stated and numerous his editorial suggestions may be. His tireless efforts result in the making of a much better book, and so if you like what you hold and read today, you should know who really deserves the credit.

He is also a genius at selecting assistants. Peter Hubbard, now deservedly promoted within HarperCollins, still offers pertinent and helpful comments; his successor, Danny Goldstein, has risen amply to the occasion, and dealt with all the trickiness of putting a book of this complexity together with adroitness, efficiency, and endless good cheer. In London I have also greatly enjoyed working with Martin Redfern, who has so skillfully guided the making of the British version of the book.

And finally, I raise a glass or two to my agents at William Morris Endeavor—in New York, the astonishingly energetic Suzanne Gluck, assisted first by Sarah Ceglarski and Elizabeth Tingue, and more recently by Caroline Donofrio and Mina Shaghaghi; and in London, the magically capable Eugenie Furniss. Bless you all, and thank you.

SW
Sandisfield, Massachusetts
July 2010

A GLOSSARY OF POSSIBLY
UNFAMILIAR TERMS

Aiguille   A needlelike spire of rock; usually applied to mountains or undersea ranges.

Benthic   Describes creatures and plants that live at the bottom of the sea.

Berber   Refers to the indigenous peoples of North Africa west of the Nile; most Moroccans are Arab-Berber. Hence: the Barbary Coast.

Bilge   The broadest internal part of the bottom of a ship, where all the leakage and swill collect.

Bollard   A stout post on a wharf, to which a ship may be tied. See also *hawser*.

Bouillon   A strong meat broth, often served hot to passengers on the boat deck of a ship on passage through cold waters.

Bridgewing   The extended and usually open part of a ship's bridge, where officers may view the entire length of the vessel, and from which dockside commands may be given.

Brig   A square-rigged sailing vessel, usually with two masts.

Cant   To steer a slow-moving ship into a dock, usually at journey's end.

Careen   To turn a beached ship over onto its side so that its hull may be cleaned or repaired.

Castile   One of the ancient monarchies, in the central north of the Iberian peninsula, that eventually made up the Kingdom of Spain.

Cog   A small sailing boat, usually with a single mast, often used for short-distance Baltic trading.

Compline   The final evening service held in a religious house.

Cordillera   A chain of mountains; most commonly applied to the Andes and the Rocky Mountains.

Coriolis force   Named after a nineteenth-century French mathematician, this force is the effect the earth's rotation has on winds and ocean currents.

Curragh   A small and simply built Irish boat, originally made of reeds and skins, still in use today in parts of rural Ireland.

Cwm   The Welsh word for a valley or depression on a hillside. The Western Cwm on the flanks of Mt. Everest is the best-known example outside Wales.

Cyanobacteria   Blue-green algae, named for their dominant color, that derive their energy from photosynthesis.

Dory   A small rowing boat, with high prows and stern and little freeboard, much used by nineteenth-century American whalers.

Einkorn wheat   A type of wild wheat considered to have been the first crop grown in an organized fashion by early agriculturists in the Fertile Crescent.

Ell   An obsolete measure of length, based on doubling the average forearm; in England it is about forty-five inches.

Erg   A Berber word for the great areas of sand dunes found in the central Sahara (a word which itself means desert in Berber).

Flense   To remove the skin and blubber from a dead whale.

Forepeak   The small and oddly angled internal space at the very tip of a boat; usually where the paint is stored, or, in naval vessels, the laundrymen.

Freeboard   The height from the waterline to the lowest part of a ship's deck. A vessel with little freeboard risks being swamped.

Fynbos   An aggregation of very rich vegetation—more than six thousand endemic species—that grows in a small area of South Africa's Western Cape Province that has been blessed with a Mediterranean climate.

Gesso   A material, made from gypsum or plaster of paris, used to prepare a surface, such as a wall or a canvas, for painting.

Godown   A Malay word for a warehouse, now commonly employed across India and South Asia.

Growler   A half-hidden small iceberg, an often lethal menace to shipping.

Gyre   In this context, a spinning pattern of wind, current, or climate.

Hammada   A Saharan desert area where the wind has blown away the sand, leaving only rocks and scrub.

Hard   A sloping surface of stone or concrete leading down into the water, up which boats may be hauled, down which they may be launched, or on which they may be cleaned or repaired.

Harmattan   A dry easterly wind, freighted with red desert dust, that blows in the winter months across western Africa.

Hawser   A thick and very strong rope, made either of hemp or steel cable, used to secure ships to the quayside.

Henge   A medieval formation, usually circular, of stones surrounding ceremonial structures or burial chambers. Stonehenge, in southern England, is the best-known example.

Hove-to   Of a ship at sea: intentionally stopped, temporarily drifting.

Kelp   Thick, floating seaweed, often strong enough to allow small craft to tie up alongside.

Knot   A unit of speed equivalent to one nautical mile per hour, employed to describe the velocities of ships and some kinds of aircraft.

Krill   A small ocean-borne, shrimplike crustacean, much favored as the constant diet of baleen whales.

Lee shore A coastline toward which the wind generally blows. A lee shore can be extremely dangerous for a nearby sailing vessel, especially if the craft becomes embayed and unable to escape being blown onto land.

Letterpress A type of printing, nowadays of little commercial importance, in which paper is briefly pressed by machinery against inked type, leaving both the ink and a discernible impression on the paper's surface.

Levant The countries of the eastern Mediterranean, so named because to seaborne mariners the sun appears to rise—become *lévant*, in French—from within them.

Long sea Of sea voyages, those which pass between continents, and during which the vessel is seldom in sight of land, except at the beginning and end of passage.

Longphort An Irish term that describes a sheltered and fortified base from which Viking invaders could conduct commerce and launch further raids.

Loxodrome Another name for the rhumb line, along which some sailors once made their journeys, keeping at all times at a constant angle to the lines of longitude that they crossed.

Lutefisk A dish, popular in Norway but little loved elsewhere, made from whitefish soaked for some days in caustic soda. The gelatinous whitefish is then eaten with flat bread and meatballs.

Machair Strips of low-lying grassland, often on a shallow bed of shell sand, found especially along the coastlines of western Scotland.

Main The mainland coast of the Caribbean, generally taken to refer to the Spanish possessions from which the galleon trade was first conducted.

Majuscule A large form of written or printed script, whether in capital or uncial (rounded) form.

Marlinspike A sharp metal spike whose official purpose was to pry apart the strands of a rope for splicing. An essential item in a sailor's kitbag, it had innumerable other uses, from oyster opener to weapon.

Medina The walled section—and thus usually the non-European section—of a North African town.

Minuscule A small kind of cursive script, often used by monks in the writing of lengthy documents.

Nunatak The Inuit word for an isolated mountain rising through an ice cap or beside a glacier.

Pelagic Unlike *littoral*, which refers to the seashore, and *benthic*, which refers to the sea bottom, *pelagic* pertains to the open ocean, far from land and close to the surface.

Pemmican A hard and initially unpalatable cake of compressed meat and grease used at first by Native Americans (the word is Cree), but later by travelers generally, for emergency rations.

Pinnace A small, often two-masted sailing vessel customarily employed to attend a much larger vessel and serve its needs in port or on short voyages.

Pipal A kind of fig tree, *Ficus religiosa*, known to many as the Bo-tree, found in the Indian subcontinent and sacred to Hindus and Buddhists in equal measure. It often serves as the meeting place in an Indian village.

Puncheon A large barrel or cask used for the transport of fermenting wine or spirits.

Quahog A kind of hard-shelled clam found in North America, only on the Atlantic coasts.

Quinquereme A powerful Roman galley with three banks of oars. The two uppermost are each pulled by two men, and the one below pulled by one. Thus each set of oars is rowed by a total of five men—hence the Latin name.

Rhumb line An imaginary line, often sailed along by the navigationally unsophisticated, which intersects lines of longitude at a constant angle.

Riprap Stones or other heavy material dumped into the sea to provide a strong foundation for a wharf or for a mole that provides shelter to a harbor.

Sal volatile A solution of the chemical ammonium carbonate, much used by delicate ladies of a certain age and time as a restorative for fainting or swooning fits.

Scrimshaw Though it frequently refers to the delicate engraving made by seamen on ivory tusks, it can signify any craft, such as knitting, performed by sailors to pass the time on long sea voyages.

Shallop A small sea boat, stoutly made and often armed.

Short sea The waters of bays and estuaries and narrow straits, through and across which vessels may ply without losing sight, at least for extended periods, of the nearby land.

Sloop A small and moderately fast sailing ship, often carrying guns on its upper decks.

Snood A kind of hairnet which sits at the back of a woman's head, like a cap.

Sphagnum Peat moss, which grows in especially boggy places. The word is Greek.

Stromatolite The fossil remains of a large aggregation of ancient algae or cyanobacteria, made solid by the absorption of calcareous material to form what looks like an oddly shaped and often large limestone nodule.

Subduction An important process in tectonic plate movement, in which a heavy oceanic plate collides with a lighter continental plate and is forced down below it. Often it then melts and its magma is forced upward, forming lines of volcanoes such as those in the Pacific Northwest of America.

Supercargo A ship's officer who supervises the loading, management, and eventual unloading of the cargo.

Taffrail The rail at the very stern end of a ship, to which departing passengers migrate to wave their farewells to those remaining on land.

Telltale A mechanical device of varying design that communicates the heading of a ship to crew other than those in the wheelhouse.

Touareg Nomadic desert people, a subgroup of the Berbers, who inhabit the Sahara from Algeria to Mauritania. They often wear robes of vivid blue.

Well-found A naval term signifying that a ship is amply supplied with victuals and equipment to allow it to make even the most hazardous of voyages.

Zircon A crystalline mineral, a silicate compound of zirconium, found in the world's most ancient rocks.

BIBLIOGRAPHY

Adams, Captain John. *Remarks on the Country Extending from Cape Palmas to the River Congo.* London: Whittaker, 1823.

Adkins, Roy. *Trafalgar: The Biography of a Battle.* London: Little, Brown, 2005.

Adkins, Roy, and Lesley Adkins. *The War for All the Oceans: From Nelson at the Nile to Napoleon at Waterloo.* London: Penguin, 2006.

Agnew, David. *Fishing South: The History and Management of the South Georgia Fisheries.* St. Albans: Penna Press, 2004.

Air Ministry. *Atlantic Bridge: The Official Account of RAF Transport Command's Ocean Ferry.* London: HMSO, 1945.

Amos, William H., and Stephen H. Amos. *Atlantic and Gulf Coasts.* National Audubon Society Nature Guides. New York: Knopf, 1985.

Anstey, Roger. *The Atlantic Slave Trade and British Abolition, 1760–1810.* London: Macmillan, 1975.

Archibald, Malcolm. *Across the Pond: Chapters from the Atlantic.* Latheronwheel, Caithness, UK: Whittles Publishing, 2001.

Armitage, David, and Michael J. Braddick, eds. *The British Atlantic World, 1500–1800.* Basingstoke, UK: Palgrave Macmillan, 2002.

Armstrong, Warren. *Atlantic Bridge: From Sail to Steam to Wings.* London: Frederick Muller, 1956.

Bailyn, Bernard. *Atlantic History: Concept and Contours.* Cambridge, Mass.: Harvard University Press, 2005.

———. *Voyagers to the West: A Passage in the Peopling of America on the Eve of the Revolution.* New York: Knopf, 1986.

Barty-King, Hugh. *Girdle Round the Earth: The Story of Cable and Wireless.* London: Heinemann, 1979.

Bathurst, Bella. *The Lighthouse Stevensons.* New York: HarperCollins, 1999.

Baumann, Elwood D. *The Devil's Triangle.* New York: Franklin Watts, 1976.

Belloc, Hilaire. *The Cruise of the "Nona."* London: Penguin, 1958.

Blum, Hester. *The View from the Masthead: Maritime Imagination and Antebellum American Sea Narratives*. Chapel Hill: University of North Carolina Press, 2008.

Bonsor, N. R. P. *North Atlantic Seaway: An Illustrated History of the Passenger Services Linking the Old World with the New*. Prescot, Lancashire, UK: T. Stephenson, 1955.

Bonturi, Orlando. *Brazil and the Vital South Atlantic*. Washington, D.C.: National Defense University, 1988.

Booker, Christopher. *The Real Global Warming Disaster*. London: Continuum, 2009.

Borgstrom, Georg, and Arthur Heighway, eds. *Atlantic Ocean Fisheries*. London: Fishing News (Books) Ltd., 1961.

Braudel, Fernand. *The Mediterranean and the Mediterranean World in the Age of Philip II*. New York: Harper and Row, 1973.

Breverton, Terry. *Admiral Sir Henry Morgan: King of the Buccaneers*. Gretna, La.: Pelican, 2005.

Bridges, E. Lucas. *Uttermost Part of the Earth*. London: Hodder and Stoughton, 1951.

Brinnin, John Malcolm. *The Sway of the Grand Saloon: A Social History of the North Atlantic*. New York: Delacourt, 1971.

Buckley, William F., Jr. *Atlantic High: A Celebration*. New York: Doubleday, 1982.

Butel, Paul. *The Atlantic*. Translated by Iain Grant. London: Routledge, 1999.

Buttress, Rob, and Andy duPort. *Reeds Nautical Almanac*. London: A and C Black, 2009.

Carr, J. Revell. *All Brave Sailors: The Sinking of the Anglo-Saxon, August 21, 1940*. New York: Simon & Schuster, 2004.

Carson, Rachel. *The Sea Around Us*. New York: Oxford University Press, 1951.

Chapin, Miriam. *Atlantic Canada*. Toronto: Ryerson Press, 1956.

Clover, Charles. *The End of the Line: How Overfishing Is Changing the World and What We Eat*. Berkeley: University of California Press, 2006.

Coote, John, ed. *The Faber Book of the Sea*. London: Faber, 1989.

Cordingly, David. *Under the Black Flag: The Romance and Reality of Life Among the Pirates*. New York: Random House, 1996.

Cramer, Deborah. *Great Waters: An Atlantic Passage*. New York: W. W. Norton, 2001.

———. *Ocean: Our Water, Our World*. Washington, D.C.: Smithsonian Books, 2008.

Cullen, Vicky. *Down to the Sea for Science*. Woods Hole, Mass.: Woods Hole Oceanographic Institution, 2005.

Cunliffe, Barry. *Facing the Ocean: The Atlantic and Its Peoples, 8000 BC–AD 1500*. Oxford: Oxford University Press, 2001.

Cuny, Paul J. *Lloyds Nautical Year Book*. London: Lloyds of London Press, 1991.

Danson, Edwin. *Weighing the World: The Quest to Measure the Earth*. New York: Oxford University Press, 2005.

Davies, David Twiston, ed. *The Daily Telegraph Book of Naval Obituaries*. London: Grub Street, 2004.

Davies, Hunter. *A Walk Around the West Indies*. London: Weidenfeld and Nicolson, 2000.

Dawson, Jeff. *The Dunedin Star Disaster*. London: Weidenfeld, 2005.

DePaolo, Donald J., et al. *Origin and Evolution of Earth: Research Questions for a Changing Planet*. Washington, D.C.: National Academies Press, 2008.

Dolin, Eric Jay. *Leviathan: The History of Whaling in America.* New York: W. W. Norton, 2007.

Donnelly, Ignatius. *Atlantis: The Antediluvian World.* New York: Harper and Brothers, 1949.

Durschmied, Erik. *The Weather Factor: How Nature Has Changed History.* London: Hodder, 2000.

Earle, Sylvia, and Linda Glover, eds. *Ocean: An Illustrated Atlas.* Washington, D.C.: National Geographic Society, 2009.

Eddy, Paul, and Magnus Linklater. *War in the Falklands.* New York: Harper and Row, 1982.

Ellis, Richard. *Deep Atlantic: Life, Death, and Exploration in the Abyss.* Knopf, 1996.

———. *Encyclopedia of the Sea.* New York: Knopf, 2006.

———. *Men and Whales.* New York: Knopf, 1991.

———. *Tuna: A Love Story.* New York: Knopf, 2008.

Emanuel, Kerry. *Divine Wind: The History and Science of Hurricanes.* New York: Oxford University Press, 2005.

Emmons, Frederick. *The Atlantic Liners, 1925–70.* New York: Bonanza, 1972.

Fanning, A. E. *Steady as She Goes: A History of the Compass Department of the Admiralty.* London: HMSO, 1986.

Fernández-Armesto, Felipe. *The Americas: A Hemispheric History.* New York: Modern Library, 2003.

———. *Amerigo: The Man Who Gave His Name to America.* London: Weidenfeld and Nicolson, 2006.

———. *Columbus and the Conquest of the Impossible.* London: Weidenfeld, 1974.

———. *Ideas That Changed the World.* New York: DK Publishing, 2003.

———. *Pathfinders: A Global History of Exploration.* Toronto: Viking Canada, 2006.

———, ed. *The Times Atlas of World Exploration.* London: HarperCollins, 1991.

Finamore, Daniel, ed. *Maritime History as World History.* Salem, Mass.: Peabody Essex Museum, 2004.

Forbes, Jack D. *The American Discovery of Europe.* Urbana: University of Illinois Press, 2007.

Fox, Robert. *Antarctica and the South Atlantic: Discovery, Development and Dispute.* London: BBC Books, 1985.

Fox, Stephen. *Transatlantic: Samuel Cunard, Isambard Brunel, and the Great Atlantic Steamships.* New York: HarperCollins, 2003.

Franck, Irene M., and David M. Brownstone. *To the Ends of the Earth: The Great Travel and Trade Routes of Human History.* New York: Facts on File, 1984.

Fuller, Major-General J. F. C. *Decisive Battles of the Western World and Their Influence upon History.* 3 volumes. London: Cassell, 1951.

Gaskell, T. F. *The Gulf Stream.* New York: John Day, 1973.

Gillis, John R. *Islands of the Mind: How the Human Imagination Created the Atlantic World.* New York: Palgrave Macmillan, 2004.

Gilroy, Paul. *The Black Atlantic: Modernity and Double Consciousness.* Cambridge, Mass.: Harvard University Press, 1993.

Gimlette, John. *The Theatre of Fish: Travels Through Newfoundland and Labrador.* London: Hutchinson, 2005.

Gordon, John Steele. *A Thread Across the Ocean: The Heroic Story of the Transatlantic Cable*. New York: Walker, 2002.

Graham, Gerald S. *Empire of the North Atlantic: The Maritime Struggle for North America*. Toronto: University of Toronto Press, 1950.

Gruber, Ruth. *Haven: The Dramatic Story of 1,000 World War II Refugees and How They Came to America*. New York: Three Rivers, 1983.

Guthrie, John. *Bizarre Ships of the Nineteenth Century*. London: Hutchinson, 1970.

Hall, Rear Admiral G. P. D. (Hydrographer of the Navy). *Ocean Passages for the World*. Taunton, UK: Ministry of Defence, 1973.

Hamilton-Paterson, James. *The Great Deep: The Sea and Its Thresholds*. New York: Random House, 1992.

Harris, Michael. *Lament for an Ocean: The Collapse of the Atlantic Cod Fishery*. Toronto: McClelland and Stewart, 1998.

———. *Rare Ambition: The Crosbies of Newfoundland*. Toronto: Penguin, 1992.

Hastings, Max, and Simon Jenkins. *The Battle for the Falklands*. London: Michael Joseph, 1983.

Hattendorf, John B., ed. *The Oxford Encyclopedia of Maritime History*. 4 volumes. New York: Oxford University Press, 2007.

Hattersley, Roy. *Nelson*. New York: Saturday Review Press, 1974.

Hearn, Chester G. *Tracks in the Sea: Matthew Fontaine Maury and the Mapping of the Oceans*. Camden, Maine: International Marine, 2002.

Hendrickson, Robert. *The Ocean Almanac*. London: Hutchinson, 1992.

Heyerdahl, Thor. *The Ra Expeditions*. New York: Signet, 1972.

Higgins, Jack. *Storm Warning*. New York: Holt, Rinehart and Winston, 1976.

Hoare, Philip. *Leviathan; or, The Whale*. London, Fourth Estate, 2008.

Hobhouse, Henry. *Seeds of Change: Five Plants That Transformed Mankind*. London: Macmillan, 1992.

Hughes, Richard. *In Hazard: A Sea Story*. London: Penguin, 1938.

International Hydrographic Organization. *Names and Limits of Oceans and Seas*. Monaco: International Hydrographic Organization, 2002.

Jablonski, Edward. *Atlantic Fever*. New York: Macmillan, 1972.

Jackson, E. L. *St. Helena: The Historic Island from Its Discovery to the Present Date*. London: Ward, Lock, 1903.

Jacobsen, Jørgen-Frantz. *Barbara*. Norwich, UK: Norvik Press, 1993.

Johnson, Donald S. *Phantom Islands of the Atlantic: The Legends of Seven Lands That Never Were*. New York: Walker, 1996.

Kay, F. George. *The Atlantic Ocean: Bridge Between Two Worlds*. London: Museum Press, 1954.

Keegan, John. *A History of Warfare*. New York: Knopf, 1993.

———. *The Price of Admiralty: War at Sea, from Man of War to Submarine*. London: Hutchinson, 1988.

Kemp, Peter, ed. *The Oxford Companion to Ships and the Sea*. Oxford: Oxford University Press, 1976.

Kennedy, Sr. Jean de Chantal. *Biography of a Colonial Town: Hamilton, Bermuda, 1790–1897*. Hamilton: Bermuda Book Stores, 1961.

Kennedy, Ludovic, ed. *A Book of Sea Journeys*. New York: Rawson Wade, 1981.

Kent, Rockwell. *Voyaging Southward from the Strait of Magellan*. Hanover, N.H.: University Press of New England, 1951.

Kirk, Stephen. *First in Flight: The Wright Brothers in North Carolina*. Winston-Salem, N.C.: John F. Blair Publishers, 1995.

Klein, Bernhard, and Gesa Mackenthun. *Sea Changes: Historicizing the Ocean*. New York: Routledge, 2004.

Knecht, G. Bruce. *Hooked: Pirates, Poaching, and the Perfect Fish*. New York: Rodale, 2006.

Knight, Franklin W., and Peggy K. Liss, eds. *Atlantic Port Cities: Economy, Culture, and Society in the Atlantic World, 1650–1850*. Knoxville: University of Tennessee Press, 1991.

Kopper, Philip. *The Wild Edge: Life and Lore of the Great Atlantic Beaches*. Chester, Conn.: Globe Pequot Press, 1991.

Kraus, Michael. *The Atlantic Civilization: Eighteenth-Century Origins*. Ithaca, N.Y.: Cornell University Press, 1949.

Kunzig, Robert. *Mapping the Deep: The Extraordinary Story of Ocean Science*. New York: W. W. Norton, 2000.

Kurlansky, Mark. *The Big Oyster: New York on the Half Shell*. New York: Ballantine, 2006.

———. *Cod: A Biography of the Fish That Changed The World*. New York: Walker, 1997.

———. *The Last Fish Tale: The Fate of the Atlantic and Survival in Gloucester, America's Oldest Fishing Port and Most Original Town*. New York: Ballantine, 2008.

———. *Salt: A World History*. New York: Walker, 2002.

Labaree, Benjamin W., ed. *The Atlantic World of Robert G. Albion*. Middletown, Conn.: Wesleyan University Press, 1975.

Lambert, Frank. *The Barbary Wars: American Independence in the Atlantic World*. New York: Hill and Wang, 2005.

Landes, David S. *The Wealth and Poverty of Nations: Why Some Are So Rich, and Some So Poor*. New York: W. W. Norton, 1998.

Leonard, Jonathan Norton. *Atlantic Beaches*. New York: Time-Life Books, 1972.

Lester, Toby. *The Fourth Part of the World: The Race to the Ends of the Earth, and the Epic Story of the Map That Gave America Its Name*. New York: Simon & Schuster, 2009.

Linebaugh, Peter, and Marcus Rediker. *The Many-Headed Hydra: Sailors, Slaves, Commoners, and the Hidden History of the Revolutionary Atlantic*. Boston: Beacon Press, 2000.

Lodwick, John. *The Forbidden Coast: A Journey Through the Rio de Oro*. London: Travel Book Club, 1956.

Longstreth, T. Morris. *To Nova Scotia: The Sunrise Province*. Toronto: Ryerson Press, 1935.

Lundy, Derek. *Godforsaken Sea: The True Story of a Race Through the World's Most Dangerous Waters*. New York: Random House, 1998.

Mac Donald, Laura. *Curse of the Narrows*. New York: Walker Books, 2005.

MacLean, Rory. *The Oatmeal Ark: From the Scottish Isles to a Promised Land*. London: HarperCollins, 1997.

Maddocks, Melvin. *The Atlantic Crossing*. Alexandria, Va.: Time-Life Books, 1981.

Manchester Guardian. C. P. Scott, 1846–1932: The Making of the Manchester Guardian. London: Frederick Muller, 1946.

Mann, Charles C. *1491: New Revelations of the Americas Before Columbus.* New York: Vintage, 2005.

Marcus, G. J. *The Conquest of the North Atlantic.* Woodbridge, Suffolk, UK: Boydell Press, 1980.

Marsh, John, and Lyman Anson. *Skeleton Coast.* London: Hodder and Stoughton, 1958.

Masselman, George. *The Atlantic: Sea of Darkness.* New York: McGraw-Hill, 1969.

Matthiessen, Peter. *Men's Lives.* New York: Vintage, 1986.

Maury, Matthew Fontaine. *The Physical Geography of the Sea and Its Meteorology.* Mineola, N.Y.: Dover, 2003.

McCalman, Iain. *Darwin's Armada: How Four Voyages to Australasia Won the Battle for Evolution and Changed the World.* Melbourne: Viking, 2009.

McEwen, W. A., and A. H. Lewis. *Encyclopedia of Nautical Knowledge.* Cambridge, Md.: Cornell Maritime Press, 1953.

McGrail, Seán. *Boats of the World: From the Stone Age to Medieval Times.* Oxford: Oxford University Press, 2001.

McKee, Alexander. *Against the Odds: Battles at Sea, 1591–1949.* Annapolis, Md.: Naval Institute Press, 1991.

Merediz, Eyda M. *Refracted Images: The Canary Islands Through a New World Lens.* Tempe: Arizona Center for Medieval and Renaissance Studies, 2004.

Middlebrook, Martin. *Convoy.* New York: Morrow, 1976.

Miles, Jonathan. *The Wreck of the Medusa: The Most Famous Sea Disaster of the Nineteenth Century.* New York: Atlantic Monthly Press, 2007.

Monsarrat, Nicholas. *The Cruel Sea.* London: Penguin, 1951.

Mooney, Chris. *Storm World: Hurricanes, Politics, and the Battle over Global Warming.* Orlando, Fla.: Harcourt, 2007.

Moorehead, Alan. *Darwin and the Beagle.* New York: Harper and Row, 1969.

Morison, Samuel Eliot. *Admiral of the Ocean Sea: A Life of Christopher Columbus.* Boston: Little, Brown, 1942.

——. *The European Discovery of America.* 2 volumes. New York: Oxford University Press, 1971.

Morris, J. *The Pax Britannica Series.* 3 volumes. London: Faber, 1978.

Morris, Robert D. *The Blue Death: Disease, Disaster, and the Water We Drink.* New York: HarperCollins, 2007.

Morrison, H. Robert, and Christine E. Lee. *America's Atlantic Isles.* Washington, D.C.: National Geographic, 1981.

Murphy, Dallas. *To Follow the Water: Exploring the Ocean to Discover Climate, from the Gulf Stream to the Blue Beyond.* New York: Basic Books, 2007.

Murphy, Hugh, and Derek J. Oddy. *The Mirror of the Seas. A Centenary History of the Society for Nautical Research.* Greenwich, UK: Society for Nautical Research, 2010.

Neill, Peter, ed. *American Sea Writing: A Literary Anthology.* New York: Library of America, 2000.

Nichols, Peter. *Sea Change: Alone Across the Atlantic in a Wooden Boat.* New York: Penguin, 1997.

Nicolson, Adam. *Seamanship: A Voyage Along the Wild Coasts of the British Isles*. New York: HarperCollins, 2004.

——. *Sea Room: An Island Life*. London: HarperCollins, 2001.

Nield, Ted. *Supercontinent: Ten Billion Years in the Life of Our Planet*. London: Granta Books, 2007.

O'Hanlon, Redmond. *Trawler: A Journey Through the North Atlantic*. London: Hamish Hamilton, 2003.

Oliver, Mary. *Why I Wake Early*. Boston: Beacon Press, 2004.

O'Siochain, P. A. *Aran: Islands of Legend*. Dublin: Foilsiuchain Eireann, 1962.

Outhwaite, Leonard. *The Atlantic: A History of the Ocean*. New York: Coward-McCann, 1957.

——. *Atlantic Circle: Around the Ocean with the Winds and Tides*. New York: Scribner's, 1931.

Owen, David, ed. *Seven Ages: Poetry for a Lifetime*. London: Penguin, 1992.

Parker, Bruce. *The Power of the Sea*. New York: Palgrave Macmillan, 2010.

Patterson, Kevin. *The Water in Between: A Journey at Sea*. Toronto: Vintage, 2000.

Pauly, Daniel, and Jay Maclean. *In a Perfect Ocean: The State of Fisheries and Ecosystems in the North Atlantic Ocean*. Washington, D.C.: Island Press, 2003.

Pearson, Michael. *The Indian Ocean*. London: Routledge, 2003.

Perry, Richard. *Lundy: Isle of Puffins*. London: Lindsay Drummond, 1940.

Pestana, Carla Gardina. *The English Atlantic in an Age of Revolution, 1640–1661*. Cambridge, Mass.: Harvard University Press, 2004.

Philbrick, Nathaniel. *Sea of Glory: America's Voyage of Discovery; The U.S. Exploring Expedition, 1838–1842*. New York: Viking, 2003.

Pike, Dag. *The Challenge of the Atlantic: Man's Battle with the World's Toughest Ocean*. Wellingbrough, UK: P. Stephens, 1988.

Pohl, Frederick J. *Atlantic Crossings Before Columbus*. New York: W. W. Norton, 1961.

Preston, Diana. *Lusitania: An Epic Tragedy*. New York: Berkley Books, 2002.

Price, Jacob M. *The Atlantic Frontier of the Thirteen American Colonies and States*. Aldershot, UK: Variorum, 1996.

Pritchard, H. Hesketh. *Through the Heart of Patagonia*. London: Heinemann, 1902.

Proctor, Noble S., and Patrick J. Lynch. *A Field Guide to North Atlantic Wildlife*. New Haven, Conn.: Yale University Press, 2005.

Prosser de Goodall, Rae Natalie. *Tierra del Fuego*. Ushuaia, Argentina: Ediciones Shanamaiim, 1970.

Raban, Jonathan. *Coasting*. London: Harvill, 1986.

——, ed. *The Oxford Book of the Sea*. Oxford: Oxford University Press, 1992.

Riley, Captain James. *Sufferings in Africa: The Incredible True Story of a Shipwreck, Enslavement, and Survival on the Sahara*. New York: Skyhorse, 2007.

Roberts, Alice. *The Incredible Human Journey: The Story of How We Colonised the Planet*. London: Bloomsbury, 2009.

Roberts, Callum. *The Unnatural History of the Sea*. Washington, D.C.: Island Press, 2007.

Robinson, Adrian, and Roy Millward. *The Shell Book of the British Coast*. Newton Abbott: David and Charles, 1983.

Rodgers, Daniel T. *Atlantic Crossings: Social Politics in a Progressive Age*. Cambridge, Mass.: Harvard University Press, 1998.

Rogers, John J. W., and M. Santosh. *Continents and Supercontinents*. New York: Oxford University Press, 2004.

Roland, Alex, W. Jeffrey Bolster, and Alexander Keyssar. *The Way of the Ship: America's Maritime History Reenvisioned, 1600–2000*. Hoboken, N.J.: John Wiley and Sons, 2008.

Rozwadowski, Helen M. *Fathoming the Ocean: The Discovery and Exploration of the Deep Sea*. Cambridge, Mass.: Harvard University Press, 2005.

Safina, Carl. *Eye of the Albatross: Visions of Hope and Survival*. New York: Henry Holt, 2002.

———. *Song for the Blue Ocean*. New York: Henry Holt, 1997.

Sanderson, Michael, ed. *Catalogue of the Library of the National Maritime Museum*. 7 volumes. London: HMSO, 1968.

Sandler, Martin. *Atlantic Ocean: An Illustrated History of the Ocean That Changed the World*. New York: Sterling, 2008.

Schei, Liv K., and Gunnie Moberg. *The Faroe Islands*. Edinburgh: Birlinn, 2003.

Schlee, Susan. *On Almost Any Wind: The Saga of the Oceanographic Research Vessel Atlantis*. Ithaca, N.Y.: Cornell University Press, 1978.

Schoeman, Amy. *Skeleton Coast*. Johannesburg: Southern Book Publishers, 1984.

Schwartz, Stuart, ed. *Tropical Babylons: Sugar and the Making of the Atlantic World, 1450–1680*. Chapel Hill: University of North Carolina Press, 2004.

Scientific American. *Oceans: A Scientific American Reader*. Chicago: University of Chicago Press, 2007.

Scott, R. Bruce. *Gentleman on Imperial Service: A Story of the Transpacific Telecommunications Cable*. Victoria, BC: Sono Nis Press, 1994.

Seed, Patricia. *Ceremonies of Possession in Europe's Conquest of the New World, 1492–1640*. New York: Cambridge University Press, 1995.

Segal, Aaron. *An Atlas of International Migration*. London: Hans Zell, 1993.

Severin, Tim. *The Brendan Voyage: A Leather Boat Tracks the Discovery of America by the Irish Sailor Saints*. New York: McGraw-Hill, 1978.

Sewell, Kenneth, and Jerome Preisler. *All Hands Down: The True Story of the Soviet Attack on the USS Scorpion*. New York: Simon & Schuster, 2008.

Shackleton, Keith, and Ted Stokes. *Birds of the Atlantic Ocean*. Feltham, UK: Country Life Books, 1968.

Shaw, David W. *Daring the Sea: The Dramatic True Story of the First Men to Cross the Atlantic in a Rowboat*. New York: Kensington Books, 1998.

Sherry, Frank. *Raiders and Rebels: A History of the Golden Age of Piracy*. New York: Morrow, 1986.

Simpson, Colin. *The Lusitania*. London: Little, Brown, 1972.

Slocum, Joshua. *Sailing Alone Around the World*. Teddington, UK: Echo, 2006.

Snow, Edward Rowe. *Great Atlantic Adventures*. New York: Dodd, Mead, 1970.

Snow, Richard. *A Measureless Peril: America in the Fight for the Atlantic*. New York: Scribner, 2010.

Solomon, Susan, et al., eds. *Climate Change 2007: The Physical Science Basis*. New York: Cambridge University Press, 2007.

South, Mark. *The Cure for Anything Is Salt Water: How I Threw My Life Overboard and Found Happiness at Sea*. New York: HarperCollins, 2007.

Springer, Haskell, ed. *America and the Sea: A Literary History*. Athens: University of Georgia Press, 1995.

St. Clair, William. *The Grand Slave Emporium: Cape Coast Castle and the British Slave Trade*. London: Profile Books, 2006.

Steele, Ian K. *The English Atlantic, 1675–1740: An Exploration of Communication and Community*. New York: Oxford University Press, 1986.

Stick, David. *Graveyard of the Atlantic: Shipwrecks of the North Carolina Coast*. Chapel Hill: University of North Carolina Press, 1952.

Stone, Roger D. *The Voyage of the Sanderling*. New York: Knopf, 1989.

Studnicki-Gizbert, Daviken. *A Nation Upon the Sea: Portugal's Atlantic Diaspora and the Crisis of the Spanish Empire, 1492–1640*. New York: Oxford University Press, 2007.

Thoreau, Henry David. *Cape Cod*. New York: Crowell, 1961.

Tomalin, Nicholas, and Ron Hall. *The Strange Last Voyage of Donald Crowhurst*. Camden, Maine: International Marine, 1995.

Ulanski, Stan. *The Gulf Stream: Tiny Plankton, Giant Bluefin, and the Amazing Story of the Powerful River in the Atlantic*. Chapel Hill: University of North Carolina Press, 2008.

Verne, Jules. *Lighthouse at the End of the World*. Translated by William Butcher. Lincoln: University of Nebraska Press, 2007.

Weightman, Gavin. *The Frozen Water Trade: How Ice from New England Lakes Kept the World Cool*. London: HarperCollins, 2001.

Wells, H. G. *The Outline of History: Being a Plain History of Life and Mankind*. London: Cassell, 1920.

Wertenbaker, William. *The Floor of the Sea: Maurice Ewing and the Search to Understand the Earth*. Boston: Little, Brown, 1974.

Westall, Dorris, ed. *Maine: WPA Guide*. Boston: Houghton Mifflin, 1937.

White, David Fairbank. *Bitter Ocean: The Battle of the Atlantic, 1939–1945*. New York: Simon & Schuster, 2006.

Williamson, Kenneth. *The Atlantic Islands: A Study of the Faeroe Life and Scene*. London: Collins, 1948.

INDEX

A NOTE ON THE TYPE

The typeface employed throughout this book is a modern interpretation of the classic eighteenth-century Bodoni face, and known as Filosofia. This was created in 1996 by the Bratislava-born type designer Zuzana Licko, who with her Dutch-born partner, Rudy VanderLans, astonished the typographic world during the closing decades of the twentieth century with a whirlwind of type design, largely occasioned by the invention of the Macintosh computer in 1984. Filosofia, with its slightly bulging serifs and lighter-than-classical-Bodoni vertical lines, clearly owes much to one of the most beloved of all Italian faces, but is more amiable and less wearing to the eyes when ranged over texts as lengthy and complex as that of *Atlantic*. I am proud that this book's designer felt able to employ this wonderful new typeface, and applaud with gratitude its most gifted creator.

SW

About the author

About the book

Read on

Insights,
Interviews
& More . . .

Meet Simon Winchester

Setsuko Winchester

AUTHOR, JOURNALIST, AND BROADCASTER Simon Winchester has worked as a foreign correspondent for most of his career. Before joining his first newspaper in 1967, however, he graduated from Oxford with a degree in geology and spent a year working as a geologist in the Ruwenzori Mountains in western Uganda and on oil rigs in the North Sea.

His journalistic work, mainly for *The Guardian* and the *Sunday Times*, has seen him based in Belfast; Washington, D.C.; New Delhi; New York; London; and Hong Kong, where he covered such stories as the Ulster crisis, the creation of Bangladesh, the fall of President Marcos, the Watergate affair, the Jonestown

Massacre, the assassination of Egypt's President Sadat, the death and cremation of Pol Pot, and the 1982 Falklands War. During the Falklands conflict, he was arrested and spent three months in prison in Ushuaia, Tierra del Fuego, on spying charges.

Winchester has been a freelance writer since 1987. He now works principally as an author, though he contributes to a number of American and British magazines and journals, including *Harper's*, *Smithsonian*, *National Geographic*, *The Spectator*, *Granta*, the *New York Times*, and *The Atlantic*. He was appointed Asia-Pacific editor of *Condé Nast Traveler* at its inception in 1987, and later became editor at large. His writing has won him several awards, including British Journalist of the Year.

He writes and presents television films on a variety of historical topics—including a series on the final years of colonial Hong Kong—and is a frequent contributor to the BBC radio program *From Our Own Correspondent*. Winchester also lectures widely—most recently before London's Royal Geographical Society (of which he is a fellow) and to audiences aboard the cruise liners *QE2* and *Seabourn Pride*. He is also a founding member and editor in chief of the *Sandisfield Times*, an independent newspaper that covers serious local issues, events, history, politics, and literature of the town of Sandisfield, Massachusetts.

Simon Winchester's books cover a wide range of subjects: the remnants ▶

Meet Simon Winchester *(continued)*

of the British Empire, the colonial architecture of India, aristocracy, the American Midwest, his months in an Argentine prison on spying charges, his description of a six-month walk through the Korean Peninsula, and the Pacific Ocean and the future of China. More recently he has written *The River at the Center of the World*, about China's Yangtze River; the bestselling *The Professor and the Madman*, the fascinating tale about the men who invented the *Oxford English Dictionary*; *The Fracture Zone: My Return to the Balkans*, which recounts his journey from Austria to Turkey during the 1999 Kosovo crisis; and the bestselling *The Map That Changed the World*, about the nineteenth-century geologist William Smith. His recent books *Krakatoa: The Day the World Exploded: August 27, 1883* (April 2003), *A Crack in the Edge of the World: America and the Great California Earthquake of 1906* (October 2005), and *The Man Who Loved China: The Fantastic Story of the Eccentric Scientist Who Unlocked the Mysteries of the Middle Kingdom* (May 2008) have all been *New York Times* bestsellers and have appeared on numerous best of and notable lists.

Simon Winchester lives in New York City and has a small farm in the Berkshires in Massachusetts. Mr. Winchester was made an Officer of the Order of the British Empire by Her Majesty, the Queen in 2006. He received the honor in a ceremony at Buckingham Palace. ∾

An Ocean of Memories

I HOPE I WILL BE FORGIVEN for beginning this brief postscript with what might seem a rather self-serving declaration: that no book I have written before has elicited so great, so enthusiastic, and so seemingly intimate a response from so many hundreds of my readers. There has indeed been a true tidal wave of correspondence—and while inevitably there have been some critical letters pointing out shortcomings that this edition takes pains to rectify, the greater number of communications have been kind and warm, the type of letters that any writer is happy to receive, to read, and to reread time and again.

Two connections—and the two quite separate chains of correspondence that then followed—stand out. One came as a result of a passage that is near the beginning of the book, while the other, with a pleasing symmetry, came about as a result of a story that I told very close to the book's end. Both amply illustrate the way in which the sea itself manages to create connections of great power that continue to sound and resonate for many years and decades.

The first was a contact made in person. I was at a literary festival in Cheltenham, in southern England, and I had given a talk to an audience of a couple of hundred. It was a slightly unusually structured talk, for reasons I have now forgotten. Usually I try not to give away the narrative arc of any book I have written, preferring instead ▶

5

to pick out a scattering of the more amusing or illustrative episodes, in the hope that this might leave people in the audience perhaps wanting to learn more—by buying the book itself. But on this occasion I found myself talking more fully than usual about the story itself, and specifically about the episode with which I begin the book: my own first journey across the Atlantic Ocean in the spring of 1963, when I made passage between Liverpool and Montreal aboard the Canadian Pacific liner *Empress of Britain*.

After the talk was over, I was taken to a table to sign copies of the book. There was a respectable lineup of maybe fifty people—given an audience of two hundred, a one-in-four ratio of purchasers to listeners, which those who measure such things rate as pretty good (the ratio is America being customarily much higher than in Britain). I took a book from the pile on the table, opened it at the signature page, smiled up at the first lady in line, and set to the task: "Shall I personalize this for you, in some way?"

Ten minutes in and all was going nicely, when I looked up to see standing before me a middle-aged lady, flushed and breathing heavily. "I am so sorry," she said, "but I ran all the way here. I didn't have a chance to come to your talk. But— don't get me wrong when I say this—I have to tell you that I've been looking for you for twenty-five years."

Alarm bells sounded faintly. A stalker? A madwoman? Was there a twenty-five-year-old former child standing behind her? But no, she seemed simply out of breath, not mad; and she smiled broadly.

"I have just one question," she continued. "Have you ever been on a ship called the *Empress of Britain*?"

I gulped. Indeed I had, I said: in fact, had she been at the talk, she would have heard me talking in some detail about that first, very memorable voyage.

"I knew it was you!" she exclaimed, and promptly produced from her handbag a diary, which she presented to me with a flourish. It was headed "Emigrating to Canada." She opened it up. "Take a look at the first page!"

Others standing in line behind her were now forming a

semicircle, eager to see what she had brought. "Read it," she said, and looked behind her at the gathering. "Out loud."

And so I did. "Tuesday 30th April, 1963," it began—forty-seven years previously. "The ship is nice, and I have already spotted one or two interesting males. We had lifeboat drill—I looked a real nit in my Mae West. It was rather sad saying goodbye to everyone, beginning my new life. It is raining a little. There is a boy in the next cabin. His name is Simon, and he is hitchhiking around Canada and America before going to Oxford University. He bought us a bottle of wine for dinner, and then afterwards we changed and he came back and we played cards in the bar . . ."

The lady's name was Sheila Lewer. I remembered her, faintly. Sheila Hoyle she had been back then, according to the passenger manifest. "Three girls from Blackburn" had been in the cabin next to mine, and Miss Hoyle, it turned out, had been one of them.

Her life after she disembarked at Montreal had been evidently as restless as mine: emigrating to Canada in 1963, moving on to Australia in 1972, coming back to Britain in 1974, then to Texas, then back to Britain again. But sometime in the eighties she had come across my name in a newspaper, it had triggered a distant memory, and she had started to wonder: *could it have been the lad I met on that long-ago ship?* And then, in the autumn of 2010, in her local tennis club, she saw a poster advertising my book and the talk in Cheltenham, and she had run across town to see if I was indeed the same person.

After that surprise, it's-a-small-world meeting at the festival she wrote to me again, quoting from her diary once more: "Wednesday 1st May, 1963. Jean, Marjorie, Simon, and I watched as we sailed down the Clyde. It was quite impressive. In the evening we had a drink in the bar with Simon. We were out in the open sea, and we all felt quite groggy." She had searched her attic for pictures, but could find none. Yet her memory of the voyage on the old *Empress* remains intact and impeccable, as if it had been a rite of passage for her, just as it had been for me. We have kept in touch ever since, and vow that we will for always, linked inextricably and irrevocably as ▶

we are by that first 'tween-decks meeting all those years ago, out at sea.

The second connection came by way of an e-mail, and out of nowhere. It was written in response to an episode toward the end of the book, in which I recount the wreck in 1942 of a Blue Star Line refrigerated cargo vessel, the MV *Dunedin Star*, off the notoriously treacherous Skeleton Coast of southwest Africa. Usually, survivors of shipwrecks cast up on this waterless expanse die of starvation and thirst; but by great good fortune, although two rescuers did perish attempting to reach them by sea, all of those from the *Star* were eventually rescued, overland.

A convoy of police vehicles from Windhoek, five hundred miles to the south, reached the party after a Herculean desert expedition. The stranded group had spent almost a month huddled under a shelter of canvas and driftwood, and were supplied with food and water by airdrops from Ventura bombers. There were forty-two of them—one, a babe in arms, had gone temporarily sand-blind, the only casualty. The party and their rescuers reached civilisation on Christmas Eve, to universal delight.

The out-of-nowhere e-mail, which was written to me sixty-eight years after these events, came from the party's babe in arms. His name was Sidney Palmer, and he was a retired British diplomat now living in Australia. When he wrote he hadn't read the book, but knew of my earlier writings, and knew from friends that in this account I had mentioned the wreck of the *Dunedin Star*.

"I am probably the last survivor," he wrote, "being the then eighteen-month-old baby mentioned in the account. . . . The only semi-memory I have of the event is being held in my mother's arms on the beach at night under a star-filled sky, and the sound of the wind in the tents. Whether I really remember this, or have 'inherited' it from something my mother told me, I can't tell. However, real or imagined, it lives with me to this day." When he signed his name, he added a phrase to his letter that tried, in an oh-so-British way, to deprecate its significance. "Just a footnote."

Well, of course to me it was anything but a footnote. It was instead an extraordinary and wholly unanticipated connection with history, and his subsequent writings to me then provided further vivid insights into one of the maritime world's few happily ending disaster stories.

After the rescue Sidney and his mother had been taken south to Cape Town, and then had traveled by railway train up almost the entire length of Africa to Egypt, to the port of Alexandria. From there they took a boat across the eastern Mediterranean to Turkey and to the senior Mr. Palmer, who was at the time stationed in the British Embassy in Ankara.

When he grew up the small boy evidently became fascinated by his father's life, and in time became a diplomat himself. He served in Yugoslavia, Thailand, and Burma (driving the family Ford Cortina from London to Rangoon, regretting to this day that violence in Nagaland compelled him to ship the car over its last few miles on a freighter), then in Iraq, the New Hebrides, Papua New Guinea, and Luxembourg, finally ending his career as deputy high commissioner for all of South India, based in Madras. Reading his later letters I became convinced that his life was well worthy of a book of its own, and I encouraged him. At the time of writing this he says he is composing it; "I'm only 73,000 words short of my target," his last letter said.

But that last letter from Sidney Palmer also spoke of *Atlantic*, which by then he had read and liked. He had two especially kind things to say. The first was about his eyes. "My mother never forgot the generosity of those people who gave up a portion of their meager water ration so that she could bathe my eyes, and thus hold the sand-blindness at bay. I owe my sight to them, and my life to others like them." His had been a long and full life, he added, "and whilst one could always have done better, I like to think that the gift [they gave me back in the desert] was not wasted."

The second was more poignant still. He had noticed that I had dedicated the book to the memory of Angus Campbell Macintyre, a Scotsman, the first mate of a South African tugboat, who had drowned in one of the rescue attempts and whose body was swept away into the ocean, never to be found. ▶

An Ocean of Memories *(continued)*

I had written in the book of my fears that my dedication to Macintyre's memory might be construed as merely sentimental; and that my having left a message of thanks in a bottle on his lonely gravesite Skeleton Coast might be thought similarly inappropriate. Sidney Palmer disagreed. "Allow me, on behalf of all those on board the *Dunedin Star* on her last voyage, to thank you for saying 'thank you' to Angus Macintyre. No generous gesture, however 'sentimental,' is wasted. It serves its own purpose. You hoped that your message in a bottle and the dedication of your book to him would 'serve as something, to someone.' Well, it served as something of great significance to me."

And then he showed me a photograph of the one relic that his late mother had brought with her from the wreck. It stands today in the Palmers' dining room, in their house beside a lake in New South Wales. It is a silver teapot, engraved with the shield of the Blue Star Line, to which the *Dunedin Star* belonged. "Somehow my mother ended up with this teapot. She said she was given it as a souvenir, but when, and by whom, and why anyone should be bothering about a teapot at such a time I have no idea, and with the indifference of youth, I never asked."

Since publication of the book a year ago there have been many other letters—scores of reminiscences of ocean voyages long ago, revelations about the sea's hidden geographical mysteries, some rather good poems, an invitation to drink Nebraskan wine beside the river in Montana that is most distant from the ocean, a note from the granddaughter of the woman who was the model for the great female statue sculpted to represent *America* outside the Customs House in lower Manhattan, another from the grandson of the founder of the Nicolson's Gin distillery that played so important a role in the eventual founding of the state of Israel, a letter informing me (to my consummate delight) that a passage from my book had been chosen as the answer in a *Wall Street Journal* acrostic puzzle, and several missives politely insisting that I investigate the claim that America had been discovered by and named after a Welshman, Richard

Ameryk. (I looked into the matter, and replied that I thought the notion somewhat far-fetched.)

As I write these words the picture of the old silver teapot from the shipwreck is before me—a talisman for an Atlantic Ocean that has itself become a talisman for so many around the world. The Atlantic, no longer a barrier, no longer even a bridge or a passageway, has become for me and for all these legions of writers something much, much more. For us it has become a force of nature, perhaps: a thing endowed with magical powers to forge connections entirely unexpected between peoples hitherto quite unaware—a grey and heaving sea that manages to bind so many peoples so closely together, both with each other and, through their writings and their sentiments, with the delighted and humbled author of this book.

Simon Winchester
May 2011 ᔦ

Have You Read?
More by Simon Winchester

**THE MAN WHO LOVED CHINA:
THE FANTASTIC STORY OF THE ECCENTRIC
SCIENTIST WHO UNLOCKED THE MYSTERIES
OF THE MIDDLE KINGDOM**

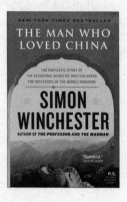

In sumptuous and illuminating detail, bestselling author Simon Winchester brings to life the extraordinary story of Joseph Needham, the brilliant Cambridge scientist, freethinking intellectual, and practicing nudist who unlocked the most closely held secrets of China, once the world's most technologically advanced country. Epic and intimate, *The Man Who Loved China* tells the sweeping story of China through Needham's remarkable life. Here is an unforgettable tale of what makes men, nations, and, indeed, mankind itself great—related by one of the world's inimitable storytellers.

"Captivating. . . . Fans of Mr. Winchester know that he can make the most obscure topic seem as if it should be on the front page."   —*Wall Street Journal*

"Needham's story is phenomenal. . . . The man was fearless in love and life. . . . Winchester [leaves] readers crying out 'more, more.' "   —*USA Today*

Unleashed by ancient geologic forces, a magnitude 8.25 earthquake rocked San Francisco in the early hours of April 18, 1906. Less than a minute later, the city lay in ruins. Bestselling author Simon Winchester brings his inimitable storytelling abilities to this extraordinary event, exploring the legendary earthquake and fires that spread horror across San Francisco and northern California in 1906 as well as the startling impact on American history and, just as important, what science recently has revealed about the fascinating subterranean processes that produced the earthquake—and almost certainly will cause it to strike again.

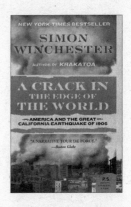

"In this brawny page-turner, bestselling writer Winchester has crafted a magnificent testament to the power of planet Earth and the efforts of humankind to understand her."
—*Publishers Weekly* (starred review)

"A narrative tour de force through more than 4,000 years of geological history that takes Winchester from the eastern edge of the North Atlantic tectonic plate in Iceland to San Francisco. . . . Without slighting the human suffering of the victims of earthquakes, tsunamis, and other natural disasters, Winchester places their tragedies in an almost cosmic context."   —*Boston Globe*

**KRAKATOA: THE DAY THE WORLD
EXPLODED: AUGUST 27, 1883**

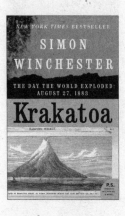

Simon Winchester details the legendary
annihilation of the volcano Krakatoa
in 1883 and its lasting, world-changing
effects, including the creation of an
immense tsunami, the release of dust
that swirled around the world for
years, and the triggering of a wave
of murderous anti-Western militancy
by fundamentalist Muslims in Java.

"One of the best books ever written
about the history and significance of
a natural disaster."
　　　　　　　—New York Times Book Review

"A real-life story bigger than any
Hollywood blockbuster."
　　　　　　　　　　—Entertainment Weekly

**THE MAP THAT CHANGED THE WORLD:
WILLIAM SMITH AND THE BIRTH OF
MODERN GEOLOGY**

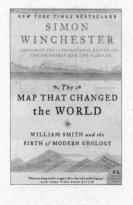

William Smith, the orphaned son of an
English country blacksmith, dug canals
for a living. From one particular dig
emerged his obsession with creating
the world's first geological map.
Known today as the father of modern
geology, Smith spent twenty-two years
researching and creating his epochal
and remarkably beautiful hand-painted
map. Instead of receiving accolades and
honors, however, he ended up the victim
of plagiarism; Smith landed in debtors'
prison and was virtually homeless for
ten years more. In 1831 this quiet genius

finally received the Geological Society of London's highest award and was offered a lifetime pension by King William IV.

"Winchester brings Smith's struggle to life in clear and beautiful language."
—*New York Times Book Review*

OUTPOSTS: JOURNEYS TO THE SURVIVING RELICS OF THE BRITISH EMPIRE

Simon Winchester, struck by a sudden need to discover exactly what was left of the British Empire, set out across the globe to visit the far-flung islands that are all that remain of what once made Britain great. He traveled thousands of miles to capture a last glint of imperial glory.

"A brilliant and delightful addition to the long and distinguished shelf of British literary odysseys."
—*Christopher Buckley, Washington Post Book World*

Have You Read? *(continued)*

THE PROFESSOR AND THE MADMAN: A TALE OF MURDER, INSANITY, AND THE MAKING OF THE *OXFORD ENGLISH DICTIONARY*

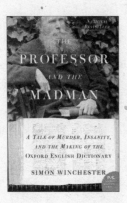

The bestselling tale of madness, genius, and the incredible obsessions of two remarkable men that led to the making of the *Oxford English Dictionary*—one of the most ambitious projects ever undertaken. As word definitions were collected, the overseeing committee led by Professor James Murray discovered that one man, Dr. W. C. Minor, had submitted more than ten thousand. When the committee insisted on honoring him, a shocking truth came to light: Dr. Minor, an American Civil War veteran, was also an inmate at an asylum for the criminally insane.

"An extraordinary tale, and Simon Winchester could not have told it better. . . . A splendid book."
—*The Economist*

"The linguistic detective story of the decade."   —William Safire,
New York Times Magazine